Aspects of Bioinformatics and Molecular Computing

Aspects of Bioinformatics and Molecular Computing

Editor: Harlan Kennedy

MURPHY & MOORE
www.murphy-moorepublishing.com

www.murphy-moorepublishing.com

MURPHY & MOORE

Cataloging-in-Publication Data

Aspects of bioinformatics and molecular computing / edited by Harlan Kennedy.
p. cm.
Includes bibliographical references and index.
ISBN 978-1-63987-734-8
1. Bioinformatics. 2. Computer simulation. 3. Computational biology.
4. Systems biology. I. Kennedy, Harlan.
QH324.2 .B56 2023
570.285--dc23

Murphy & Moore Publishing
1 Rockefeller Plaza,
New York City,
NY 10020, USA

ISBN 978-1-63987-734-8

Contents

Preface

Molecular computing and bioinformatics are two important interrelated disciplines that are concerned with the study of molecules and computers. Molecular computing is an emerging branch of computing that makes use of DNA, biochemistry, and molecular biology hardware in place of the traditional electronic computing to build computer programs. Experiments are also ongoing to replace the silicon chips used in computers with test tubes, liquids and even living cells. Molecular computing is becoming popular due to its ability to assemble more circuitry onto a microchip than can be contained in silicon chips at lower costs. Bioinformatics is a field of study that makes use of the tools of computation and analysis for developing methods and tools for understanding large and complex biological data. There are various bioinformatics tools such as visualization, pattern recognition, machine learning algorithms, and data mining. These tools help to compare, analyze and interpret genetic and genomic data. This book provides comprehensive insights into molecular computing and bioinformatics. As these fields are emerging at a rapid pace, its contents will help the readers understand their modern concepts and applications.

The information contained in this book is the result of intensive hard work done by researchers in this field. All due efforts have been made to make this book serve as a complete guiding source for students and researchers. The topics in this book have been comprehensively explained to help readers understand the growing trends in the field.

I would like to thank the entire group of writers who made sincere efforts in this book and my family who supported me in my efforts of working on this book. I take this opportunity to thank all those who have been a guiding force throughout my life.

Editor

1

DNC4mC-Deep: Identification and Analysis of DNA N4-Methylcytosine Sites based on Different Encoding Schemes by using Deep Learning

Abdul Wahab [1,†], Omid Mahmoudi [1,†], Jeehong Kim [2,*] and Kil To Chong [3,4,*]

1 Department of Electronics and Information Engineering, Jeonbuk National University, Jeonju 54896, Korea; me.wahabqayyum@gmail.com (A.W.); omidmahmoudi75@jbnu.ac.kr (O.M.)
2 Department of New & Renewable Energy, VISION College of Jeonju, Jeonju 55069, Korea
3 Department of Electronics Engineering, Jeonbuk National University, Jeonju 54896, Korea
4 Advance Electronics & Information Research Center, Jeonbuk National University, Jeonju 54896, Korea
* Correspondence: jeehong@jvision.ac.kr (J.K.); kitchong@jbnu.ac.kr (K.T.C.)
† These authors equally contributed to this work.

Abstract: N4-methylcytosine as one kind of modification of DNA has a critical role which alters genetic performance such as protein interactions, conformation, stability in DNA as well as the regulation of gene expression same cell developmental and genomic imprinting. Some different 4mC site identifiers have been proposed for various species. Herein, we proposed a computational model, DNC4mC-Deep, including six encoding techniques plus a deep learning model to predict 4mC sites in the genome of *F. vesca*, *R. chinensis*, and Cross-species dataset. It was demonstrated by the 10-fold cross-validation test to get superior performance. The DNC4mC-Deep obtained 0.829 and 0.929 of MCC on *F. vesca* and *R. chinensis* training dataset, respectively, and 0.814 on cross-species. This means the proposed method outperforms the state-of-the-art predictors at least 0.284 and 0.265 on *F. vesca* and *R. chinensis* training dataset in turn. Furthermore, the DNC4mC-Deep achieved 0.635 and 0.565 of MCC on *F. vesca* and *R. chinensis* independent dataset, respectively, and 0.562 on cross-species which shows it can achieve the best performance to predict 4mC sites as compared to the state-of-the-art predictor.

Keywords: N4-methylcytosine; rosaceae genome; DNA encoding methods; computational biology; deep learning; bioinformatics

---

1. Introduction

Dynamic DNA modifications, such as methylation and demethylation have an essential role in the regulation of gene expression. DNA methylation as a heritable epigenetic marker is one type of chemical modification of DNA, which alters genetic performance without altering the DNA sequence [1,2]. Several researches have shown that it has the ability to change DNA protein interactions, DNA conformation, DNA stability, and chromatin structure. Meanwhile, it can regulate some different functions including cell developmental, genomic imprinting, and gene expressions [3,4]. N4-methylcytosine (4mC), 5-Methylcytosine (5mC), and N6-methyladenine (6mA) as three common methylations by specific methyltransferase enzymes occur in both prokaryotes and eukaryotes [5–7].

In prokaryotes, the host DNA from exogenous pathogenic DNA can be identified by 6mA and 4mC [8], and also 4mC regulates DNA replication and its errors [9,10]. Meanwhile, 4mC as part of a restriction-modification (R-M) system prevents restriction enzymes from degrading host DNA [11]. In eukaryotes, 5mC has a crucial role in transposon suppression, gene imprinting, and regulation. By high sensitivity techniques, 6mA and 4mC can only be detected in eukaryotes [12].

The 5mC, as the most well-explored and common type of cytosine methylation plays a significant role in several biological processes [13] and can be caused by cancer, diabetes, and also some neurological diseases [14–16]. The 4mC as effective methylation protects its own DNA from the restriction of enzyme-mediated degradation. It has an important role in controlling some various processes including cell cycle, gene expression levels, differentiating self and non-self-DNA, DNA replication, and correcting DNA replication errors [9,17].

Some extensive experimental studies have been performed to detect 4mC sites in the whole genome such as 4mC-Tet-assisted bisulfite sequencing, methylation-precise PCR, mass spectrometry, and Single-Molecule of Real-Time (SMRT) sequencing [18–21]. The aforementioned experimental approaches are laborious and expensive work when performing genome-wide testing. Therefore, it is necessary to develop a computational method for identifying 4mC sites.

Lately, several 4mC sites identifiers [22,23] have been proposed for different species such as *C. elegans, D. melanogaster, A. thaliana, E. coli, G. subterraneus, G. pickeringii*. The i4mC-ROSE [24] as the first computational tool for predicting 4mC sites within the *Rosaceae* genomes has been proposed to identify the 4mC sites in the genomes of *F. vesca* [25] and *R. chinensis* [26]. It generated six probabilistic scores by using six encoding methods; random forest (RF), algorithms with k-space spectral nucleotide composition (KSNC), electron-ion interaction pseudopotentials (EIIP), k-mer composition (Kmer), binary encoding (BE), dinucleotide physicochemical properties (DPCP), and trinucleotide physicochemical properties (TPCP) that cover various aspects of DNA sequence information. Then, those scores were combined with a linear regression model for enhanced prediction performance [24]. The 4mcDeep-CBI [27] as a deep learning framework has been proposed to predict the 4mC sites in an expanded dataset of *Caenorhabditis elegans* (*C. elegans*). 3-CNN and BLSTM were used to extract deep information and to obtain advanced features.

In this work, a novel predictor, DNC4mC-Deep, has been established for the identification of 4mC sites in the genome of *F. vesca*, *R. chinensis*, and cross-species which is newly prepared. The overall framework of our work summarized as; Firstly we used the six encoding techniques named 2Kmer [28], 3Kmer [29], binary encoding (BE) [30,31], nucleotide chemical property (NCP) [32], nucleotide chemical property, and nucleotide frequency (NCPNF) [32], and multivariate mutual information (MMI) [33,34]. Then, we made a deep learning model by using the Convolution Neural Network (CNN). We applied a grid search algorithm to obtain the optimal model with tuned hyperparameters. All six encoding schemes were input separately in the optimal selected model and recorded the results of each encoding scheme and used the K-fold cross-validation method with the value of K as 10. To evaluate and analyze the results of the model on each encoding scheme, we used the performance evaluation metrics. We also presented two different applications; the first one is silico mutagenesis [35] representation using heat maps, and the second is distinguishing the most significant portions of a sequence using saliency maps [36]. After getting the results from the model by all six different feature encoding methods, we ended up with that Dinucleotide composition (DNC) is outperforms from all six encoding schemes and the state-of-the-art model. In comparison to the state-of-the-art model, DNC4mC-Deep successfully achieves 0.635, 0.565, and 0.562 of MCC on *F. vesca*, *R. chinensis*, and cross-species independent dataset, respectively.

2. Materials and Methods

2.1. Benchmark Datasets

The benchmark dataset of DNA 4mC obtained from Md. Mehedi Hassan et al. [24]. It contains the *F. vesca* and *R. chinensis* genome. To prepare the high-quality dataset they have applied the sequences with ModQV score greater than 20, whereas the remaining sequences were excluded. To solve the homology bias problem, the CD-HIT-EST [37] software was used to exclude redundant sequences with a cut-off of 0.65. All sequences contain a central cytosine (C) nucleotide with a length of 41 base pairs (bp).

In both datasets, *F. vesca* and *R. chinensis* genome were considered 75% and 25% samples from all data as the training and the independent dataset. The training dataset consists of 4854 and 2337 positive DNA sequences as 4mC samples for *F. vesca*, and *R. chinensis*, severally. The negative DNA sequences, such as non-4mC, consists of 4854 and 2337 samples for *F. vesca*, and *R. chinensis* genome. Furthermore, the independent dataset included 1617 for both positive and negative DNA sequences of *F. vesca* genome whereas for *R. chinensis* positive and negative DNA contains 779 samples.

Moreover, we made the cross-species as a new benchmark dataset from the two above datasets. To avoid the redundancy in the original datasets we used CD-HIT-EST with different threshold values. The recent dataset was also divided into the training and the independent dataset with the same proportion (75% and 25% samples) where we obtained the cross-species dataset with the most attentive threshold at 0.80 containing 7190 and 5874 positive and negative DNA sequences, respectively, on the training dataset. Meanwhile, we assumed 2394 positive and 2234 negative DNA sequences on the independent dataset. The length of each sample is 41nt. Details of the benchmark datasets are shown in Table 1.

Table 1. Benchmark datasets demonstration.

Species	Dataset	Training Dataset	Total	Independent Dataset	Total
F. vesca	4mC samples non-4mC samples	4854 4854	9708	1617 1617	3234
R. chinensis	4mC samples non-4mC samples	2337 2337	4674	779 779	1558
Cross-species	4mC sampels non-4mC samples	7190 5874	13,064	2394 2234	4628

2.2. Feature Encoding Methods

Feature encoding has a vital role in the construction of the model [38]. A DNA sequence is represented as a fixed length of feature vectors which can be classified by deep learning algorithms. In this article, six various types of feature encoding methods, binary encoding [39], DNC (2kmer), TNC (3kmer) [40–43], Multivariate Mutual Information (MMI) [44], Nucleotide Chemical Property (NCP) and Nucleotide Chemical Property and Nucleotide Frequency (NCPNF) [28,29,45–47] were employed to formulate methylcytosine samples.

2.2.1. Binary Encoding (BE)

Binary encoding is a simple and effective feature algorithm converts each nucleotide into a binary vector as follows: A (1, 0, 0, 0, 0), C (0, 1, 0, 0, 0), G (0, 0, 1, 0, 0), T (0, 0, 0, 1, 0) and N (0, 0, 0, 0, 0). A DNA sequence with m nucleotides can be represented into a vector of $5 \times m$ features [30,31].

2.2.2. Kmer

Kmer is a common feature encoding algorithm that has been widely used in various prediction works [28,29,45,48,49]. A DNA sample is expressed as V = N1, N2, N3, ... NL, where L denotes the length of the sequence and Ni is one of the regular nucleotides A, C, G, T, and N. In this work, di-nucleotide composition (DNC) and tri-nucleotide composition (TNC) were considered. In DNC all samples of 41 nt produce 40 components with the equation of $L - k + 1$. The DNC scheme generated a 25 (5^2) dimensional feature. Whereas in TNC samples of 41 nt generated 39 elements with the equation of $L - k + 2$. The TNC form into a 125 (5^3)-dimensional vector. In both equations, the L denotes the length of the sequence and k represents the value of Kmer as an integer.

2.2.3. Nucleotide Chemical Property (NCP)

The four nucleic acids have different chemical properties [50]. In terms of ring structures, A and G each contain two rings, whereas C and T contain only one. Regarding secondary structures, A and T form weak hydrogen bonds, whereas C and G form strong hydrogen bonds. In terms of chemical functionality, A and C can be classified into the amino group, while G and T can be classified into the keto group. The cluster of four nucleotides was shown in Table 2.

Table 2. Cluster of nucleotides based on chemical properties.

Chemical Property	Class	Nucleotides
Ring structure	Two ring One ring	A, G C, T
Hydrogen bond	Strong Weak	C, G A, T
Functional group	Amino Keto	A, C G, T

Three coordinates x, y, and z were used to represent ring structure, the hydrogen bond, the chemical functionality, respectively, and the value of 0 and 1 was assigned to each one. The feature extraction algorithm can be formulated as follows:

$$x_i = \begin{cases} 1 & if s_i \epsilon \{A, G\} \\ 0 & if s_i \epsilon \{C, T\} \end{cases}, y_i = \begin{cases} 1 & if s_i \epsilon \{A, T\} \\ 0 & if s_i \epsilon \{C, G\} \end{cases}, z_i = \begin{cases} 1 & if s_i \epsilon \{A, C\} \\ 0 & if s_i \epsilon \{G, T\} \end{cases}$$

where $n(s_i)$ represents A, C, G, T, and N nucleotide, which can be converted by the coordinates (1, 1, 1), (0, 0, 1), (1, 0, 0), (0, 1, 0), and (0, 0, 0), respectively.

We also tried the nucleotide chemical properties (NCP) with the frequencies of each nucleotide (NF) position in a sample. The method was got from Chen et al. [32] for both encoding schemes. We integrated the NCP and NF to represent a matrix with 41 columns and 5 rows for each sample of the DNA sequence. Each DNA base of the sequence was designated as a column of the matrix and for each column, their initial three components were characterized as the nucleotide chemical property and the last one represented as a nucleotide frequency which we denoted as NCPNF.

2.2.4. Multivariate Mutual Information (MMI)

MMI has been used in many works [33,34,51] to extract features of the nucleotides sequence. We used the MMI based feature encoding algorithm which was proposed by Pan et al. [52]. First of all, they modified a two-tuple and three-tuple nucleotides set as follows:

$$\begin{gathered} T_2 = \{AA, AC, AG, AT, AN, CC, CG, CT, CN, GG, GT, GN, TT, TN, NN\} \\ T_3 = \{AAA, AAC, AAG, AAT, AAN, ACC, ACG, ACT, ACN, AGG, AGT, \\ AGN, ATT, ATN, ANN, CCC, CCG, CCT, CCN, CGG, CGT, CGN, \\ CTT, CTN, CNN, GGG, GGT, GGN, GTT, GTN, TTT, TTN, NNN\} \end{gathered}$$

Then, the mutual information for the elements was calculated as a frequency of nucleotides in the sequence with respect to 2-tuple and 3-tuple. We extracted 55 MMI features.

3. The Proposed Deep Learning Model

In this study, an efficient deep learning model based on CNN was proposed for the identification of 4mC sites in the genome of *F. vesca*, *R. chinensis* and cross-species. CNN does not require manually extracted features like a conventional supervised learning processes. The immense advantage of a CNN, it can extract the features by itself automatically for the classification process. Additionally, a handy crafted feature can also be fed to CNN to build a heterogeneous model. A CNN has a big impact on various fields of natural language processing, image processing [53–56] and computational biology [57,58]. To get an optimum model we applied grid search and during learning the CNN, six hyperparameters were tuned. The ranges within each hyper-parameter was tuned to are listed in Table 3.

Table 3. Hyper-parameters tuning demonstration.

Parameters	Range
Convolution layers	[1, 2, 3, 4, 5]
Filters in convolution Layer	[8, 12, 16, 22, 32, 42, 64, 128]
Filter size	[2, 3, 4, 5, 6, 7, 8, 10, 12, 14]
Pool-size in Maxpooling	[2, 4]
Stride length in Maxpooling	[2, 4]
Dropout values	[0.2, 0.25, 0.3, 0.35, 0.4]

After getting the best model from the grid search, we used six different encoding schemes (DNC, TNC, BE, NCP, NCPNF, MMI) for the input of the CNN model. Each encoding technique converted into vectorization of the input sequence and used the same CNN model for training and testing also verified the robustness from the independent dataset. All the feature encoding approaches had a different impact on a single model.

In the proposed model, initially, two blocks used with the same number of layers but different values of parameters. Each block contains one convolution layer Conv1D (f, k, s) where parameter f is the number of filters, k is the kernel-size, and s represents the stride value are equal to 32, 5 and 1, respectively on both blocks. The convolution layer utilizes its ability to fetch the features by self-regulating for the input sequence of positive and negative 4mC samples. As a parameter of the convolution layer, we used L2 regularization and bias regularization to make sure that the model has no overfitting problem. We set the values for both regularizations with 0.0001 for the two Conv1D of blocks. As an activation function, an exponential linear unit (ELU) is used. Each Conv1D was followed by a group normalization layer (GN) as GroupNormalization (g) where g is a number of groups, to decrease the outcomes of convolution layers. Group normalization divides channels into groups and normalizes the feature within each group. The number of groups was set to 4 on both blocks of GN. To reduce the redundancy of the features after GN layers, we employed a max-pooling layer in each block as MaxPooling1D (l, r) where l denotes pool-size and r is the stride were set as 4 and 2, respectively. The max-pooling layer helps to reduce the dimensionality of the features from former layers. The outputs of the max-pooling layers were passed through dropout layers, Dropout (d) with a probability of 0.25 as a value of d on both blocks for the prevention of overfitting during the training. Dropout helps to switch off the effects of a few hidden nodes by adjusting the output of nodes to zero at training.

After both blocks, to unstack the output, a flatten function was used to squash the feature vectors from the previous layers. Right after a flatten layer, a fully connected (FC) dense layer used as Dense (n) with the number of n neurons which was set as 32 and also used the L2 regularization parameter for bias and weights with the value of 0.0001. ELU activation function used in the FC layer. At last, a FC layer was applied and used sigmoid function for the binary classification. Sigmoid is used to squeezes the values between the range of 0 and 1 to represent the probability of having 4mC and non-4mC sites. Figure 1 shows the complete architecture of the presented model.

The DNC4mC-Deep was carried out on the Keras Framework [59]. In DNC4mC-Deep we used stochastic gradient descent (SGD) optimizer with a momentum of 0.95 and the learning rate is set as 0.005. For the loss function, binary cross-entropy was used. On the fit function, we set the 100 for the epoch and 32 for the batch size. The checkpoint was used on call back function for saving the models and their best weights whereas early stopping was also implemented to halt the prediction accuracy at the time when validation stops improving. The patience level was set to 30 in early stopping.

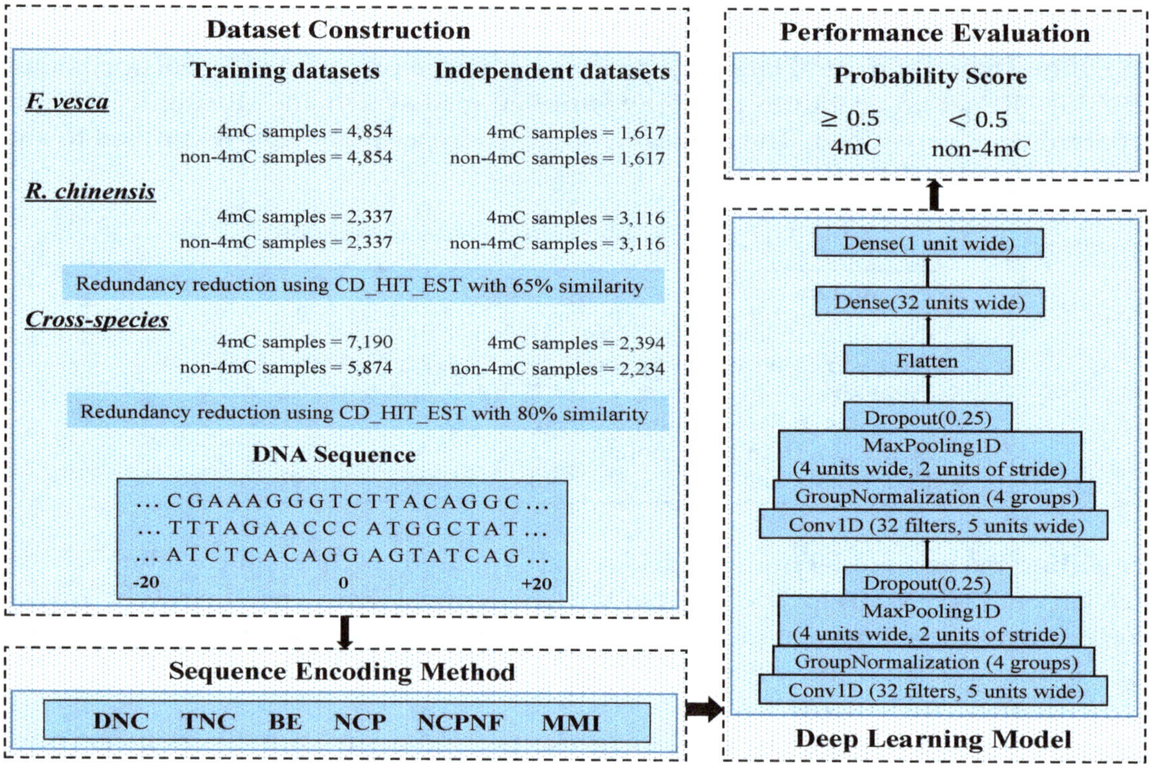

Figure 1. A complete structure of DNC4mC-Deep.

4. Performance Evaluation Metrics

The performance of the prediction model can be measured by using k-fold cross-validation. In DNC4mC-Deep we used 10 fold cross-validation to achieve the foremost prediction calculation. Cross-validation is a resampling technique which provides a precise performance estimation for the predictive model. It intermixes the entire dataset and divides into a k number of clusters, where each cluster contains eight folds for training, one fold for validation, and one for testing. The model was trained and tested k times, recorded performance each time, and concised by taking the mean score for the performance evaluation. The most common criteria which is used to evaluate the performance of the predicted models are four metrics; Mathew's correlation coefficient (MCC), accuracy (ACC), sensitivity (Sn) and specificity (Sp) with the following mathematical formulations [60–62].

$$MCC = \frac{TP * TN - FP * FN}{\sqrt{(TP + FP) * (TP + FN) * (TN + FP) * (TN + FN)}} \tag{1}$$

$$ACC = \frac{TP + TN}{TP + TN + FP + FN} \tag{2}$$

$$SN = \frac{TP}{TP + FN} \tag{3}$$

$$SP = \frac{TN}{TN + FP} \tag{4}$$

where TP and TN as true positive and true negative indicate the correct numbers of predicted samples for 4mCs and non-4mCs, respectively. Meanwhile, FP and FN as false positive and false negative represent the false numbers of predicted samples for 4mCs and non-4mCs, respectively. Besides, the receiver operating characteristics curve (ROC) and area under the ROC curve (AUC) were also used to show the performance of proposed model.

5. Results and Discussion

Six different encoding methods, namely DNC, TNC, NCP, BE, NCPNF, and MMI were used on various feature encodings for identification of the best classifier for the 4mC site prediction.

5.1. *Performance Evaluation of Various Feature Methods on the Training Datasets*

By comparing the effectiveness of the proposed methods with i4mC-ROSE model which used the same datasets, the DNC scheme yielded MCC, ACC, Sn and Sp of 0.829, 0.914, 0.926, and 0.903, respectively as the best performances for *F. vesca* dataset (Figure 2). Similarly, it achieved 0.828 for MCC, 0.914 for ACC, 0.919 for Sn and 0.910 for Sp as the maximum value on the *R. chinensis* dataset (Figure 3). The detailed performances of DNC as the best encoding method for ten different models on the *R. chinensis* dataset are given in Supplementary File 1. The TNC scheme yielded the highest value for Sp of 0.909 on the *F.vesca* dataset. Table 4, summarized the prediction performances by each six encoding methods and existing state-of-the-art model on *F. vesca* and *R. chinensis* datasets.

Furthermore, the performance evaluation by six different encoding models on the cross-species dataset is shown in Figure 4. The DNC scheme yielded the highest value for all those metrics except Sp which the TNC scheme achieved the highest value of 0.882 shown in Table 5.

Table 4. Performance evaluation of six encoding methods with state-of-the-art model on training benchmark dataset for *F. vesca* and *R. chinensis* species.

Dataset	Method	MCC	ACC	Sn	Sp	AUC
Fragaria Vesca	DNC	0.829	0.914	0.926	0.903	0.96
	TNC	0.825	0.912	0.916	0.909	0.96
	NCP	0.797	0.898	0.922	0.874	0.95
	BE	0.760	0.879	0.905	0.854	094
	NCPNF	0.782	0.891	0.907	0.874	0.95
	MMI	0.659	0.829	0.864	0.794	0.90
	i4mC-ROSE	0.545	0.767	0.635	0.899	0.88
Rosa Chinensis	DNC	0.828	0.914	0.919	0.910	0.96
	TNC	0.811	0.906	0.906	0.906	0.96
	NCP	0.811	0.906	0.901	0.910	0.96
	BE	0.805	0.903	0.891	0.914	0.95
	NCPNF	0.794	0.897	0.892	0.901	0.95
	MMI	0.691	0.846	0.833	0.858	0.92
	i4mC-ROSE	0.563	0.784	0.668	0.900	0.89

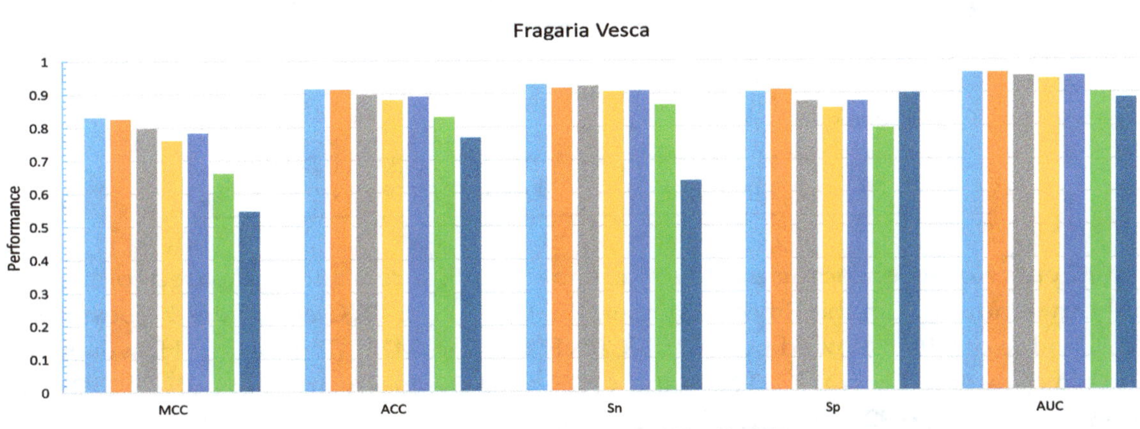

Figure 2. Grapical demonstration of performance comparison between six encoding methods and state-of-the-art model on training *Fragaria Vesca* dataset.

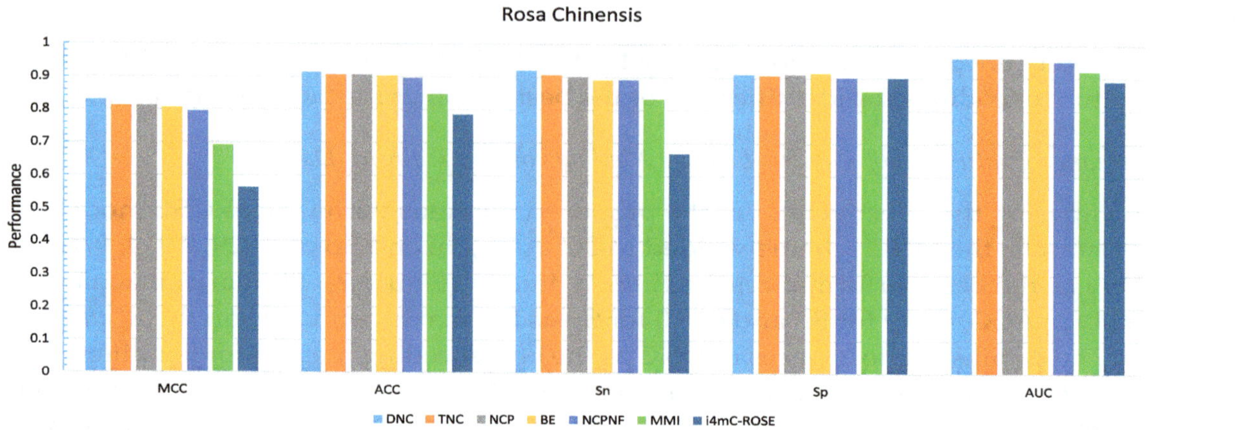

Figure 3. Grapical demonstration of performance comparison between six encoding methods and state-of-the-art model on training *Rosa Chinensis* dateset.

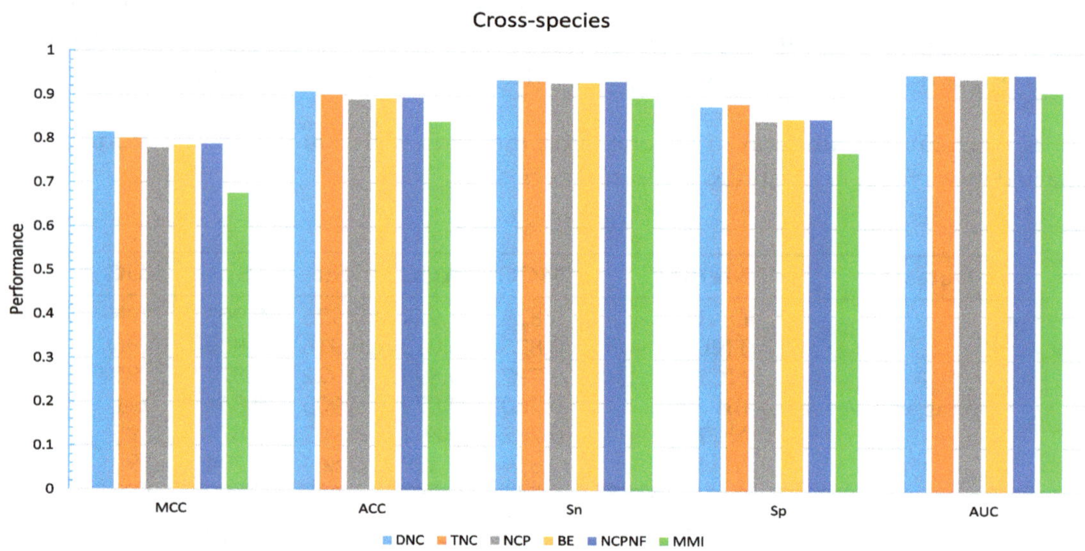

Figure 4. Grapical demonstration of performance comparison between six encoding methods on the training cross-species dataset.

Table 5. Performance evaluation of six encoding methods on training benchmark dataset for cross-species.

Dataset	Method	MCC	ACC	Sn	Sp	AUC
Cross-species	DNC	0.814	0.908	0.935	0.876	0.95
	TNC	0.800	0.901	0.933	0.882	0.95
	NCP	0.779	0.890	0.929	0.843	0.94
	BE	0.785	0.893	0.930	0.848	0.95
	NCPNF	0.788	0.895	0.933	0.848	0.95
	MMI	0.676	0.840	0.896	0.772	0.91

The ROC curve of six encoding models was shown in Figure 5 and compared to the i4mC-ROSE model for both genomes. On the *F. vesca* dataset, the DNC and TNC achieved the best performance with an AUC value of 0.96 followed by NCP, NCPNF, BE, and MMI (Figure 5a). However, the highest AUC value was presented by DNC, TNC, and NCP of 0.96 equally and next BE, NCPNF, and MMI provided 0.95, 0.95, and 0.92, respectively on *R. chinensis* dataset (Figure 5b). Besides, DNC, TNC, BE, NCNF, and NCP all have the highest value of 0.95 on training benchmark dataset cross-species (Figure 6).

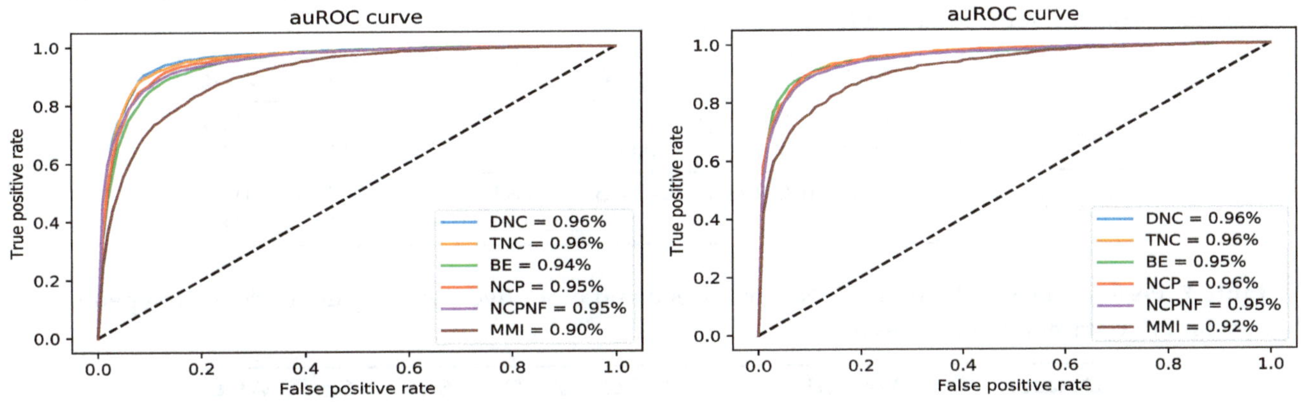

(a) Fragaria Vesca **(b)** Rosa Chinensis

Figure 5. auRoc curves of six encoding methods for the proposed model on two training benchmark dataset.

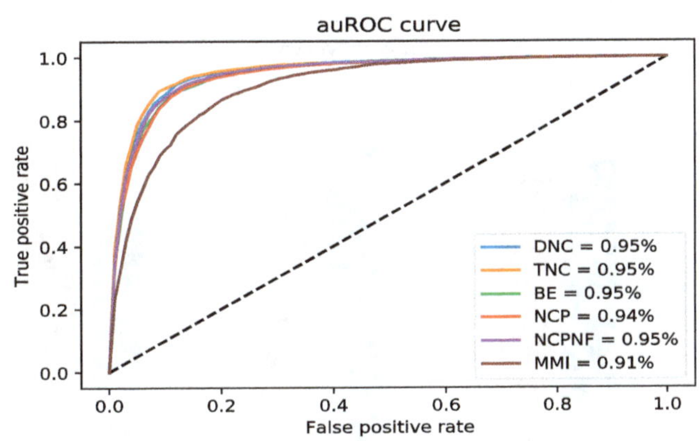

Figure 6. auRoc curves of six encoding methods for the proposed model on new cross-species training benchmark dataset.

5.2. Performance Evaluation of Various Encoding Methods on the Independent Datasets

We considered DNC as an encoder to characterize our proposed model, DNC4mC-Deep, due to its consistent performance on training datasets. It means we used the DNC4mC-Deep term instead of DNC scheme on the independent datasets. As represented in Table 6, the DNC4mC-Deep scheme achieved MCC, ACC, Sn and Sp of 0.635, 0.815, 0.878, and 0.753, respectively on *F. vesca* dataset (Figure 7). However, it yielded 0.565 MCC, 0.783 ACC, 0.801 Sn and 0.765 Sp on *R. chinensis* dataset (Figure 8). It can be seen clearly, comparing with the i4mC-ROSE method, the performance of the proposed predictor outperformed on both datasets. Additionally, as can be seen in Table 7, we compared the performance of six different encoding schemes on the cross-species dataset. The DNC4mC-Deep yielded the highest values for MCC, ACC, Sp, and AUC of 0.562, 0.780, 0.706, and 0.85, respectively. However, the NCPNF provided 0.871 Sn as the highest value Figure 9. Furthermore, we reached to 0.89, 0.87, and 0.85 of ROC for *F. vesca*, *R. chinensis*, and cross-species datasets which are depicted in Figure 10.

Table 6. Performance evaluation between the DNC4mC-Deep and state-of-the-art model on independent benchmark dataset for *F. vesca* and *R. chinensis* species.

Dataset	Method	MCC	ACC	Sn	Sp	AUC
Fragaria Vesca	DNC4mC-Deep	0.635	0.815	0.878	0.753	0.89
	i4mC-ROSE	0.601	0.797	0.721	0.873	0.89
Rosa Chinensis	DNC4mC-Deep	0.565	0.783	0.801	0.765	0.87
	i4mC-ROSE	0.535	0.759	0.636	0.881	0.86

Table 7. Performance evaluation of DNC4mC-Deep and other five encoding methods on independent benchmark dataset for cross-species.

Dataset	Method	MCC	ACC	Sn	Sp	AUC
Cross-species	DNC4mC-Deep	0.562	0.780	0.849	0.706	0.85
	TNC	0.542	0.769	0.854	0.678	0.84
	NCP	0.530	0.764	0.828	0.696	0.84
	BE	0.546	0.770	0.867	0.666	0.84
	NCPNF	0.560	0.777	0.871	0.677	0.85
	MMI	0.512	0.753	0.858	0.640	0.83

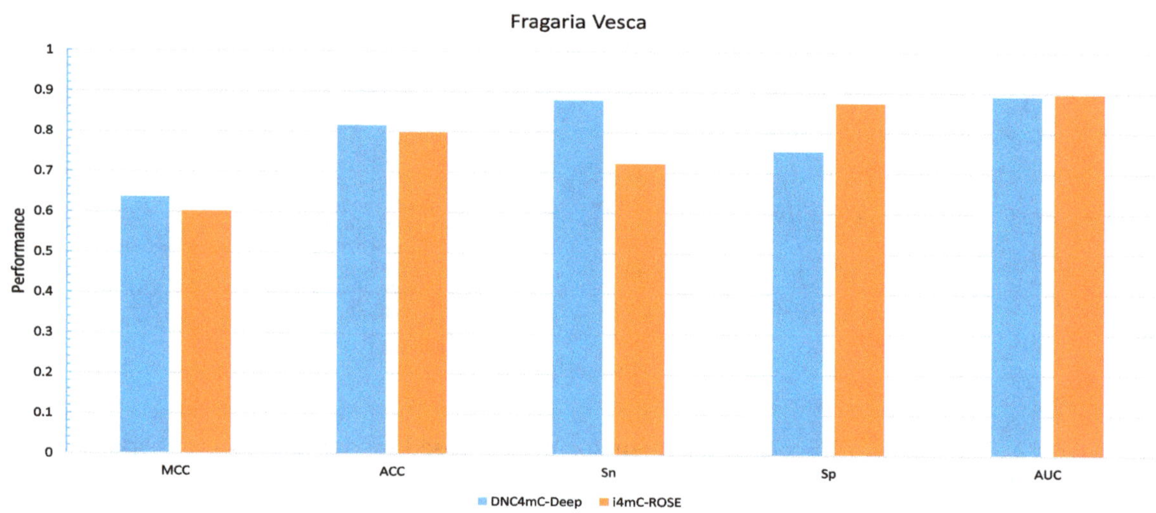

Figure 7. Grapical illustration of performance comparison between DNC4mC-Deep and state-of-the-art model on the independent Fragaria Vesca dataset.

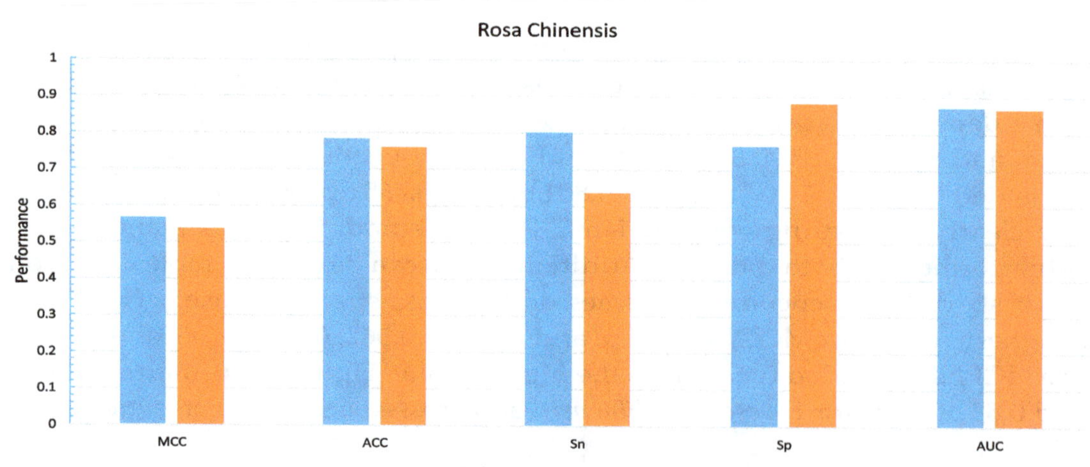

Figure 8. Grapical illustration of performance comparison between DNC4mC-Deep and state-of-the-art model on the independent *Rosa Chinensis* dataset.

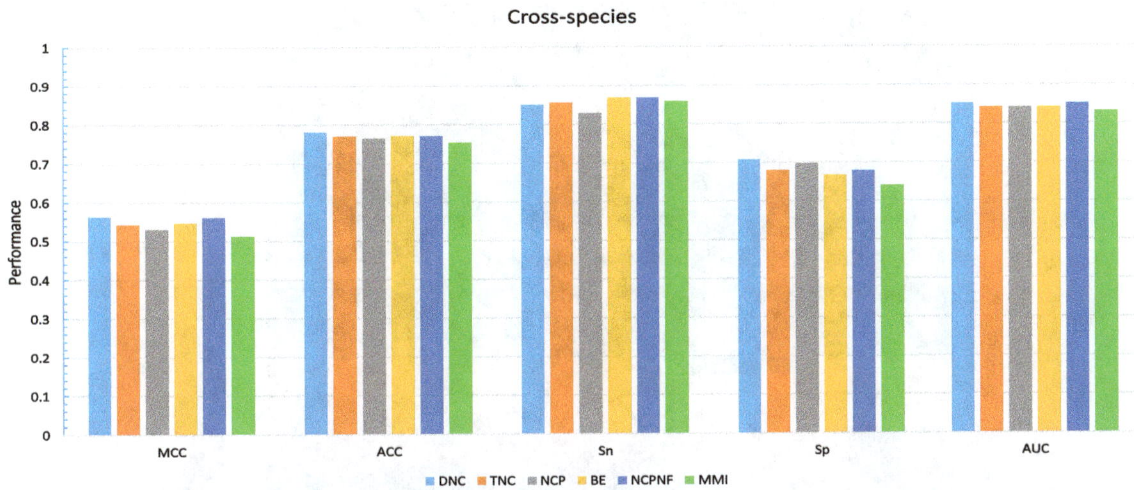

Figure 9. Grapical illustration of performance comparison between six encoding methods on the independent cross-species dataset.

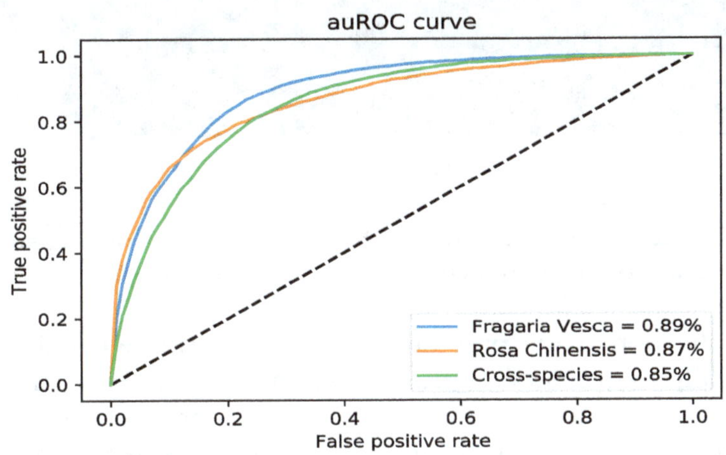

Figure 10. auRoc curves of three independent benchmark dataset on DNC4mC-Deep model.

5.3. *Interpreting Applications of Deep Learning Models*

Deep learning has an ability to accomplish the state-of-the-art results but it is further challenging to construe the algorithms as a standard statistical model. In the presented work, we demonstrated two applications to understand why those deep learning models perform well apart from others and analyze their prediction by presenting the various visualization methods.

The first most authenticated and reliable method to interpret a CNN model for computational biology is silico mutagenesis which is used in several research works [35,63,64]. We computationally mutated the nucleotides by mutating each nucleotide of a single sequence with a fixed length of five nucleotides A, C, G, T, and N. During this systematic approach, the model recomposes the output of every mutation and stores the output as an absolute difference. Next, the average of mutated predictive results of the whole dataset was taken.

A heat map was used to show the mutated modifications. CNN has the capability to visualize each convolution filter as a heat map or weight matrix. Figure 11, depicts the visualization of the mutation on *F. vesca* dataset as a local feature while learning the model. In the center of the sequences, the impact of mutation is more impactful on the final predictions because of C nucleotide which is representing the methylcytosine modification, the alteration of C nucleotide can lead to different types of gene modification. In contrast, the other sides of the heat map show the low effect of a mutation on the prediction which indicates the alteration of nucleotides cannot affect the outcome of N4-methylcytosine identification.

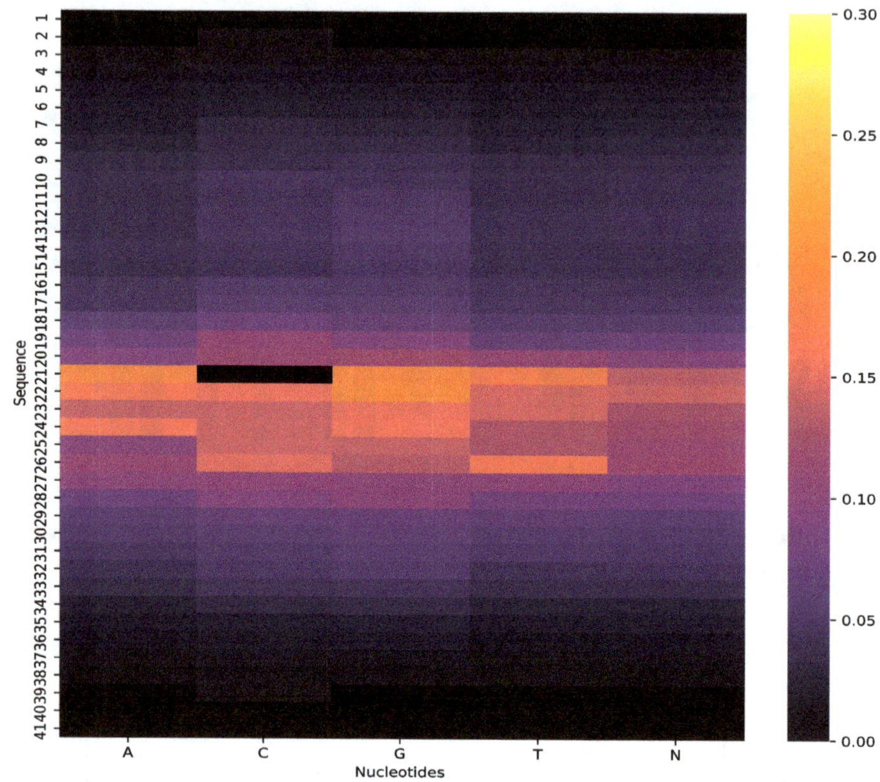

Figure 11. Heatmap visualization of silico mutation, where center of sequence with C nucleotide represents the highest effect on final prediction of Fragaria Vesca dataset.

There is another application to interpret the CNN model for knowing about the important features in the sequence, which help to gradients of the model for the final prediction. Saliency maps are the opted option to know about the most influential parts of the sequences for the classification because many researchers used in their works [36,65,66]. To visualize the effect of each position, we performed a pointwise product of the saliency map with the binary encoded sequence to acquire the derivative values for the original nucleotide letters of the sequences (A, C, G, T, and N). Samples were divided into 2-mer components across all sequences by $L - k + 1$ formulation. In Figure 12, we can see the impact of di-nucleotide characters at each position on the output score of the whole *F. vesca* dataset. At the center of the bases, di-nucleotide motif CC has high magnitude value which represents the most important feature motif in the sequences for the prediction of the CNN model. Motif CC also indicates the N4-methycytosine modification which is our considerable problem for the prediction.

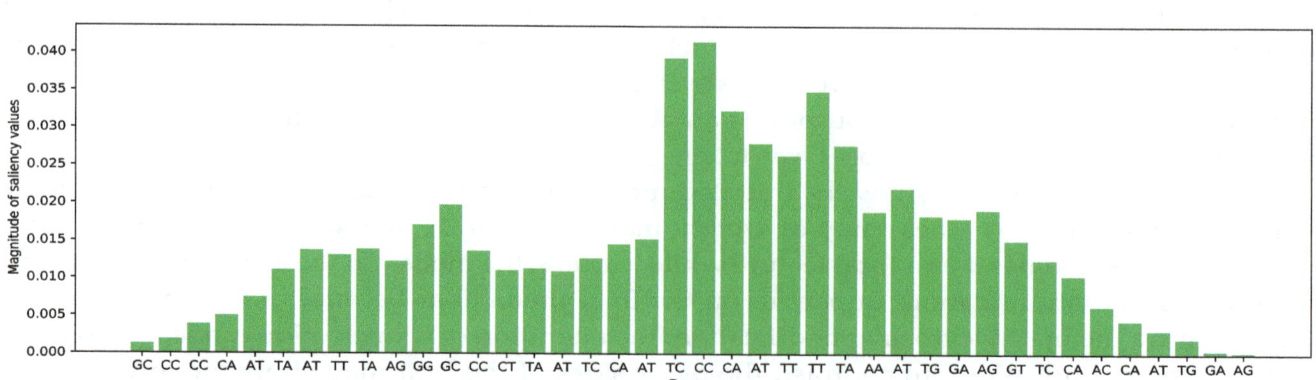

Figure 12. Saliency map of each di-nucleotide influence on model's output in the Fragaria Vesca dataset.

6. Conclusions

In this work, we presented an influential computational model named as DNC4mC-Deep to identify the N4-methylcytosine sites. There are two benchmark datasets related to the Rosaceae genome used Fragaria Vesca and Rosa Chinensis, from those two datasets we constructed a new benchmark dataset: cross-species. We used six different types of feature encoding schemes to input DNA sequence and fed to the CNN model one after another. The CNN based predictor was derived after applying the grid search algorithm. The results obtained from each encoding technique, we concluded that dinucleotide composition (DNC) outperforms and is most imperative for the strong performance of deep learning algorithms to predict 4mC sites. However, to compare with the state-of-the-art models, the CNN model with DNC encoding scheme shows the utmost effective performance and indicates the high capability of prediction. We used different evaluation metrics such as MCC, ACC, Sn, Sp, and AUC, to acquire the efficiency of the proposed predictor. Finally, we interpreted our deep learning model from two techniques: silico mutagenesis and saliency map. DNC4mC-Deep can make a high impact on the biologist to identify the N4-methylcytosine sites and can be used in brain development abnormalities. In the future, we will extend the work to prepare some new datasets and make computational models related to deep learning. Meanwhile, we established the webserver http://home.jbnu.ac.kr/NSCL/DNC4mC-Deep.htm, for users to achieve their desired results easily.

Author Contributions: Conceptualization, A.W., O.M., J.K. and K.T.C.; Methodology, A.W. and O.M., J.K.; Software, A.W.; Validation, A.W., O.M., J.K. and K.T.C.; Investigation, A.W., O.M., J.K. and K.T.C.; Writing—original draft preparation: A.W. and O.M.; Writing, review and editing, A.W., O.M., J.K. and K.T.C.; Supervision, K.T.C. All authors have read and agreed to the published version of the manuscript.

References

1. Rathi, P.; Maurer, S.; Summerer, D. Selective recognition of N 4-methylcytosine in DNA by engineered transcription-activator-like effectors. *Philos. Trans. R. Soc. B: Biol. Sci.* **2018**, *373*, 20170078. [CrossRef]
2. Jeltsch, A.; Jurkowska, R.Z. New concepts in DNA methylation. *Trends Biochem. Sci.* **2014**, *39*, 310–318. [CrossRef]
3. Liang, Z.; Shen, L.; Cui, X.; Bao, S.; Geng, Y.; Yu, G.; Liang, F.; Xie, S.; Lu, T.; Gu, X.; et al. DNA N6-adenine methylation in Arabidopsis thaliana. *Dev. Cell* **2018**, *45*, 406–416. [CrossRef]
4. Chatterjee, A.; Eccles, M.R. *DNA Methylation and Epigenomics: New Technologies and Emerging Concepts*; Springer: Berlin, Germany, 2015; Volume 103.
5. Fu, Y.; Luo, G.Z.; Chen, K.; Deng, X.; Yu, M.; Han, D.; Hao, Z.; Liu, J.; Lu, X.; Doré, L.C.; et al. N6-methyldeoxyadenosine marks active transcription start sites in Chlamydomonas. *Cell* **2015**, *161*, 879–892. [CrossRef] [PubMed]
6. Blow, M.J.; Clark, T.A.; Daum, C.G.; Deutschbauer, A.M.; Fomenkov, A.; Fries, R.; Froula, J.; Kang, D.D.; Malmstrom, R.R.; Morgan, R.D.; et al. The epigenomic landscape of prokaryotes. *PLoS Genet.* **2016**, *12*, e1005854. [CrossRef]
7. Chen, W.; Yang, H.; Feng, P.; Ding, H.; Lin, H. iDNA4mC: Identifying DNA N4-methylcytosine sites based on nucleotide chemical properties. *Bioinformatics* **2017**, *33*, 3518–3523. [CrossRef]
8. Heyn, H.; Esteller, M. An adenine code for DNA: A second life for N6-methyladenine. *Cell* **2015**, *161*, 710–713. [CrossRef]
9. Cheng, X. DNA modification by methyltransferases. *Curr. Opin. Struct. Biol.* **1995**, *5*, 4–10. [CrossRef]
10. Wei, L.; Luan, S.; Nagai, L.A.E.; Su, R.; Zou, Q. Exploring sequence-based features for the improved prediction of DNA N4-methylcytosine sites in multiple species. *Bioinformatics* **2019**, *35*, 1326–1333. [CrossRef]
11. Schweizer, H.P. Bacterial genetics: Past achievements, present state of the field, and future challenges. *Biotechniques* **2008**, *44*, 633–641. [CrossRef]

12. Suzuki, M.M.; Bird, A. DNA methylation landscapes: Provocative insights from epigenomics. *Nat. Rev. Genet.* **2008**, *9*, 465–476. [CrossRef]
13. Robertson, K.D. DNA methylation and human disease. *Nat. Rev. Genet.* **2005**, *6*, 597–610. [CrossRef] [PubMed]
14. Jones, P.A. Functions of DNA methylation: Islands, start sites, gene bodies and beyond. *Nat. Rev. Genet.* **2012**, *13*, 484–492. [CrossRef] [PubMed]
15. Yao, B.; Jin, P. Cytosine modifications in neurodevelopment and diseases. *Cell. Mol. Life Sci.* **2014**, *71*, 405–418. [CrossRef] [PubMed]
16. Ling, C.; Groop, L. Epigenetics: A molecular link between environmental factors and type 2 diabetes. *Diabetes* **2009**, *58*, 2718–2725. [CrossRef] [PubMed]
17. Chen, K.; Zhao, B.S.; He, C. Nucleic acid modifications in regulation of gene expression. *Cell Chem. Biol.* **2016**, *23*, 74–85. [CrossRef]
18. Doherty, R.; Couldrey, C. Exploring genome wide bisulfite sequencing for DNA methylation analysis in livestock: A technical assessment. *Front. Genet.* **2014**, *5*, 126. [CrossRef]
19. Flusberg, B.A.; Webster, D.R.; Lee, J.H.; Travers, K.J.; Olivares, E.C.; Clark, T.A.; Korlach, J.; Turner, S.W. Direct detection of DNA methylation during single-molecule, real-time sequencing. *Nat. Methods* **2010**, *7*, 461. [CrossRef]
20. Boch, J.; Bonas, U. Xanthomonas AvrBs3 family-type III effectors: Discovery and function. *Annu. Rev. Phytopathol.* **2010**, *48*, 419–436. [CrossRef]
21. Buryanov, Y.I.; Shevchuk, T. DNA methyltransferases and structural-functional specificity of eukaryotic DNA modification. *Biochemistry* **2005**, *70*, 730–742. [CrossRef]
22. Liu, Q.; Chen, J.; Wang, Y.; Li, S.; Jia, C.; Song, J.; Li, F. DeepTorrent: A deep learning-based approach for predicting DNA N4-methylcytosine sites. *Brief. Bioinform.* **2020**. [CrossRef] [PubMed]
23. Khanal, J.; Nazari, I.; Tayara, H.; Chong, K.T. 4mCCNN: Identification of N4-methylcytosine sites in prokaryotes using convolutional neural network. *IEEE Access* **2019**, *7*, 145455–145461. [CrossRef]
24. Hasan, M.M.; Manavalan, B.; Khatun, M.S.; Kurata, H. i4mC-ROSE, a bioinformatics tool for the identification of DNA N4-methylcytosine sites in the Rosaceae genome. *Int. J. Biol. Macromol.* **2019**, *157*, 752–758. [CrossRef] [PubMed]
25. Edger, P.P.; VanBuren, R.; Colle, M.; Poorten, T.J.; Wai, C.M.; Niederhuth, C.E.; Alger, E.I.; Ou, S.; Acharya, C.B.; Wang, J.; et al. Single-molecule sequencing and optical mapping yields an improved genome of woodland strawberry (Fragaria vesca) with chromosome-scale contiguity. *Gigascience* **2018**, *7*, gix124. [CrossRef]
26. Raymond, O.; Gouzy, J.; Just, J.; Badouin, H.; Verdenaud, M.; Lemainque, A.; Vergne, P.; Moja, S.; Choisne, N.; Pont, C.; et al. The Rosa genome provides new insights into the domestication of modern roses. *Nat. Genet.* **2018**, *50*, 772–777. [CrossRef] [PubMed]
27. Zeng, F.; Fang, G.; Yao, L. A deep neural network for identifying DNA N4-methylcytosine sites. *Front. Genet.* **2020**, *11*, 209. [CrossRef]
28. Fu, J.; Tang, J.; Wang, Y.; Cui, X.; Yang, Q.; Hong, J.; Li, X.; Li, S.; Chen, Y.; Xue, W.; et al. Discovery of the consistently well-performed analysis chain for SWATH-MS based pharmacoproteomic quantification. *Front. Pharmacol.* **2018**, *9*, 681. [CrossRef]
29. He, W.; Jia, C.; Zou, Q. 4mCPred: Machine learning methods for DNA N4-methylcytosine sites prediction. *Bioinformatics* **2019**, *35*, 593–601. [CrossRef]
30. Hao, L.; Dao, F.Y.; Guan, Z.X.; Zhang, D.; Tan, J.X.; Zhang, Y.; Chen, W.; Lin, H. iDNA6mA-Rice: A computational tool for detecting N6-methyladenine sites in rice. *Front. Genet.* **2019**, *10*, 793.
31. Chen, Z.; Zhao, P.; Li, F.; Marquez-Lago, T.T.; Leier, A.; Revote, J.; Zhu, Y.; Powell, D.R.; Akutsu, T.; Webb, G.I.; et al. iLearn: An integrated platform and meta-learner for feature engineering, machine-learning analysis and modeling of DNA, RNA and protein sequence data. *Brief. Bioinform.* **2020**, *21*, 1047–1057. [CrossRef]
32. Feng, P.; Yang, H.; Ding, H.; Lin, H.; Chen, W.; Chou, K.C. iDNA6mA-PseKNC: Identifying DNA N6-methyladenosine sites by incorporating nucleotide physicochemical properties into PseKNC. *Genomics* **2019** *111*, 96–102. [CrossRef] [PubMed]

33. Nanni, L.; Lumini, A. An ensemble of K-local hyperplanes for predicting protein–protein interactions. *Bioinformatics* **2006**, *22*, 1207–1210. [CrossRef] [PubMed]
34. Cao, J.; Xiong, L. Protein sequence classification with improved extreme learning machine algorithms. *BioMed Res. Int.* **2014**, *2014*, 103054 . [CrossRef] [PubMed]
35. Raimondi, D.; Orlando, G.; Tabaro, F.; Lenaerts, T.; Rooman, M.; Moreau, Y.; Vranken, W.F. Large-scale in-silico statistical mutagenesis analysis sheds light on the deleteriousness landscape of the human proteome. *Sci. Rep.* **2018**, *8*, 1–11. [CrossRef] [PubMed]
36. Simonyan, K.; Vedaldi, A.; Zisserman, A. Deep inside convolutional networks: Visualising image classification models and saliency maps. *arXiv* **2013**, arXiv:1312.6034.
37. Fu, L.; Niu, B.; Zhu, Z.; Wu, S.; Li, W. CD-HIT: Accelerated for clustering the next-generation sequencing data. *Bioinformatics* **2012**, *28*, 3150–3152. [CrossRef]
38. Wang, Y.; Yang, S.; Zhao, J.; Du, W.; Liang, Y.; Wang, C.; Zhou, F.; Tian, Y.; Ma, Q. Using machine learning to measure relatedness between genes: A multi-features model. *Sci. Rep.* **2019**, *9*, 1–15. [CrossRef]
39. Xu, Z.C.; Feng, P.M.; Yang, H.; Qiu, W.R.; Chen, W.; Lin, H. iRNAD: A computational tool for identifying D modification sites in RNA sequence. *Bioinformatics* **2019**, *35*, 4922–4929. [CrossRef]
40. Liu, B.; Liu, F.; Wang, X.; Chen, J.; Fang, L.; Chou, K.C. Pse-in-One: A web server for generating various modes of pseudo components of DNA, RNA, and protein sequences. *Nucleic Acids Res.* **2015**, *43*, W65–W71. [CrossRef]
41. He, J.; Fang, T.; Zhang, Z.; Huang, B.; Zhu, X.; Xiong, Y. PseUI: Pseudouridine sites identification based on RNA sequence information. *BMC Bioinform.* **2018**, *19*, 306. [CrossRef]
42. Su, Z.D.; Huang, Y.; Zhang, Z.Y.; Zhao, Y.W.; Wang, D.; Chen, W.; Chou, K.C.; Lin, H. iLoc-lncRNA: Predict the subcellular location of lncRNAs by incorporating octamer composition into general PseKNC. *Bioinformatics* **2018**, *34*, 4196–4204. [CrossRef] [PubMed]
43. Zhu, X.; He, J.; Zhao, S.; Tao, W.; Xiong, Y.; Bi, S. A comprehensive comparison and analysis of computational predictors for RNA N6-methyladenosine sites of Saccharomyces cerevisiae. *Brief. Funct. Genom.* **2019**, *18*, 367–376. [CrossRef] [PubMed]
44. Pan, Y.; Wang, Z.; Zhan, W.; Deng, L. Computational identification of binding energy hot spots in protein–rna complexes using an ensemble approach. *Bioinformatics* **2018**, *34*, 1473–1480. [CrossRef] [PubMed]
45. Wei, L.; Chen, H.; Su, R. M6APred-EL: A sequence-based predictor for identifying N6-methyladenosine sites using ensemble learning. *Mol. Ther.-Nucleic Acids* **2018**, *12*, 635–644. [CrossRef]
46. Xue, W.; Yang, F.; Wang, P.; Zheng, G.; Chen, Y.; Yao, X.; Zhu, F. What contributes to serotonin–norepinephrine reuptake inhibitors' dual-targeting mechanism? The key role of transmembrane domain 6 in human serotonin and norepinephrine transporters revealed by molecular dynamics simulation. *ACS Chem. Neurosci.* **2018**, *9*, 1128–1140. [CrossRef]
47. Tan, J.X.; Lv, H.; Wang, F.; Dao, F.Y.; Chen, W.; Ding, H. A survey for predicting enzyme family classes using machine learning methods. *Curr. Drug Targets* **2019**, *20*, 540–550. [CrossRef]
48. Yang, H.; Yang, W.; Dao, F.Y.; Lv, H.; Ding, H.; Chen, W.; Lin, H. A comparison and assessment of computational method for identifying recombination hotspots in Saccharomyces cerevisiae. *Brief. Bioinform.* **2019**. [CrossRef]
49. Zhou, Y.; Zeng, P.; Li, Y.H.; Zhang, Z.; Cui, Q. SRAMP: Prediction of mammalian N6-methyladenosine (m6A) sites based on sequence-derived features. *Nucleic Acids Res.* **2016**, *44*, e91. [CrossRef]
50. Jeong, B.S.; Bari, A.G.; Reaz, M.R.; Jeon, S.; Lim, C.G.; Choi, H.J. Codon-based encoding for DNA sequence analysis. *Methods* **2014**, *67*, 373–379. [CrossRef]
51. Cerf, N.J.; Adami, C. Information theory of quantum entanglement and measurement. *Phys. D Nonlinear Phenom.* **1998**, *120*, 62–81. [CrossRef]
52. Pan, G.; Jiang, L.; Tang, J.; Guo, F. A novel computational method for detecting DNA methylation sites with DNA sequence information and physicochemical properties. *Int. J. Mol. Sci.* **2018**, *19*, 511. [CrossRef] [PubMed]
53. ur Rehman, M.; Khan, S.H.; Rizvi, S.D.; Abbas, Z.; Zafar, A. Classification of skin lesion by interference of segmentation and convolotion neural network. In Proceedings of the 2018 2nd International Conference on Engineering Innovation (ICEI), Bangkok, Thailand, 5–6 July 2018; pp. 81–85.

54. Khan, S.H.; Abbas, Z.; Rizvi, S.D. Classification of Diabetic Retinopathy Images Based on Customised CNN Architecture. In Proceedings of the 2019 Amity International Conference on Artificial Intelligence (AICAI), Dubai, UAE, 4–6 February 2019; pp. 244–248.
55. Nizami, I.F.; Majid, M.; ur Rehman, M.; Anwar, S.M.; Nasim, A.; Khurshid, K. No-reference image quality assessment using bag-of-features with feature selection. *Multimed. Tools Appl.* **2020**, *79*, 7811–7836. [CrossRef]
56. Ilyas, T.; Khan, A.; Umraiz, M.; Kim, H. SEEK: A Framework of Superpixel Learning with CNN Features for Unsupervised Segmentation. *Electronics* **2020**, *9*, 383. [CrossRef]
57. Wahab, A.; Ali, S.D.; Tayara, H.; Chong, K.T. iIM-CNN: Intelligent identifier of 6mA sites on different species by using convolution neural network. *IEEE Access* **2019**, *7*, 178577–178583. [CrossRef]
58. Mahmoudi, O.; Wahab, A.; Chong, K.T. iMethyl-Deep: N6 Methyladenosine Identification of Yeast Genome with Automatic Feature Extraction Technique by Using Deep Learning Algorithm. *Genes* **2020**, *11*, 529. [CrossRef]
59. Chollet, F. Keras: Deep learning library for theano and tensorflow. *Io/k* **2015**, *7*, T1.
60. Khanal, J.; Tayara, H.; Chong, K.T. Identifying Enhancers and Their Strength by the Integration of Word Embedding and Convolution Neural Network. *IEEE Access* **2020**, *8*, 58369–58376. [CrossRef]
61. Tayara, H.; Chong, K.T. Improving the quantification of DNA sequences using evolutionary information based on deep learning. *Cells* **2019**, *8*, 1635. [CrossRef]
62. Tahir, M.; Tayara, H.; Chong, K.T. Convolutional neural networks for discrimination of RNA pseudouridine sites. *IBRO Rep.* **2019**, *6*, S552. [CrossRef]
63. Zou, J.; Huss, M.; Abid, A.; Mohammadi, P.; Torkamani, A.; Telenti, A. A primer on deep learning in genomics. *Nat. Genet.* **2019**, *51*, 12–18. [CrossRef]
64. McCafferty, C.L.; Sergeev, Y.V. Global computational mutagenesis provides a critical stability framework in protein structures. *PLoS ONE* **2017**, *12*, e0189064. [CrossRef] [PubMed]
65. Lanchantin, J.; Singh, R.; Wang, B.; Qi, Y. Deep motif dashboard: Visualizing and understanding genomic sequences using deep neural networks. In *Pacific Symposium on Biocomputing 2017*; World Scientific: Singapore, 2017; pp. 254–265.
66. Baehrens, D.; Schroeter, T.; Harmeling, S.; Kawanabe, M.; Hansen, K.; Müller, K.R. How to explain individual classification decisions. *J. Mach. Learn. Res.* **2010**, *11*, 1803–1831.

2

How 'Protein-Docking' Translates into the New Emerging Field of Docking Small Molecules to Nucleic Acids?

Francesca Tessaro [1,2] and **Leonardo Scapozza** [1,2,*]

1 Pharmaceutical Biochemistry, School of Pharmaceutical Sciences, University of Geneva CMU, Rue Michel-Servet 1, 1211 Geneva 4, Switzerland; francesca.tessaro@unige.ch

2 Institute of Pharmaceutical Sciences of Western Switzerland, University of Geneva, 1211 Geneva, Switzerland

* Correspondence: leonardo.scapozza@unige.ch

Abstract: In this review, we retraced the '40-year evolution' of molecular docking algorithms. Over the course of the years, their development allowed to progress from the so-called 'rigid-docking' searching methods to the more sophisticated 'semi-flexible' and 'flexible docking' algorithms. Together with the advancement of computing architecture and power, molecular docking's applications also exponentially increased, from a single-ligand binding calculation to large screening and polypharmacology profiles. Recently targeting nucleic acids with small molecules has emerged as a valuable therapeutic strategy especially for cancer treatment, along with bacterial and viral infections. For example, therapeutic intervention at the mRNA level allows to overcome the problematic of undruggable proteins without modifying the genome. Despite the promising therapeutic potential of nucleic acids, molecular docking programs have been optimized mostly for proteins. Here, we have analyzed literature data on nucleic acid to benchmark some of the widely used docking programs. Finally, the comparison between proteins and nucleic acid targets docking highlighted similarity and differences, which are intrinsically related to their chemical and structural nature.

Keywords: molecular docking; nucleic acids; structure-based computational methods

---

1. Introduction

Nowadays computational methods are routinely applied in drug discovery campaigns and its positive impact in research is recognized in the scientific community. Computer-aided drug discovery (CADD) raised in the years 1970s and its continuous progress, together with the advancement in computer technologies, made its usage an indispensable tool. CADD methods are usually divided into two main groups: ligand-based and structure-based. The first approach builds structure-activity models based on ligand structure information, whereas the second method takes advantage of the three-dimensional structure of the biomolecular target to investigate ligand binding. Among the structure-based methods, molecular docking is a well-known tool for the identification of ligand-target recognition. In this review, we will explore the evolution of molecular docking from a single ligand-protein calculation to more recent and sophisticated applications with a particular emphasis on docking applied to nucleic acids (NA) as emerging therapeutic targets.

2. Nature of the Biomolecular Target for Structure-Based Methods

2.1. Experimentally Determined Structure

The protein data bank (PDB) (https://www.rcsb.org/) is a repository of experimental macromolecular structures, which is constantly updated. By May 2020, the PDB counted ~163'633 experimental structures of proteins, RNAs, and DNAs belonging to different organisms. The majority of the structures are solved by X-ray crystallography (~145'437) (https://www.rcsb.org/stats/summary) but also nuclear magnetic resonance (NMR) and electron microscopy (EM) techniques strongly contribute in increasing the collection. The introduction of novel protocols and methods in x-ray crystallography enormously ameliorate structure resolution if compared to the first solved myoglobin structure in 1960 [1]. Nevertheless, we are lacking structural information for many biological macromolecules. Other techniques like NMR are preferred for small systems, particularly suitable for nucleic acids as small RNA or DNA sequences where structure dynamic information might result to be crucial for the investigation of the biological event. However, Cryo-EM technology has evolved from the method for studying relatively large systems as cells organelles or macromolecular complexes but with a substantial loss on resolution, into the most recent high resolution Cryo-EM allowing to reach high resolution (better than 3 Å) for structures as small as 64 kDa [2]. This method also has the advantage of using a smaller amount of material compared to X-ray or NMR techniques.

2.2. Computational Methods for Structure Prediction

The PDB represents the largest source of structural data of biomolecules suitable for structure-based computational tools, and indeed for molecular docking. Despite the increasing number of deposited structures in the PDB, we are far from having structural information for all biological relevant proteins. The number of non-redundant protein sequences is estimated to be 152 million from the National Centre for Biotechnology Information (https://www.ncbi.nlm.nih.gov/refseq/). This represents an enormous gap between sequence annotation and available 3D structures. In addition, the rate at which new protein sequences are defined far exceeds the rate at which protein structures are experimentally determined. Hence, in absence of experimental structures, computational predictions emerged as a valuable alternative for structure-based studies. These computational methods can be divided into two main categories: 'template-based' and 'free-modelling'. Among them, template-based methods, homology modelling in particular, results in being the most accurate for binding site identification and drug design purposes [3]. It has been estimated that with the use of homology representation, the human proteome can reach 70% of its coverage at ≥ 30% of sequence identity, which represent around 95% of the human drug targets used for therapeutic intervention [4].

Homology modelling (or comparative modelling) consists in building target models based on sequence similarity with a homologous template, which presents a known 3D structure. The general assumption of this methodology is that evolutionary related biomolecules with a similar sequence tend to have also similar structure. Normally, in the case of proteins, sequence identity with the template greater than ~30% is considered the threshold for a model suitable for structure-based studies, whereas for drug design purpose a more stringent 50% identity is required for a reliable model [5]. Sequence alignment and model validation are both crucial steps in obtaining a reliable model. Ramachandran plot, for example, geometrically evaluates backbone outliers reporting the torsion angles, ϕ and ψ, which should reside in the allowed region of the graph [1].

Homology modelling has been used for many applications in structure-based studies from target identification, site-directed mutagenesis, binding site inspection to druggability analysis. Structural models obtained by homology represents the starting point for other computational methods notably molecular docking and dynamic simulations. One example where homology models were used as a starting source of information is represented by a structure-based investigation of a novel class of aspartic proteases: *Toxoplasma gondii* Aspartyl protease 3 and *Plasmodium Falciparum* Plasmepsin IX/X, where with the application of molecular docking we could unravel the atomistic mode of action of a nanomolar inhibitor (49c) [6].

3. "40 Years of Protein Docking Highlights and Advancements"

After 40 years since the first application of docking, huge advancements have been made both in searching algorithms and in scoring functions, allowing its usage in a broad-range of biomolecular systems. Here, we want to highlights some of the important milestones, which contributed to the history and development of molecular docking theory and applications.

3.1. *An Evolutionary Perspective: From Rigid to Flexible Docking*

Along the course of the years, ameliorations of the docking performances mainly focused on balancing the dual aspect of predicted pose correctness vs. speed of the searching algorithm required for the calculation. The development of docking algorithms was also favored by the introduction of modern computers and parallelization, which determined upscale improvements on the speed of the calculations. Complexity of the algorithms can be classified accordingly to the number of degrees of freedom they neglect (Figure 1).

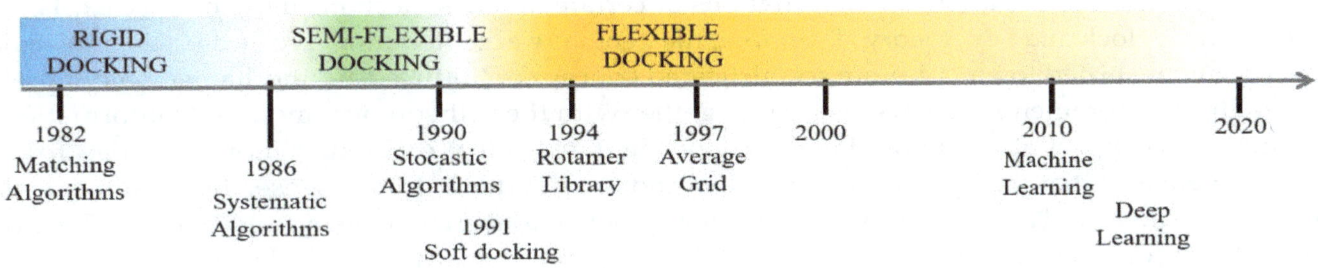

Figure 1. Evolutionary timeline highlighting key events contributing to the development of molecular docking algorithms, starting from rigid to flexible docking.

In 1982, DOCK program [7] was introduced and designed to find molecules matching by shape complementarity. Molecular and geometrical shape algorithms also called matching algorithms were firstly applied in rigid-docking calculations: where the two molecules treated as rigid bodies, explore only the six translational and rotational degrees of freedom. In the matching algorithms, the binding site is represented as a collection of spheres of different radii, which allow mapping the ligand to the sphere centers and finding matching sets [7]. Rigid docking has an appreciable speed and shows to be very effective in database exploration [8]. However, one of the major drawbacks of rigid docking is the fact of not considering ligand-flexibility and on relying only on pre-calculated conformations of the ligands. This method refers to the primordial concept of molecular recognition: the 'lock and key' model introduced by Fischer [9]. Because of the limited degree of freedom, virtually reducing the complexity of the system, rigid docking was successfully applied also for protein-protein and protein-peptide docking [10].

Ligand conformational freedom starts to be introduced with ligand-flexible docking methods, also referred as semi-flexible docking. Systematic algorithms, for example, generate all possible ligand conformations in order to find the best ligand-binding match. Fragmentations and incremental construction (IC) methods belong to the stochastic algorithms. With this algorithm each ligand is divided in small fragments and then rigidly docked into the binding site [11]. Only when the anchor fragment is established the rest of the molecule is joined and energy minimized in the binding site. This method works well with small to medium sized molecules, while with bigger molecules the number of generated fragments increase and thus might create difficulties in pose prediction. On the one hand, a systematic search of the ligand conformation is more accurate, on the other, it requires high computational resources and time. For instance, stochastic algorithms are a good trade-off between speed and chance to obtain the correct binding pose. Among them, Monte Carlo algorithm is a well-known stochastic method: at each iteration the ligand goes through a random bond-rotation, rigid-body translation or rotation. The binding pose is evaluated based on molecular-mechanic energy calculation and then the standard Metropolis Monte Carlo method based on the Boltzmann constant is applied to accept or reject the pose for the next iteration. A particular advantage of Monte Carlo method is the possibility to explore diverse energetic local minima conformations using higher temperature values, and therefore overcome the problem of the energy barriers [12]. Tabu method was introduced few years later with the aim of improving Monte Carlo conformational search while preventing the exploration of already sampled zones.

Genetic algorithms [13,14] also belong to the stochastic methods and they took inspiration from the biological evolution of the Darwin's theory. Each gene represents a degree of freedom of the ligand and thus the chromosomes describe the ligand conformations. As happens in nature, mutations and crossovers occur and define new populations of conformers to which a fitness score is associated. Only the docking pose with good scoring will be selected and moved to the next iterations.

Phenomena such as allosteric modulations or certain kinetics mechanisms cannot be explained with the early 'lock and key' theory. Therefore, the generally accepted model regarding the molecular binding event shifted toward a more sophisticated concept of 'induced-fit' mechanism, more prone to explain the aforementioned events [15]. Together with these theoretical models, 'conformational ensemble' theory was also proposed [16], supporting the idea that different conformers exists for the same protein and that the ligand preferentially binds one of them. In this sense, the ideal molecular docking engine would allow a conformational degree of freedom on both ligand and protein. Therefore, accounting for protein flexibility is fundamental for explaining many biological processes around the ligand-binding event [17]. Molecular dynamic simulations in this sense can perfectly describe the dynamic behavior of a macromolecular system as function of time. However, the major drawback on its application is the computational cost required for analyzing in an exhaustive way the energetic landscape of the systems. Indeed, for large systems the energetic barriers that separate the possible binding modes might have results too high to be overcome, which might lead to an inadequate conformational sampling.

Before the advent of molecular dynamic, various methods tried to address the protein flexibility. For example, 'soft docking' [18] allows a partial overlap of the ligand-protein atoms by decreasing the van der Waals repulsion energy term employed by the scoring function. On one hand, this method results in being computationally very efficient and suitable for testing large libraries, on the other hand protein atoms are roughly fixed, approximating small movements. Just a few year later, Leach [19] introduced rotamer libraries composed by alternative conformations of amino acids side-chains. Here, the use of rotamer libraries speeds up the sampling search and allows avoiding ligand-pose minimization, which makes this approach appealing for its velocity. However, all these strategies considered a single protein conformation. Thus, large movements with the exception of side chains rotations are omitted.

The 'conformational ensemble' docking approach recalls the homonymous theory [16], where the target exists as a collection of conformers derived by molecular dynamics or Monte Carlo simulations. Based on this theory, the ligand binding event can occur either by recognizing its preferred protein conformation or by a mutual conformational rearrangement accommodating the ligand in its binding site as induce-fit effect [15]. An alternative way to consider the ensemble of conformers has been introduced by Knegtel [20], where an energy and geometry weighted average grid is used for describing the receptor binding site. This approach has been further improved not by using an average grid but by merging the conserved moieties while creating alternative conformations for the remaining parts [21]. Today the awareness in the dynamic behavior of proteins is widely accepted and embraced by the scientific community. Sampling strategies for an exhaustive search of protein conformations have been explored during the course of the years [22,23]. Finally, the last decade of molecular docking advancements has seen the powerful integration of machine-learning methods, and even more with deep learning, which hold huge promises for computational drug discovery research [24].

3.2. *Molecular Docking as a Crossing Tool for Multiple Scopes*

From its original application molecular docking was conceived as in silico tool for investigating target-ligand interaction especially in support of drug discovery campaigns either in the screening phase, de novo design or ligand optimization. Advancements in either algorithms, scoring functions, computer architectures (GPUs), parallelization, and artificial intelligence contribute to reconfirm the crucial role of molecular docking in science (Figure 2) [25]. Indeed, it is now employed and integrated in a variety of discovery tasks not only for large library screening, but also as reverse docking for target profiling in drug repositioning, polypharmacology, and prediction of adverse effects [25,26], as shown in Figure 2.

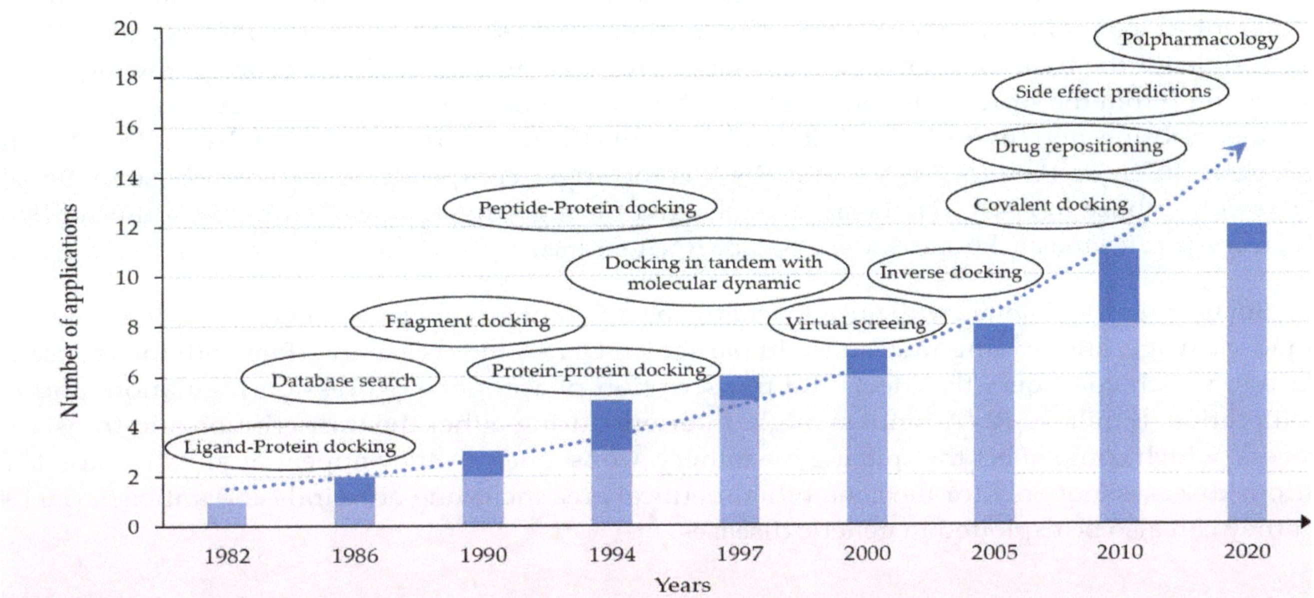

Figure 2. Increase of molecular docking applications. The dark blue compartment shows the new applications that are summed up with the existing ones.

Recent works highlight how large screening protocols involving millions of compounds will positively influence early drug discovery programs and will allow the identification of novel chemotypes belonging to their relative target [27]. The trustworthiness of molecular docking depends also on the accuracy of the used scoring function. Recent reviews have described the strengths and weaknesses of the classical scoring function as well as the more recent advancement in scoring function development [28,29]. Here we would like to highlight that artificial intelligence and statistical analysis emerged as tools to improve binding affinity and scoring functions taking advantage of the ever-growing publicly available databases [30]. One of latest applications is covalent docking or 'reactive docking', where results are very encouraging despite the challenging task of handling the ligand bond formation often using classical molecular mechanics force fields [31–33]. In this sense, new methodologies and scoring functions need to be developed to ameliorate the scoring accuracy without affecting the computational cost and hindering its application for large screening [34]. Moreover, as mentioned above, one of the major drawbacks of molecular docking is the reduced ability to handle protein flexibility. Even with the implementation of several strategies [18–20] it does not allow extended conformational changes of the protein structure especially at the backbone level. However, to overcome these limitations, docking is often integrated with molecular dynamics and thermodynamic calculations for both exploring the conformational ensemble and for further pose refinement [22,23,35,36]. Finally, molecular docking does not apply only to small molecules recognizing proteins but rather its broad usage allows to dock different types of macromolecules as protein-protein, peptide-protein, and not ultimately nucleic acids.

4. Nucleic Acids as Emerging Therapeutic Targets

Just 1.5% of the human genome encodes for proteins, only a 0.05% has been successfully drugged with existing small molecules [ ]. If the latest number is considered, it appears clear that limitations on the chemical coverage within the human genome relies mainly on the difficulty to target the so-called 'undruggable' proteins. Disease-associated proteins represents a small fraction of the whole human genome. Therefore, if we consider that nucleic acids can be potentially modulated by small molecules then the therapeutic landscape of macromolecular targets will exponentially increase, as exemplify by the consistent fraction (70%) of the human genome transcribed into non-coding RNA [4]. DNA and RNA play essential roles in many biological processes and represent an important class of drug targets [37]. They can be distinguished by their nucleosides and sugar composition, which results into a different functional role. Indeed, while DNA is long-term storage of genetic information, RNA is responsible for the transfer of genetic information from the nucleus to the ribosome allowing protein production. From the structural point of view, DNA is composed by four nucleotide bases: adenosine, guanine, cytosine, and thymidine; while RNA presents the uracil base instead of thymidine. Another structural difference between DNA and RNA is the sugar composition: 2-deoxyribose in DNA is replaced by ribose in RNA. Each base is connected through their sugar by a phosphodiester bond, which constitutes the backbone of the DNA or RNA strand.

Small molecules interact with nucleic acids using different mechanisms: intercalation, cross-linkage, strand-cleavage, and reading-molecules. In particular, DNA-binders can interfere with the replication process, which consequently affects the transcription phase, gene expression regulation, and cell proliferation. Similarly, RNA-binders might interfere during either the transcription and translation process, which could affect the splicing machinery. Consequently, they appear as suitable targets for various diseases, not only for the chemotherapeutical area including antivirals and antibacterial [38], but they can also be exploited in genetic diseases.

4.1. DNA-Targeting for Cancer and Antimicrobial Therapy

DNA is considered as one of the main molecular targets in chemotherapy [39]. The discovery of first alkylating agents go back to chemical warfare during World War II, which spawned the modern era of cancer therapy [40]. Unfortunately, its non-specific interaction with the DNA cause off-target effects and high toxicity, partially limiting its use in clinic today. Also, the majority of first DNA-binders were causing double strands break which were beneficial for blocking the tumor progression, but a as drawback, this damage could lead to novel mutations and cause secondary tumors, which would appear 10–15 years later. Indeed, much more effort has been put into the development of DNA-binders without genotoxic effects. This resulted in molecules classified as groove binders, intercalators, and covalent binders. One notable example of DNA target is the oncogene c-Myc, which is over-activated in many types of cancer, the stabilization of its G-quadruplex structure inhibits the telomerase activity [41] (Figure 3A). The blockage of cell proliferation has raised interest also as therapeutic strategy for anti-bacterial treatment. Today, the potential of using DNA as a target to overcome bacterial resistance is underestimated; different strategies to increase the therapeutic index are used as the increase of the selective uptake of the drug by the microbes or designing compounds with sequence selectivity [42]. Examples of small molecules binding DNA for cancer and anti-infective therapy are reviewed in other works [39,40,42–46].

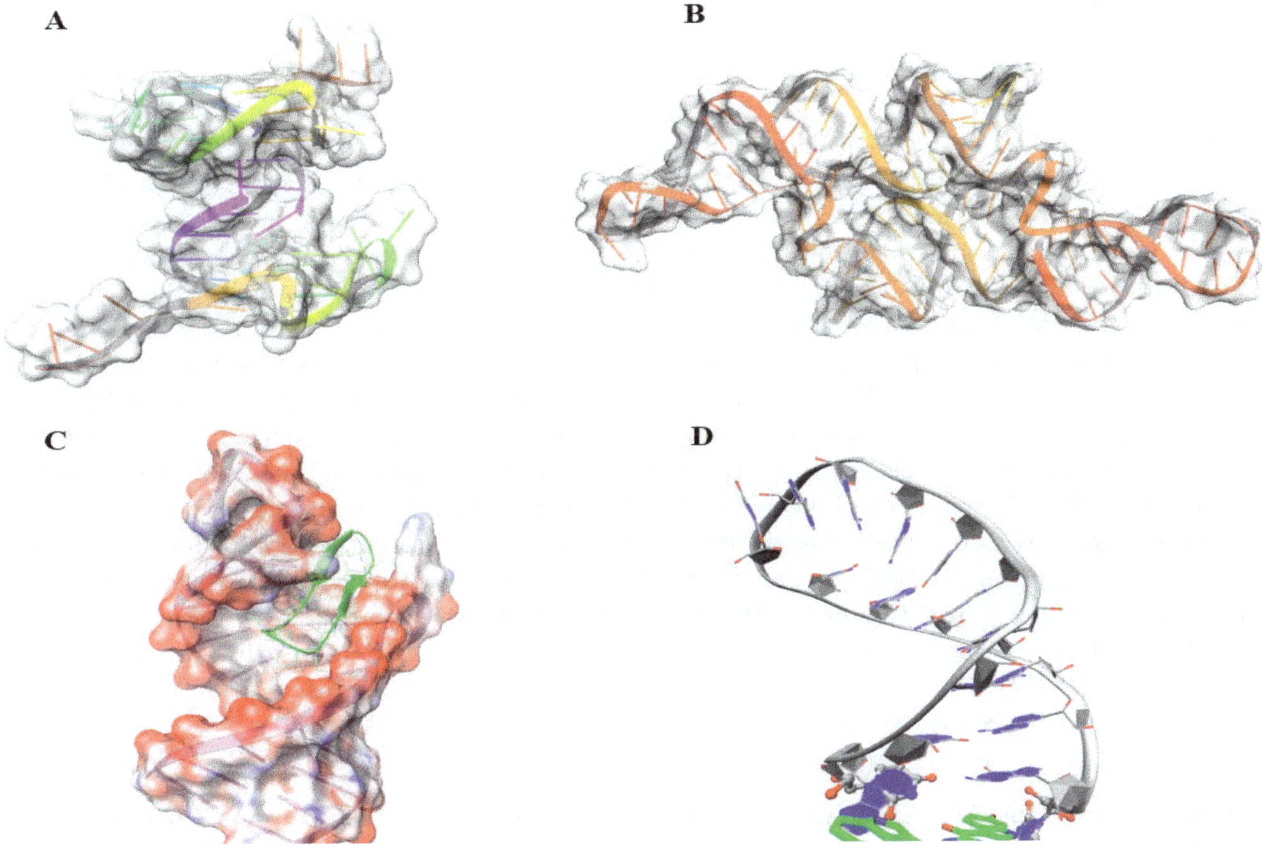

Figure 3. Example of nucleic acid targets. (**A**) Structure of the G-quadruplex oncogene c-Myc shown with transparent surface and backbone as cartoon style (PDB code: 6AU4). (**B**) Example of riboswitch from Glutamine II Riboswitch (PDB code: 6QN3), represented with transparent surface and the backbone as cartoon style. (**C**) Representation of the HIV trans-activation response (TAR) RNA bound to a high affinity macrocycle (PDB code: 6D2U). The electrostatic surface highlights the charge distribution, whereas the ligand bound is shown in licorice (green). (**D**) Structure of TSL2 hairpin bound with the small molecule splicing modulator PK4C9 (green). The RNA structure is shown in grey cartoon style and the ligand as licorice [47].

4.2. RNA as Antiviral and Antibacterial Target

The transcription and translation process can be regulated also at the RNA level and its targeting has appeared to be a suitable approach for anti-bacterial and anti-viral therapy. Riboswitches are a regulatory part of the mRNA, which control gene expression and consequently protein biosynthesis (Figure 3B). Natural ligands binding to riboswitches modulate protein gene products by upregulating or downregulating the translation process or by altering mRNA stability. The therapeutic potential of this regulatory mRNA has been widely explored as an anti-bacterial target, especially because they show high selectivity for their bound-ligand [48]. The ribosomal RNA (rRNA) is another validated target for anti-bacterial therapy, which is directly involved in the translation process. It is composed by two subunits: 30S is responsible of the correct position and reading of the tRNA and subunit 50S executing the peptide linkage. Aminoglycosides are a known class of antibiotic inducing mistranslation of the polypeptide chain and thus interfering with protein biosynthesis [49]. RNA-targeting has also raised interest. HIV trans-activation response (TAR) RNA is considered as a valid target for HIV infection and can bind macrocycle molecules that have anti-HIV activity (Figure 3C) [50].

4.3. mRNA Triggering Splicing Machinery

The splicing event consists in introns removal from the pre-mRNA sequence generating the mature form of the mRNA, which will then constitute the template for protein translation. The spliceosome machinery is composed by multiple proteins specifically recognizing RNA motifs. A sophisticated mechanism of interactions among the different constituents precisely rules this process [51]. Indeed, any mutations or errors during this transformation might lead to malfunction that is directly translated at a protein level (i.e., truncated proteins, misfolding, over or down expressions). Targeting splicing is an astute intervention because it occurs at the early stage of the gene expression without altering the genome [52]. Many strategies have now been developed for splicing modifications, which are used to increase specific alternative spliced isoforms or to correct aberrant gene expression caused by gene mutation altering the splicing [52]. For example, spinal muscular atrophy (SMA), a life-threatening neurodegenerative disease, is the most common genetic cause of infant mortality. Mutations at the survival motor neuron 1 (SMN1) gene directly affect the production levels of the SMN protein. A second copy of the gene, SMN2, exists, but because of a different alternative splicing pattern, it generates low SMN protein levels, which correlate with the severity of the disease. It has been demonstrated that triggering the splicing of the SMN2 gene determines exon7 (E7) inclusion and restores the lacking levels of SMN protein. Thus, this is a valid strategy for SMA treatment [53–55]. Recently we have reported the first target-based study of small molecule splicing modifiers binding the RNA terminal stem loop 2 (TSL2) [47]. PK4C9 was identified as a promising TSL2-binder hit, able to modify the splicing and restore SMN protein levels in SMA patient cells (Figure 3D) [47].

5. Current 'Protein Docking Algorithms' Applied to Nucleic Acids: Challenges, Solutions, and Pitfalls

Since the time of the development of molecular docking, the majority of the therapeutic targets investigated were proteins, the refinement of the docking programs was focusing mainly on these. A number of docking programs were modified and retrained in order to be applied also to nucleic acids (NA), whereas others were newly designed. Huge advances have been made to ameliorate algorithm search and scoring functions including different aspects as target flexibility, ligand geometry, and solvent effect. However, molecular docking applied to nucleic acids presents its own set of challenges and certain features might sometimes differ from the ones used for protein docking.

5.1. Challenges of Nucleic Acid Docking

5.1.1. Structural and Active Site Features:

Similar to proteins, nucleic acids present their own set of geometrical and structural features, which characterize and distinguish them. DNA is mostly found as a double strand helix, for which three forms exists A, B, and Z. RNA can be found also as a double strand helix for which the A-form is the most present at physiological pH, whereas DNA privileges the B-form. This double helix forms two shallow cavities, the major and minor groove, which travel all along the length of the sequence. Interestingly, it has been seen that the DNA major groove is wider and accessible to large molecules as proteins and peptides. Whereas, DNA minor groove results in being narrow and more suitable to bind small molecules upon water displacements. Contrary, RNA major groove is more prone to interact with small molecules because highly electrostatic and narrower, while the RNA minor groove binds proteins via Van der Waals forces. Moreover, RNA has the characteristic to fold into different secondary structures as bulges, hairpin stem loops, G-quadruplex etc. Overall, similar to proteins, nucleic acids tertiary structures can be distinguished by specific and sophisticated geometrical motifs that form suitable drug-binding pockets able to recognized unique small molecules [56].

5.1.2. Charge Distribution Effect on Ligand-Nucleic Acids Interactions:

If from a structural point of view nucleic acids results in being suitable for target-based drug discovery, the high prevalence of charges and the metal ions poses a major challenge, especially when scoring the free energy of the ligand-NA binding. This is clearly evidenced by the comparison of the electrostatic potential surface of a single nucleic acid base with the one of a neutral or charged amino acid (Figure 4).

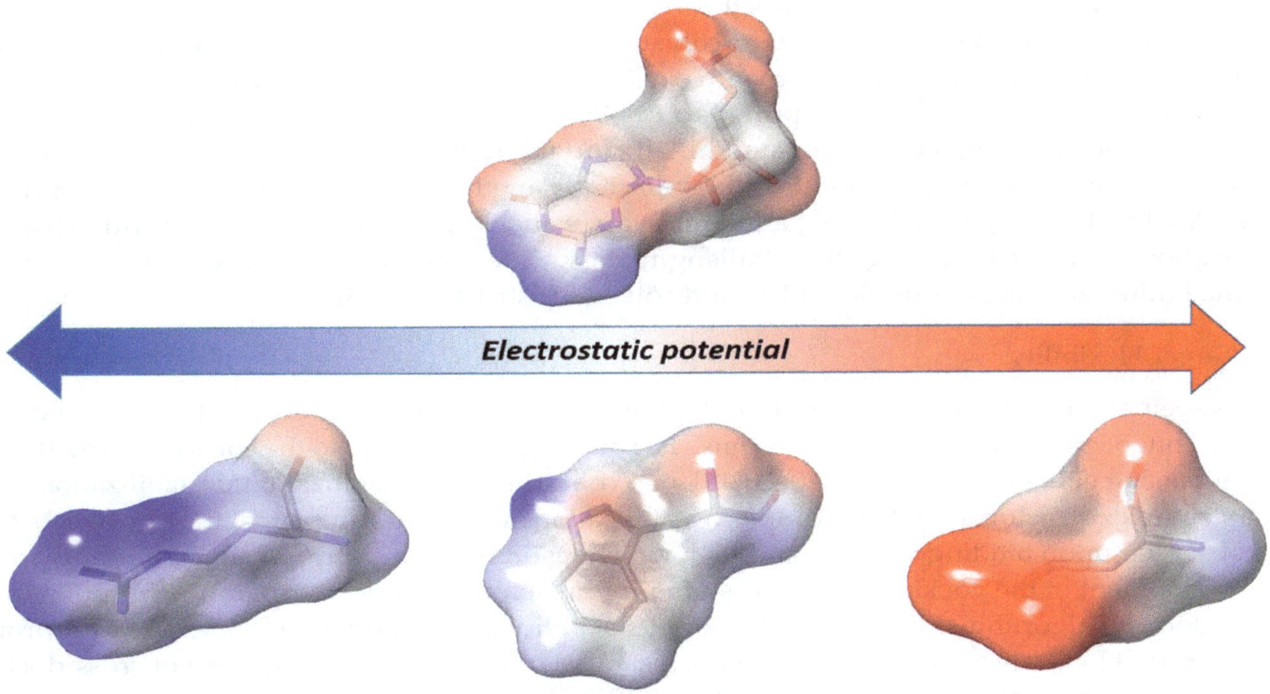

Figure 4. Comparison of the different charge distribution represented by electrostatic surface between polar amino acids (from left: arginine, tryptophan, and glutamate) and a nucleotide as guanosine monophosphate. The color ramp indicates in red the negative electrostatic potential energy values and in blue the positive values.

In a solution, the negative phosphates are normally neutralized by counterions as Na^+. Their mobile nature results indeed difficult to handle with molecular docking and often the strategy of choice is to add to the phosphate a +1 partial charge, implicitly considering the counterions [57]. The high prevalence of charges necessitates the development of dedicated scoring functions, able to describe unique ligand-NAs interactions. One of the major issues, when docking or screening positively charged ligands or amine-rich aminoglycosides is the ability to discriminate specific interactions, given the prevalence of backbone negative charges. Ribodock is an empirical scoring function developed for describing peculiar RNA-ligand interactions [58]. In particular, it accounts for a new term ($S_{posC\text{-}acc}$), which describes a peculiar ligand-RNA or protein-RNA interaction occurring between the positive charge of a guanidine moiety (i.e., arginine residue) and the partially negatively charged carbonyl oxygen counterpart present in the nucleobases (i.e., guanine). The nature of this interaction is still debated, as it might be considered electrostatic instead of π-stacking [59].

5.1.3. Metal Ions and Solvation:

The presence of the phosphate negative charges brings along a multitude of metal ions and water molecules strongly interacting to each other. Therefore, to disrupt the hydration shell ligands needs high polarity when considering the polyanionic character of the nucleic acid molecules. Site-bound ions together with waters are essential for stabilizing and shaping nucleic acids structure as shown for G-quadruplex metal-network [60] Analogous to proteins, water molecules play a major role in stabilizing the network of target-ligand interactions. Often, they are taking part in the ligand binding mechanisms. A particular challenge is represented by the water polarization effect caused by the frequent presence of charges and ions, which might reflect their effect at several layers of distances. Therefore, the role of water needs to be taken into consideration when advancing towards the ligand-optimization phase. A first step in this direction has been made in the Autodock program, where hydration is implemented as a new potential term for 'virtually' displaceable water molecules. This permits a substantial improvement in retrieving the native pose with RMSD values ranging below the 1.41 Å [59]. DOCK 6 introduces implicit solvation models, by means of the AMBER Generalized Born (GB) and Poisson Boltzmann (PB) in its solvation module describing the electrostatic contribution of waters and ions [61]. The combination with explicit water and counterions reaches accuracy values comparable to DrugScore^{RNA} when applied to RNA [61]. The latter is a knowledge-based, scoring function specifically designed to simulate RNA-ligand interactions. It is built on the most frequent interactions observed in crystal complexes [62]. Recently, methods like the 3D-RISM methods [60] or the SPLASH'EM scoring function [63] allowing waters and cations site-prediction during structure optimization aimed at addressing this challenging task. The accuracy of these methods still suffers from the limited number of nucleic acid-ligand solved 3D structures [63].

5.1.4. Target Flexibility

As seen already for proteins, a not-negligible aspect is the intrinsic flexibility of nucleic acids structure, often ignored by docking algorithms. The binding of the ligand to the corresponding RNA or DNA is causing either an induced-fit effect [15], or a conformational stabilization/destabilization of the tertiary complex structure [16]. For example, a stabilization of the conformation of the riboswitches has been observed upon binding of the ligand [64], which results in being trapped in the binding pocket. The development of the docking programs suited for nucleic acids follows a parallel evolution to the protein docking algorithms. In reply to the target-flexibility issue, similar strategies used for protein were proposed [65]. These strategies showed substantial improvement when different cross-docking scoring were applied [66]. In this context, the MORDOR program allows flexibility on both ligand and nucleic acids by applying molecular mechanics minimization with a restrained conformational search based on the X-ray or NMR experimental structure [67]. Despite the high accuracy, this program is not recommended for screening large libraries due to the elevated computing time needed. Full flexibility of both molecules has been proposed using a Monte Carlo algorithm for ab initio ligand-DNA docking without introducing any bias on prior binding site selection [68]. Monte Carlo variables define the conformations of both ligand and DNA relative to each other and to the internal flexibility of the two molecules [68]. Another strategy to overcome the flexibility issue during the binding process is to perform docking using a conformational target ensemble that intrinsically accounts for flexibility. Our study on the PK4C9 binding to the RNA TSL2 hairpin demonstrates the importance of considering two different conformational populations, respectively tri-loop and penta-loop, where the latter preferentially accommodate the ligand [47]. Elastic potential grids have been also introduced to address target flexibility in DrugScore^{RNA} scoring function [69], as previously seen for protein-ligand docking [70]. They consisted in adapting a pre-calculated 3D grid of potential field values of an initial RNA conformation to other conformations by moving the grid intersection points in space, while keeping the potential field values constant [69]. While the use of elastic potential grids allows to consider backbone and base motion simultaneously, movements as rotational flip motions and changing in the interaction types determine the key limitations of this approach. In addition, it does not allow to model highly negative charge pockets for accommodating positively charged ligands [69].

5.2. Benchmark of Docking Programs Applied to Nucleic Acids

Finally, we want to rationalize the advances and performances of docking programs applied to nucleic acids along the history, which refers to works retrieved from the literature [57,71–74]. However, we have to warn the reader that literature on docking programs applied to nucleic acids is less abundant compared to protein docking and because of the small data set used, sometimes, an appropriate statistical comparison results in being difficult. Indeed, our purpose is to emphasize the evolution and the ameliorations introduced in nucleic acids docking, rather than comparing different program's performances. A recapitulating chart is made with the intent of a figurative appreciation of the different studies (Figure 5).

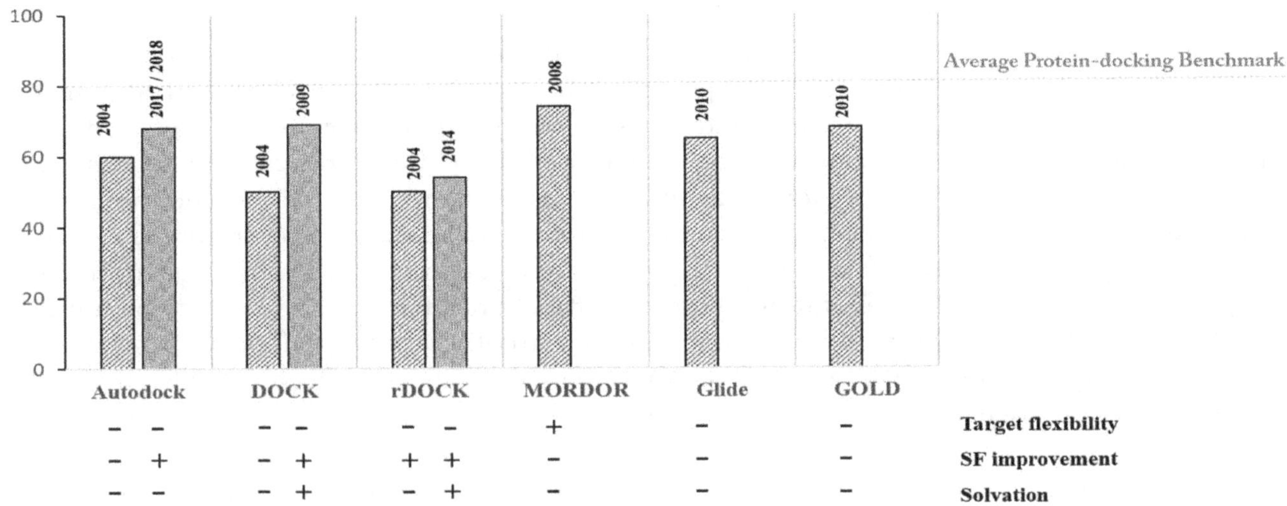

Figure 5. Benchmark comparison of some docking programs applied to or developed for nucleic acids. The average value of protein-docking benchmarks (80%), which has been retrieved from comparative assessments of published benchmarks [75–79], is used as indicator for comparison.

The first benchmark studies date back to 2004, the original DOCK and Autodock programs were tested against a dataset of 16 RNA complexes. Both programs perform well with rigid/aromatic ligands with an average accuracy (RMSD values < 2.5 Å compared to the experimental structure) reaching the 78% pose prediction accuracy (not reported in Figure 5). Autodock outperformed DOCK when working with weak binders and aminoglycosides. Considering the overall dataset, the accuracy of the two programs ranges between 50 to 60% pose prediction accuracy [57]. In the same period, Ribodock (rDOCK in the Figure 5), a novel scoring function developed for nucleic acids, was tested in a small dataset of 10 complexes. It was able to achieve 50% of success rate in predicting the correct binding mode within 2.0 Å RMSD [58]. A slight increase in accuracy (54%) was achieved later with improvements in the solvation term of the scoring function [78]. Nine aminoglycosides were successfully docked into hydrated and flexible RNAs using Autodock with some implementation aimed at considering solvation. These results underlie the importance of including key water molecules during the calculation and the use of ensembles of conformations to assess target flexibility [59]. The MORDOR program, introduced in 2008, successfully predicted 74% of the experimental binding modes within 2.5 Å [67]. The obtained accuracy amelioration is directly linked with the improvement in handling target flexibility and the consequent induced-fit effect. Ameliorations have been implemented in DOCK version 6.0, where the use of explicit waters, sodium counterions, and rescoring with solvent models (see above) contribute to a success rate ranging from 47% to 80% depending on ligand complexity expressed by the number of ligand rotatable bonds [61]. In a comprehensive study, the performance of two widely used programs, Glide and GOLD, was evaluated over a data set of 60 RNA complexes [74]. The observed improvement in accuracy up to 65% (Glide) and 68% (GOLD) relates with the use of multiple input conformations of the ligands and of an ensemble of target conformations (specifically from NMR structures). Later studies highlight divergences in Autodock calculations between the predicted highest score geometry vs. the experimental one [72]. Indeed, combining predicted higher scores and high docking frequencies would generate better results, closer to the experimental geometries. To this purpose, a novel scoring function SPA-LN was developed for

Autodock improving both prediction of binding affinity and specificity with accuracy values around 76% (a less stringent 3Å RMSD cut-off was considered) [80]. Recently, the introduction of a vibrational entropy term in the Autodock scoring function further improved the accuracy in successfully predicting the binding geometry of the ligand-DNA complex [81].

6. Conclusions

In this work, we have retraced through the evolution of docking programs either in terms of searching algorithms allowing the passage from rigid to flexible docking, and also in terms of applications ranging from a single-ligand docking to the prediction of a drug's side effects, polypharmacology until nucleic acids docking. The role of this computational method has evolved, placing itself as a crossing-node tool for medicinal chemistry campaigns. Moreover, the increasing number of experimental structures deposited in the PDB and NDB databases contribute to its progress, enlarging its usage for the exploration of novel targets for new therapeutics. Despite the effort made on developing novel algorithms, ameliorating the scoring functions, docking still faces some limitations such as target flexibility. In this sense, the combination with other computational tools as molecular dynamic and sampling methods allows to overcome this limit. The advancements also in computer architectures and parallelization enable now to perform the calculation in reasonable time.

In this review, we have analyzed the use of molecular docking applied to nucleic acids, since they are emerging as valuable therapeutic targets. Many protein docking programs are also applied to nucleic acids, although, when compared to protein docking their accuracy in binding mode prediction is lower. Similar to proteins, nucleic acids flexibility, the consequent induced-fit effect, or conformational selection during the ligand-binding event are partially ignored by the docking programs. As for docking to protein, this might negatively affect the docking accuracy. Unique for nucleic acids is the electrostatic distribution of the charges, which is associated with the solvation effect of water molecules and ions distribution. New scoring functions specifically designed for nucleic acids are emerging. They include tailored terms for better describing the systems (i.e., solvation, electrostatic, entropy, etc.). Nevertheless, a major difficult task linked to the presence of metal ions and the continuous fast charge exchange between water molecules and ions remains to be solved.

In conclusion, this review reports on how molecular docking and its applications have evolved over time, the strategies adopted for partially overcoming the problems linked to ligand-macromolecule recognition, and it points out the remaining challenges to perfectly recapitulate virtually the complexity of intermolecular interactions.

Author Contributions: Conceptualization and writing—original draft preparation, F.T. and L.S. All authors have read and agreed to the published version of the manuscript.

Acknowledgments: We thank Gervasio Francesco for his critical reading of the manuscript.

References

1. Kendrew, J.C.; Dickerson, R.E.; Strandberg, B.E.; Hart, R.G.; Davies, D.R.; Phillips, D.C.; Shore, V. Structure of myoglobin: A three-dimensional Fourier synthesis at 2 Å. resolution. *Nature* **1960**, *185*, 422. [CrossRef] [PubMed]
2. Herzik, M.A.; Wu, M.; Lander, G.C. High-resolution structure determination of sub-100 kDa complexes using conventional cryo-EM. *Nat. Commun.* **2019**, *10*, 1032. [CrossRef] [PubMed]
3. Moult, J.; Fidelis, K.; Kryshtafovych, A.; Schwede, T.; Tramontano, A. Critical assessment of methods of protein structure prediction (CASP)-Round XII. *Proteins* **2018**, *86* (Suppl. 1), 7–15. [CrossRef] [PubMed]

4. Somody, J.C.; MacKinnon, S.S.; Windemuth, A. Structural coverage of the proteome for pharmaceutical applications. *Drug Discov. Today* **2017**, *22*, 1792–1799. [CrossRef] [PubMed]
5. Hillisch, A.; Pineda, L.F.; Hilgenfeld, R. Utility of homology models in the drug discovery process. *Drug Discov. Today* **2004**, *9*, 659–669. [CrossRef]
6. Mukherjee, B.; Tessaro, F.; Vahokoski, J.; Kursula, I.; Marq, J.B.; Scapozza, L.; Soldati-Favre, D. Modeling and resistant alleles explain the selectivity of antimalarial compound 49c towards apicomplexan aspartyl proteases. *EMBO J.* **2018**, *37*, e98047. [CrossRef]
7. Kuntz, I.D.; Blaney, J.M.; Oatley, S.J.; Langridge, R.; Ferrin, T.E. A geometric approach to macromolecule-ligand interactions. *J. Mol. Biol.* **1982**, *161*, 269–288. [CrossRef]
8. Shoichet, B.K.; Kuntz, I.D.; Bodian, D.L. Molecular docking using shape descriptors. *J. Comput. Chem.* **1992**, *13*, 380–397. [CrossRef]
9. Fischer, E. Einfluss der Configuration auf die Wirkung der Enzyme. II. *Ber. Der Dtsch. Chem. Ges.* **1894**, *27*, 3479–3483. [CrossRef]
10. Stoddard, B.L.; Koshland, D.E., Jr. Prediction of the structure of a receptor-protein complex using a binary docking method. *Nature* **1992**, *358*, 774–776. [CrossRef]
11. DesJarlais, R.L.; Sheridan, R.P.; Dixon, J.S.; Kuntz, I.D.; Venkataraghavan, R. Docking flexible ligands to macromolecular receptors by molecular shape. *J. Med. Chem.* **1986**, *29*, 2149–2153. [CrossRef] [PubMed]
12. Hart, T.N.; Read, R.J. A multiple-start Monte Carlo docking method. *Proteins* **1992**, *13*, 206–222. [CrossRef] [PubMed]
13. Oshiro, C.M.; Kuntz, I.D.; Dixon, J.S. Flexible ligand docking using a genetic algorithm. *J. Comput. Aided Mol. Des.* **1995**, *9*, 113–130. [CrossRef] [PubMed]
14. Jones, G.; Willett, P.; Glen, R.C. Molecular recognition of receptor sites using a genetic algorithm with a description of desolvation. *J. Mol. Biol.* **1995**, *245*, 43–53. [CrossRef]
15. Koshland, D.E. Application of a Theory of Enzyme Specificity to Protein Synthesis. *Proc. Natl. Acad. Sci. USA* **1958**, *44*, 98–104. [CrossRef]
16. Monod, J.; Wyman, J.; Changeux, J.-P. On the nature of allosteric transitions: A plausible model. *J. Mol. Biol.* **1965**, *12*, 88–118. [CrossRef]
17. Teague, S.J. Implications of protein flexibility for drug discovery. *Nat. Rev. Drug Discov.* **2003**, *2*, 527–541. [CrossRef]
18. Jiang, F.; Kim, S.H. "Soft docking": Matching of molecular surface cubes. *J. Mol. Biol.* **1991**, *219*, 79–102. [CrossRef]
19. Leach, A.R. Ligand docking to proteins with discrete side-chain flexibility. *J. Mol. Biol.* **1994**, *235*, 345–356. [CrossRef]
20. Knegtel, R.M.A.; Kuntz, I.D.; Oshiro, C.M. Molecular docking to ensembles of protein structures11Edited by B. Honig. *J. Mol. Biol.* **1997**, *266*, 424–440. [CrossRef]
21. Claussen, H.; Buning, C.; Rarey, M.; Lengauer, T. FlexE: Efficient molecular docking considering protein structure variations. *J. Mol. Biol.* **2001**, *308*, 377–395. [CrossRef]
22. Salmaso, V.; Moro, S. Bridging Molecular Docking to Molecular Dynamics in Exploring Ligand-Protein Recognition Process: An Overview. *Front. Pharmacol.* **2018**, *9*, 923. [CrossRef] [PubMed]
23. Amaro, R.E.; Baudry, J.; Chodera, J.; Demir, O.; McCammon, J.A.; Miao, Y.; Smith, J.C. Ensemble Docking in Drug Discovery. *Biophys. J.* **2018**, *114*, 2271–2278. [CrossRef]
24. Torres, P.H.M.; Sodero, A.C.R.; Jofily, P.; Silva, F.P., Jr. Key Topics in Molecular Docking for Drug Design. *Int. J. Mol. Sci.* **2019**, *20*, 4574. [CrossRef] [PubMed]
25. Pinzi, L.; Rastelli, G. Molecular Docking: Shifting Paradigms in Drug Discovery. *Int. J. Mol. Sci.* **2019**, *20*, 4331. [CrossRef] [PubMed]
26. Westermaier, Y.; Barril, X.; Scapozza, L. Virtual screening: An in silico tool for interlacing the chemical universe with the proteome. *Methods* **2015**, *71*, 44–57. [CrossRef] [PubMed]
27. Lyu, J.; Wang, S.; Balius, T.E.; Singh, I.; Levit, A.; Moroz, Y.S.; O'Meara, M.J.; Che, T.; Algaa, E.; Tolmachova, K.; et al. Ultra-large library docking for discovering new chemotypes. *Nature* **2019**, *566*, 224–229. [CrossRef]
28. Wenying, S.; Xuanyi, L.; Hequan, Y.; Kejiang, L. Convolutional Neural Network Based Virtual Screening. *Curr. Med. Chem.* **2020**, *27*, 1–15.
29. Li, J.; Fu, A.; Zhang, L. An Overview of Scoring Functions Used for Protein-Ligand Interactions in Molecular Docking. *Interdiscip. Sci.* **2019**, *11*, 320–328. [CrossRef]

30. Lavecchia, A. Machine-learning approaches in drug discovery: Methods and applications. *Drug Discov. Today* **2015**, *20*, 318–331. [CrossRef]
31. Gupta, M.; Sharma, R.; Kumar, A. Docking techniques in pharmacology: How much promising? *Comput. Biol. Chem.* **2018**, *76*, 210–217. [CrossRef] [PubMed]
32. Kumalo, H.M.; Bhakat, S.; Soliman, M.E.S. Theory and Applications of Covalent Docking in Drug Discovery: Merits and Pitfalls. *Molecules* **2015**, *20*, 1984–2000. [CrossRef] [PubMed]
33. Bianco, G.; Forli, S.; Goodsell, D.S.; Olson, A.J. Covalent docking using autodock: Two-point attractor and flexible side chain methods. *Protein Sci.* **2016**, *25*, 295–301. [CrossRef] [PubMed]
34. Cavasotto, C.N.; Adler, N.S.; Aucar, M.G. Quantum Chemical Approaches in Structure-Based Virtual Screening and Lead Optimization. *Front. Chem.* **2018**, *6*, 188. [CrossRef]
35. De Vivo, M.; Masetti, M.; Bottegoni, G.; Cavalli, A. Role of Molecular Dynamics and Related Methods in Drug Discovery. *J. Med. Chem.* **2016**, *59*, 4035–4061. [CrossRef]
36. Totrov, M.; Abagyan, R. Flexible ligand docking to multiple receptor conformations: A practical alternative. *Curr. Opin. Struct. Biol.* **2008**, *18*, 178–184. [CrossRef]
37. Warner, K.D.; Hajdin, C.E.; Weeks, K.M. Principles for targeting RNA with drug-like small molecules. *Nat. Rev. Drug Discov.* **2018**, *17*, 547–558. [CrossRef]
38. Hurley, L.H.; Boyd, F.L. DNA as a target for drug action. *Trends Pharmacol. Sci.* **1988**, *9*, 402–407. [CrossRef]
39. Wang, M.; Yu, Y.; Liang, C.; Lu, A.; Zhang, G. Recent Advances in Developing Small Molecules Targeting Nucleic Acid. *Int. J. Mol. Sci.* **2016**, *17*, 779. [CrossRef]
40. Ali, A.; Bhattacharya, S. DNA binders in clinical trials and chemotherapy. *Bioorg. Med. Chem.* **2014**, *22*, 4506–4521. [CrossRef]
41. Kohn, K.W. Beyond DNA cross-linking: History and prospects of DNA-targeted cancer treatment–fifteenth Bruce F. Cain Memorial Award Lecture. *Cancer Res.* **1996**, *56*, 5533–5546. [PubMed]
42. Dutta, D.; Debnath, M.; Müller, D.; Paul, R.; Das, T.; Bessi, I.; Schwalbe, H.; Dash, J. Cell penetrating thiazole peptides inhibit c-MYC expression via site-specific targeting of c-MYC G-quadruplex. *Nucleic Acids Res.* **2018**, *46*, 5355–5365. [CrossRef] [PubMed]
43. Bolhuis, A.; Aldrich-Wright, J.R. DNA as a target for antimicrobials. *Bioorg. Chem.* **2014**, *55*, 51–59. [CrossRef] [PubMed]
44. Hurley, L.H. DNA and its associated processes as targets for cancer therapy. *Nat. Rev. Cancer* **2002**, *2*, 188–200. [CrossRef] [PubMed]
45. Cimino-Reale, G.; Zaffaroni, N.; Folini, M. Emerging Role of G-quadruplex DNA as Target in Anticancer Therapy. *Curr. Pharm. Des.* **2016**, *22*, 6612–6624. [CrossRef] [PubMed]
46. Barrett, M.P.; Gemmell, C.G.; Suckling, C.J. Minor groove binders as anti-infective agents. *Pharm. Ther.* **2013**, *139*, 12–23. [CrossRef]
47. Wilson, W.D.; Tanious, F.A.; Mathis, A.; Tevis, D.; Hall, J.E.; Boykin, D.W. Antiparasitic compounds that target DNA. *Biochimie* **2008**, *90*, 999–1014. [CrossRef]
48. Garcia-Lopez, A.; Tessaro, F.; Jonker, H.R.A.; Wacker, A.; Richter, C.; Comte, A.; Berntenis, N.; Schmucki, R.; Hatje, K.; Petermann, O.; et al. Targeting RNA structure in SMN2 reverses spinal muscular atrophy molecular phenotypes. *Nat. Commun.* **2018**, *9*, 2032. [CrossRef]
49. McCown, P.J.; Corbino, K.A.; Stav, S.; Sherlock, M.E.; Breaker, R.R. Riboswitch diversity and distribution. *RNA* **2017**, *23*, 995–1011. [CrossRef]
50. Krause, K.M.; Serio, A.W.; Kane, T.R.; Connolly, L.E. Aminoglycosides: An Overview. *Cold Spring Harb. Perspect. Med.* **2016**, *6*, a027029. [CrossRef]
51. Shortridge, M.D.; Wille, P.T.; Jones, A.N.; Davidson, A.; Bogdanovic, J.; Arts, E.; Karn, J.; Robinson, J.A.; Varani, G. An ultra-high affinity ligand of HIV-1 TAR reveals the RNA structure recognized by P-TEFb. *Nucleic Acids Res.* **2019**, *47*, 1523–1531. [CrossRef] [PubMed]
52. Will, C.L.; Lührmann, R. Spliceosome structure and function. *Cold Spring Harb. Perspect. Biol.* **2011**, *3*, a003707. [CrossRef] [PubMed]
53. Havens, M.A.; Duelli, D.M.; Hastings, M.L. Targeting RNA splicing for disease therapy. *Wires Rna* **2013**, *4*, 247–266. [CrossRef] [PubMed]
54. Marquis, J.; Meyer, K.; Angehrn, L.; Kämpfer, S.S.; Rothen-Rutishauser, B.; Schümperli, D. Spinal muscular atrophy: SMN2 pre-mRNA splicing corrected by a U7 snRNA derivative carrying a splicing enhancer sequence. *Mol. Ther.* **2007**, *15*, 1479–1486. [CrossRef]

55. Naryshkin, N.A.; Weetall, M.; Dakka, A.; Narasimhan, J.; Zhao, X.; Feng, Z.; Ling, K.K.Y.; Karp, G.M.; Qi, H.; Woll, M.G.; et al. Motor neuron disease. SMN2 splicing modifiers improve motor function and longevity in mice with spinal muscular atrophy. *Science* **2014**, *345*, 688–693. [CrossRef]
56. Campagne, S.; Boigner, S.; Rüdisser, S.; Moursy, A.; Gillioz, L.; Knörlein, A.; Hall, J.; Ratni, H.; Cléry, A.; Allain, F.H.T. Structural basis of a small molecule targeting RNA for a specific splicing correction. *Nat. Chem. Biol.* **2019**, *15*, 1191–1198. [CrossRef]
57. Detering, C.; Varani, G. Validation of automated docking programs for docking and database screening against RNA drug targets. *J. Med. Chem.* **2004**, *47*, 4188–4201. [CrossRef]
58. Morley, S.D.; Afshar, M. Validation of an empirical RNA-ligand scoring function for fast flexible docking using RiboDock®. *J. Comput. Aided Mol. Des.* **2004**, *18*, 189–208. [CrossRef]
59. Giambaşu, G.M.; Case, D.A.; York, D.M. Predicting Site-Binding Modes of Ions and Water to Nucleic Acids Using Molecular Solvation Theory. *J. Am. Chem. Soc.* **2019**, *141*, 2435–2445. [CrossRef]
60. Moitessier, N.; Westhof, E.; Hanessian, S. Docking of Aminoglycosides to Hydrated and Flexible RNA. *J. Med. Chem.* **2006**, *49*, 1023–1033. [CrossRef]
61. Lang, P.T.; Brozell, S.R.; Mukherjee, S.; Pettersen, E.F.; Meng, E.C.; Thomas, V.; Rizzo, R.C.; Case, D.A.; James, T.L.; Kuntz, I.D. DOCK 6: Combining techniques to model RNA–small molecule complexes. *RNA* **2009**, *15*, 1219–1230. [CrossRef] [PubMed]
62. Pfeffer, P.; Gohlke, H. DrugScoreRNAKnowledge-Based Scoring Function to Predict RNA−Ligand Interactions. *J. Chem. Inf. Model.* **2007**, *47*, 1868–1876. [CrossRef] [PubMed]
63. Wei, W.; Luo, J.; Waldispühl, J.; Moitessier, N. Predicting Positions of Bridging Water Molecules in Nucleic Acid–Ligand Complexes. *J. Chem. Inf. Model.* **2019**, *59*, 2941–2951. [CrossRef] [PubMed]
64. Stagno, J.R.; Liu, Y.; Bhandari, Y.R.; Conrad, C.E.; Panja, S.; Swain, M.; Fan, L.; Nelson, G.; Li, C.; Wendel, D.R.; et al. Structures of riboswitch RNA reaction states by mix-and-inject XFEL serial crystallography. *Nature* **2017**, *541*, 242–246. [CrossRef]
65. Chen, Q.; Shafer, R.H.; Kuntz, I.D. Structure-Based Discovery of Ligands Targeted to the RNA Double Helix. *Biochemistry* **1997**, *36*, 11402–11407. [CrossRef] [PubMed]
66. Lind, K.E.; Du, Z.; Fujinaga, K.; Peterlin, B.M.; James, T.L. Structure-Based Computational Database Screening, In Vitro Assay, and NMR Assessment of Compounds that Target TAR RNA. *Chem. Biol.* **2002**, *9*, 185–193. [CrossRef]
67. Guilbert, C.; James, T.L. Docking to RNA via Root-Mean-Square-Deviation-Driven Energy Minimization with Flexible Ligands and Flexible Targets. *J. Chem. Inf. Model.* **2008**, *48*, 1257–1268. [CrossRef]
68. Rohs, R.; Bloch, I.; Sklenar, H.; Shakked, Z. Molecular flexibility in ab initio drug docking to DNA: Binding-site and binding-mode transitions in all-atom Monte Carlo simulations. *Nucleic Acids Res.* **2005**, *33*, 7048–7057. [CrossRef]
69. Krüger, D.M.; Bergs, J.; Kazemi, S.; Gohlke, H. Target Flexibility in RNA−Ligand Docking Modeled by Elastic Potential Grids. *ACS Med. Chem. Lett.* **2011**, *2*, 489–493. [CrossRef]
70. Kazemi, S.; Krüger, D.M.; Sirockin, F.; Gohlke, H. Elastic potential grids: Accurate and efficient representation of intermolecular interactions for fully flexible docking. *ChemMedChem* **2009**, *4*, 1264–1268. [CrossRef]
71. Luo, J.; Wei, W.; Waldispuhl, J.; Moitessier, N. Challenges and current status of computational methods for docking small molecules to nucleic acids. *Eur. J. Med. Chem.* **2019**, *168*, 414–425. [CrossRef] [PubMed]
72. Deligkaris, C.; Ascone, A.T.; Sweeney, K.J.; Greene, A.J.Q. Validation of a computational docking methodology to identify the non-covalent binding site of ligands to DNA. *Mol. Biosyst.* **2014**, *10*, 2106–2125. [CrossRef] [PubMed]
73. Chen, L.; Calin, G.A.; Zhang, S. Novel insights of structure-based modeling for RNA-targeted drug discovery. *J. Chem. Inf. Model.* **2012**, *52*, 2741–2753. [CrossRef] [PubMed]
74. Li, Y.; Shen, J.; Sun, X.; Li, W.; Liu, G.; Tang, Y. Accuracy Assessment of Protein-Based Docking Programs against RNA Targets. *J. Chem. Inf. Model.* **2010**, *50*, 1134–1146. [CrossRef] [PubMed]
75. Su, M.; Yang, Q.; Du, Y.; Feng, G.; Liu, Z.; Li, Y.; Wang, R. Comparative Assessment of Scoring Functions: The CASF-2016 Update. *J. Chem. Inf. Model.* **2019**, *59*, 895–913. [CrossRef] [PubMed]
76. Gaillard, T. Evaluation of AutoDock and AutoDock Vina on the CASF-2013 Benchmark. *J. Chem. Inf. Model.* **2018**, *58*, 1697–1706. [CrossRef]

77. Wang, Z.; Sun, H.; Yao, X.; Li, D.; Xu, L.; Li, Y.; Tian, S.; Hou, T. Comprehensive evaluation of ten docking programs on a diverse set of protein-ligand complexes: The prediction accuracy of sampling power and scoring power. *Phys. Chem. Chem. Phys.* **2016**, *18*, 12964–12975. [CrossRef]
78. Ruiz-Carmona, S.; Alvarez-Garcia, D.; Foloppe, N.; Garmendia-Doval, A.B.; Juhos, S.; Schmidtke, P.; Barril, X.; Hubbard, R.E.; Morley, S.D. rDock: A Fast, Versatile and Open Source Program for Docking Ligands to Proteins and Nucleic Acids. *PLoS Comput. Biol.* **2014**, *10*, e1003571. [CrossRef]
79. Hartshorn, M.J.; Verdonk, M.L.; Chessari, G.; Brewerton, S.C.; Mooij, W.T.; Mortenson, P.N.; Murray, C.W. Diverse, high-quality test set for the validation of protein-ligand docking performance. *J. Med. Chem.* **2007**, *50*, 726–741. [CrossRef]
80. Yan, Z.; Wang, J. SPA-LN: A scoring function of ligand-nucleic acid interactions via optimizing both specificity and affinity. *Nucleic Acids Res.* **2017**, *45*, e110. [CrossRef]
81. McElfresh, G.W.; Deligkaris, C. A vibrational entropy term for DNA docking with autodock. *Comput. Biol. Chem.* **2018**, *74*, 286–293. [CrossRef] [PubMed]

3

Identification of D Modification Sites by Integrating Heterogeneous Features in *Saccharomyces cerevisiae*

Pengmian Feng [1,5,*,†], Zhaochun Xu [2,3,†], Hui Yang [2], Hao Lv [2], Hui Ding [2] and Li Liu [4,*]

1 Innovative Institute of Chinese Medicine and Pharmacy, Chengdu University of Traditional Chinese Medicine, Chengdu 611730, China
2 Key Laboratory for Neuro-Information of Ministry of Education, School of Life Science and Technology, Center for Informational Biology, University of Electronic Science and Technology of China, Chengdu 610054, China; jdzxuzhaochun@163.com (Z.X.); huiyang0325@163.com (H.Y.); 13208188368@163.com (H.L.); hding@uestc.edu.cn (H.D.)
3 Computer Department, Jingdezhen Ceramic Institute, Jingdezhen 333403, China
4 Laboratory of Theoretical Biophysics, School of Physical Science and Technology, Inner Mongolia University, Hohhot 010021, China
5 School of Public Health, North China University of Science and Technology, Tangshan 063000, China
* Correspondence: fengpengmian@gmail.com (P.F.); liliu2010imu@163.com (L.L.);

† These authors contributed equally to this work.

Academic Editors: Xiangxiang Zeng, Alfonso Rodríguez-Patón and Quan Zou

Abstract: As an abundant post-transcriptional modification, dihydrouridine (D) has been found in transfer RNA (tRNA) from bacteria, eukaryotes, and archaea. Nonetheless, knowledge of the exact biochemical roles of dihydrouridine in mediating tRNA function is still limited. Accurate identification of the position of D sites is essential for understanding their functions. Therefore, it is desirable to develop novel methods to identify D sites. In this study, an ensemble classifier was proposed for the detection of D modification sites in the *Saccharomyces cerevisiae* transcriptome by using heterogeneous features. The jackknife test results demonstrate that the proposed predictor is promising for the identification of D modification sites. It is anticipated that the proposed method can be widely used for identifying D modification sites in tRNA.

Keywords: dihydrouridine; nucleotide physicochemical property; pseudo dinucleotide composition; RNA secondary structure; ensemble classifier

---

1. Introduction

To date, more than 100 kinds of post-transcriptional modifications have been identified in transfer RNAs (tRNAs). It has been demonstrated that these modifications are involved in all core aspects of tRNA function [1]. Among them, dihydrouridine (D) is a prevalent tRNA modification, which has been found in the three domains of life [2].

The D modification is formed by a dihydrouridine synthase [3]. Unlike uridine (U), the ring of D is not aromatic, which precludes its interactions with other bases in tRNA by stacking interactions [4,5]. By destabilizing the tRNA structure, D can enhance the conformational flexibility of tRNA [6]. Therefore, it is concluded that the flexibility and even the folding of tRNA could be affected by D modification [4,7].

Recent studies have also shown that tRNA lacking D degrades significantly faster, suggesting that D modification can protect tRNAs from degradation [1,8]. Despite the abundant occurrence of D modification, our knowledge about its roles in mediating tRNA biological functions is still limited.

Therefore, it is urgent to develop novel methods to describe the distribution of D modification sites. Since it is cost ineffective and labor intensive to detect D modification sites by using experimental techniques, it is necessary to develop theoretical methods for the detection of D modification.

Therefore, in the present study, an ensemble classifier was proposed for the detection of D modification sites in the *Saccharomyces cerevisiae* transcriptome, in which the nucleotide physicochemical property, pseudo dinucleotide composition, and secondary structure component were employed to train the basic predictors, respectively. In the jackknife test, the ensemble classifier obtained an accuracy of 83.09% for identifying D modification sites. This result demonstrated the superiority of the proposed method for identifying D modification sites in the *S. cerevisiae* transcriptome.

2. Results

2.1. Performances of Different Features

In order to demonstrate the effectiveness of the different kinds of features for identifying D sites, we first built support vector machine (SVM) predictors based on each kind of sequence encoding schemes (i.e., nucleotide physicochemical property, pseudo dinucleotide composition, or secondary structure component). Their jackknife test results for identifying D sites in the *S. cerevisiae* transcriptome are reported in Table 1. Although the nucleotide-physicochemical-property-based predictor (NPCP-SVM) obtained the highest accuracy (Acc) for identifying D sites, its sensitivity (Sn) was only 67.65%, indicating that it still could not accurately identify the real D sites. For the predictors based on pseudo dinucleotide composition and secondary structure component (namely PseDNC-SVM and SSC-SVM), their accuracies (Acc) were only 75.74% and 72.79% with the atthews correlation coefficients (MCC) of 0.5 and 0.45, respectively. Taken together, these results indicate that the performances of the aforementioned three predictors were not fully satisfactory. Therefore, there is still scope to improve the performance for identifying D sites.

Table 1. Performances of different methods for identifying dihydrouridine (D) sites.

Methods	Sn (%)	Sp (%)	Acc (%)	MCC
NPCP-SVM	67.65	100	83.82	0.59
PseDNC-SVM	73.53	77.94	75.74	0.50
SSC-SVM	70.59	75.00	72.79	0.45
Ensemble SVM	76.47	89.71	83.09	0.62

2.2. Improving Predictive Performance Using Ensemble Learning

Several recent works have demonstrated that the ensemble learning scheme can improve the performance of predictors [9–13]. In order to improve the performance of identifying D sites, we constructed an ensemble predictor based on SVM by using different kinds of features. Therefore, three basic SVM-based predictors were built by using nucleotide physicochemical property, pseudo dinucleotide composition, and secondary structure component, respectively. Figure 1 shows the prediction process with the ensemble classifier. The three predictors were integrated as an ensemble predictor via a voting strategy (see Materials and Methods). By combining the results of the three predictors together, a sequence in the benchmark dataset was predicted as a D-site-containing sequence if its prediction probabilities yielded by more than two predictors were all greater than 0.5.

The jackknife test results of the ensemble predictor for identifying D sites in *S. cerevisiae* transcriptome are also listed in Table 1. It was found that the sensitivity of the ensemble predictor was improved to 76.47%. Although its specificity and accuracy was a little lower than NPCP-SVM, the MCC of the ensemble predictor was 0.62, which was higher than that of any single SVM-based predictor, indicating the ensemble predictor was much more stable than NPCP-SVM, PseDNC-SVM, and SSC-SVM for the detection of D modification sites.

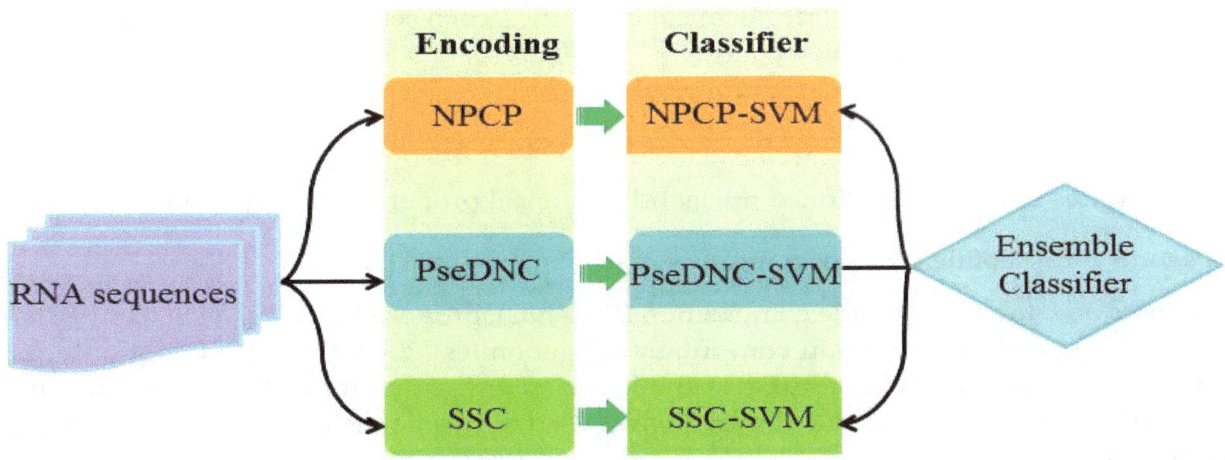

Figure 1. The flaw chart of the ensemble classifiers. NPCP-SVM stands for nucleotide-physicochemical-property-based predictor; PseDNC-SVM stands for pseudo-dinucleotide-composition-based predictor; SSC-SVM stands for secondary-structure-based predictor.

3. Materials and Methods

3.1. Benchmark Dataset

The original 208 positive samples (D-site-containing sequences) were fetched from the RMBase database [14]. All of these sequences in RMBase were 41 nt long with the D site in the center. Preliminary tests indicated that the best prediction results were achieved when the sequence was 41 nt long. In order to avoid redundancy, sequences with more than 80% sequence similarity were removed using the CD-HIT program [15]. Accordingly, we obtained 68 D-site-containing sequences from the *S. cerevisiae* transcriptome.

Negative samples were obtained by selecting 41-nt-long sequences that satisfied the following rules: (1) uridine is the center of the sequence, and (2) no dihydrouridine modification of the centered uridine has been identified experimentally. Accordingly, we could obtain a huge number of negative samples, from which we randomly picked 68 samples to form the negative subset for the purpose of using a balance benchmark dataset to train the model. In summary, our benchmark dataset comprised 68 D-site-containing sequences and 68 false D-site-containing sequences from the *S. cerevisiae* transcriptome, which is available at https://github.com/chenweiimu/D-Pred.

3.2. Sequence Encoding Scheme

3.2.1. Nucleotide Physicochemical Property (NPCP)

Adenosine (A), cytosine (C), guanine (G), and uridine (U) have different chemical properties [16,17]. In terms of ring structures, A and G are purines containing two rings, whereas C and U are pyrimidines containing one ring. When forming secondary structures, C and G form strong hydrogen bonds, whereas A and U form weak hydrogen bonds. In terms of amino/keto bases, A and C belong to the amino group, while G and U belong to the keto group [16,17].

In order to encode RNA sequences using these properties, the (x, y, z) coordinates were used to describe the chemical properties of the four nucleotides, and a value of 0 or 1 was assigned to (x, y, z), respectively. If x, y, and z coordinates stand for the ring structure, the hydrogen bond, and the amino/keto bases, A, C, G, and U can be represented by (1, 1, 1), (0, 0, 1), (1, 0, 0), and (0, 1, 0), respectively.

Accordingly, by using nucleotide chemical properties, each sequence could be encoded by a 123 (3 × 41)-dimensional vector, as given bellow:

$$\mathbf{R}_1 = \begin{bmatrix} \varepsilon_1 & \varepsilon_2 & \varepsilon_3 & \cdots & \varepsilon_i & \cdots & \varepsilon_{123} \end{bmatrix}^{\mathbf{T}} \quad (1)$$

where ε_i indicates the abovementioned nucleotide chemical properties, and its value is 0 or 1.

3.2.2. Pseudo Dinucleotide Composition

The pseudo *k*-tuple nucleotide composition (PseKNC), proposed by Chen et al. [18,19], has been successfully and widely applied in computational genomics [20–22]. PseKNC not only includes local sequence order information but also the global sequence pattern [23]. In the current study, the pseudo dinucleotide composition (PseDNC) was used to encode the RNA sequences and is defined as follows [18,19]:

$$\mathbf{R} = \begin{bmatrix} d_1 & d_2 & \cdots & d_{16} & d_{16+1} & \cdots & d_{16+\lambda} \end{bmatrix}^{\mathbf{T}} \quad (2)$$

where

$$d_u = \begin{cases} \frac{f_u}{\sum_{i=1}^{16} f_i + w\sum_{j=1}^{\lambda} \theta_j} & (1 \le u \le 16) \\ \frac{w\theta_{u-16}}{\sum_{i=1}^{16} f_i + w\sum_{j=1}^{\lambda} \theta_j} & (16 < u \le 16+\lambda) \end{cases}. \quad (3)$$

In Equation (3), f_u $(u = 1, 2, \cdots, 16)$ is the normalized occurrence frequency of the u-th nonoverlapping dinucleotide in the RNA sequence, and

$$\theta_j = \frac{1}{L-j-1}\sum_{i=1}^{L-j-1} C_{i,\ i+j} \quad (j = 1, 2, \cdots, \lambda;\ \lambda < L) \quad (4)$$

where θ_j is the j-tier correlation factor that reflects the sequence order correlation between all the j-th most contiguous dinucleotides. The coupling factor $C_{i,\ i+j}$ is defined as

$$C_{i,\ i+j} = \frac{1}{\mu}\sum_{g=1}^{\mu} \left[P_g(D_i) - P_g(D_{i+j})\right]^2 \quad (5)$$

where μ is the number of RNA physicochemical properties considered, $P_g(D_i)$ is the normalized numerical value of the g-th ($g = 1, 2, 3, \ldots, \mu$) RNA local structural property for the dinucleotide R_iR_{i+1} at position i, and $P_g(D_{i+j})$ is the corresponding value for the dinucleotide $R_{i+j}R_{i+j+1}$ at position $i + j$.

Inspired by a recent study [24], the three RNA physicochemical properties, namely, enthalpy [25], entropy [25], and free energy [26], were used to define PseDNC. Thus, in Equation (4), μ is equal to 3. The normalized numerical values of the three physicochemical properties of the 16 different RNA dinucleotides were obtained from our previous work [24].

The two parameters w and λ were optimized in the following ranges [0, 1] and [1, 10] with steps of 0.1 and 1, respectively. In the current work, the optimal values for w and λ were 0.5 and 4, respectively. Hence, the RNA sequence can be formulated by a (16 + 4) = 20-dimensional vector as given below:

$$\mathbf{R}_2 = \begin{bmatrix} d_1 & d_2 & \cdots & d_{16} & d_{17} & \cdots & d_{20} \end{bmatrix}^{\mathbf{T}} \quad (6)$$

3.2.3. Secondary Structure Component (SSC)

Considering the fact that RNA modification is affected by its structures [27], the RNA sequences were also encoded using the RNA secondary structures. By using the RNAfold tool (version 2.1.9) in ViennaRNA package with default parameters [28], we obtained the secondary structure status at each position, which was represented by brackets ("(" or ")") indicating paired nucleotides and by dots (".") indicating unpaired nucleotides. In the current study, we did not distinguish "(" and ")" and used "(" for both situations. For a given trinucleotide, there were eight (2^3) possible structure statuses (i.e., "(((", "((.", "(..", "(.(", ".((", ".(.", "..(", and " ... "). If the first nucleotide in the trinucleotide was further considered, there would be 32 (4 × 8) possible sequence-structure modes, which were denoted as "A-(((", "A-((.", "A-(..", ... , and "U- ... ". Therefore, a given sequence could be represented by using the following sequence-structure:

$$\mathbf{R}_3 = \left[f^{A}_{(((}, f^{A}_{((.}, f^{A}_{(..}, \ldots, f^{A}_{\ldots}, f^{C}_{(((}, \ldots, f^{U}_{\ldots}\right]^{\mathbf{T}}. \quad (7)$$

The elements in the vector of R_3 indicate the frequency of the 32 sequence-structure modes.

3.3. Support Vector Machine

SVM is a well-known machine learning method for pattern recognition and has been widely used in bioinformatics [29–35]. In the current study, the LibSVM package 3.18 (http://www.csie.ntu.edu.tw/~cjlin/libsvm/) was used to perform SVM. Due to its effectiveness and speed in training process, the radial basis kernel function (RBF) of SVM was often used to find the classification hyperplane. The regularization parameter C and kernel parameter γ of the SVM operation engine was optimized in the ranges of $[2^{-5}, 2^{15}]$ and $[2^{-15}, 2^{-5}]$ with steps of 2 and 2^{-1}, respectively. The prediction was made according to the probability score yielded from SVM. If its probability score was greater than 0.5, a uridine would be predicted as a D site, otherwise, a non-D-site.

3.4. Ensemble Classifiers

By using the NPCP, PseKNC, and SSC features, three basic classifiers were built, which voted for the final result according to the following rule [9]:

$$V_i = \sum_{k=1}^{3} f(pre(C_k), Class_i) \quad (i = 1,\ 2) \quad (8)$$

where V_i is the voting score for the sequence belonging to the $Class_i$. $f(pre(C_k), Class_i)$ is defined as

$$f(pre(C_k), Class_i) = \begin{cases} 1 & if\ pre(C_k) \in Class_i \\ 0 & if\ pre(C_k) \notin Class_i \end{cases} (i = 1,\ 2; k = 1,\ 2,\ 3). \quad (9)$$

The final prediction is determined by

$$Sgn(i) = \text{argmax}_i\{V_i\} \quad (i = 1,\ 2)\ . \quad (10)$$

$Sgn(i)$ is the argument that maximizes the voting score V_i.

3.5. Performance Evaluation

The performance of the method were evaluated by using sensitivity (Sn), specificity (Sp), accuracy (Acc), and the Matthews correlation coefficient (MCC), as given below [36–40]:

$$\begin{cases} \text{Sn} = 1 - \frac{N^{+}_{-}}{N^{+}} & 0 \le \text{Sn} \le 1 \\ \text{Sp} = 1 - \frac{N^{-}_{+}}{N^{-}} & 0 \le \text{Sp} \le 1 \\ \text{Acc} = 1 - \frac{N^{+}_{-}+N^{-}_{+}}{N^{+}+N^{-}} & 0 \le \text{Acc} \le 1 \\ \text{MCC} = \frac{1-\left(\frac{N^{+}_{-}}{N^{+}}+\frac{N^{-}_{+}}{N^{-}}\right)}{\sqrt{\left(1+\frac{N^{-}_{+}-N^{+}_{-}}{N^{+}}\right)\left(1+\frac{N^{+}_{-}-N^{-}_{+}}{N^{-}}\right)}} & -1 \le \text{MCC} \le 1 \end{cases} \quad (11)$$

where N^+ represents the total number of D-site-containing sequences, while N^+_- is the number of D-site-containing sequences incorrectly predicted to be of false D-site-containing sequences. N^- is the total number of false D-site-containing sequences, while N^-_+ the number of the false D-site-containing sequences incorrectly predicted to be of D-site-containing sequences.

3.6. Jackknife Cross-Validation

Among the three methods (i.e., independent dataset test, K-fold cross-validation test, and jackknife cross-validation), the jackknife cross-validation is deemed to be the least arbitrary, as demonstrated by in a recent review paper [41]. In the jackknife cross-validation, each sample in the training dataset is in turn singled out as an independent test sample and all the rule parameters are calculated without including the one being identified [42–46]. Accordingly, jackknife cross-validation was also used to examine the performance of the method proposed in the current study.

4. Conclusions

In this study, by integrating heterogeneous sequence-based features, a SVM-based ensemble classifier was proposed to identify D modification sites in the *S. cerevisiae* transcriptome. In this predictor, not only was the local and global sequence information included by encoding RNA sequences using PseDNC, but the nucleotide chemical properties and structures were also considered by representing RNA sequences using nucleotide physicochemical properties and predicted RNA secondary structures. The jackknife test results demonstrate that the proposed predictor is promising for the identification of D modification sites. It is anticipated that the proposed method will become an essential computational tool for identifying D modification sites in tRNA.

However, the proposed method has two flaws. The limited number of experimentally verified D modification data hindered us from extracting effective features to describe the D modification sites containing sequences. The other shortcoming is that the present method directly uses the entirety of the features, which may reduce the generalization capacity of the model and increase the computational time. Therefore, in future work, we shall make efforts to collect more D modification data and also employ the feature selection method to winnow out the optimal features.

Author Contributions: P.F. and L.L. conceived and designed the experiments; P.F., Z.X., H.Y., H.L., and H.D. performed the experiments; P.F. and L.L. wrote the paper.

References

1. Dyubankova, N.; Sochacka, E.; Kraszewska, K.; Nawrot, B.; Herdewijn, P.; Lescrinier, E. Contribution of dihydrouridine in folding of the D-arm in tRNA. *Organ. Biomol. Chem.* **2015**, *13*, 4960–4966. [CrossRef] [PubMed]
2. Sprinzl, M.; Horn, C.; Brown, M.; Ioudovitch, A.; Steinberg, S. Compilation of tRNA sequences and sequences of tRNA genes. *Nucleic Acids Res.* **1998**, *26*, 148–153. [CrossRef] [PubMed]
3. Yu, F.; Tanaka, Y.; Yamashita, K.; Suzuki, T.; Nakamura, A.; Hirano, N.; Suzuki, T.; Yao, M.; Tanaka, I. Molecular basis of dihydrouridine formation on tRNA. *Proc. Natl. Acad. Sci. USA* **2011**, *108*, 19593–19598. [CrossRef]
4. Jones, C.I.; Spencer, A.C.; Hsu, J.L.; Spremulli, L.L.; Martinis, S.A.; DeRider, M.; Agris, P.F. A counterintuitive Mg^{2+}-dependent and modification-assisted functional folding of mitochondrial tRNAs. *J. Mol. Biol.* **2006**, *362*, 771–786. [CrossRef] [PubMed]
5. Dalluge, J.J.; Hashizume, T.; Sopchik, A.E.; McCloskey, J.A.; Davis, D.R. Conformational flexibility in RNA: The role of dihydrouridine. *Nucleic Acids Res.* **1996**, *24*, 1073–1079. [CrossRef]
6. Kasprzak, J.M.; Czerwoniec, A.; Bujnicki, J.M. Molecular evolution of dihydrouridine synthases. *BMC Bioinform.* **2012**, *13*, 153. [CrossRef]

7. Whelan, F.; Jenkins, H.T.; Griffiths, S.C.; Byrne, R.T.; Dodson, E.J.; Antson, A.A. From bacterial to human dihydrouridine synthase: Automated structure determination. *Acta Crystallogr. Sect. D Biol. Crystallogr.* **2015**, *71*, 1564–1571. [CrossRef]
8. Alexandrov, A.; Chernyakov, I.; Gu, W.; Hiley, S.L.; Hughes, T.R.; Grayhack, E.J.; Phizicky, E.M. Rapid tRNA decay can result from lack of nonessential modifications. *Mol. Cell* **2006**, *21*, 87–96. [CrossRef] [PubMed]
9. Chen, W.; Xing, P.; Zou, Q. Detecting N6-methyladenosine sites from RNA transcriptomes using ensemble Support Vector Machines. *Sci. Rep.* **2017**, *7*, 40242. [CrossRef] [PubMed]
10. Jia, C.; Zuo, Y.; Zou, Q. O-GlcNAcPRED-II: An integrated classification algorithm for identifying O-GlcNAcylation sites based on fuzzy undersampling and a K-means PCA oversampling technique. *Bioinformatics* **2018**, *34*, 2029–2036. [CrossRef]
11. Zou, Q.; Guo, J.; Ju, Y.; Wu, M.; Zeng, X.; Hong, Z. Improving tRNAscan-SE Annotation Results via Ensemble Classifiers. *Mol. Inform.* **2015**, *34*, 761–770. [CrossRef] [PubMed]
12. Chen, W.; Feng, P.M.; Lin, H.; Chou, K.C. iRSpot-PseDNC: Identify recombination spots with pseudo dinucleotide composition. *Nucleic Acids Res.* **2013**, *41*, e68. [CrossRef] [PubMed]
13. Wan, S.; Duan, Y.; Zou, Q. HPSLPred: An Ensemble Multi-label Classifier for Human Protein Subcellular Location Prediction with Imbalanced Source. *Proteomics* **2017**, *17*, 1700262. [CrossRef] [PubMed]
14. Xuan, J.J.; Sun, W.J.; Lin, P.H.; Zhou, K.R.; Liu, S.; Zheng, L.L.; Qu, L.H.; Yang, J.H. RMBase v2.0: Deciphering the map of RNA modifications from epitranscriptome sequencing data. *Nucleic Acids Res.* **2018**, *46*, D327–D334. [CrossRef] [PubMed]
15. Fu, L.; Niu, B.; Zhu, Z.; Wu, S.; Li, W. CD-HIT: Accelerated for clustering the next-generation sequencing data. *Bioinformatics* **2012**, *28*, 3150–3152. [CrossRef] [PubMed]
16. Zhang, J.; Feng, P.; Lin, H.; Chen, W. Identifying RNA N(6)-Methyladenosine Sites in *Escherichia coli* Genome. *Front. Microbiol.* **2018**, *9*, 955. [CrossRef] [PubMed]
17. Feng, P.; Ding, H.; Yang, H.; Chen, W.; Lin, H.; Chou, K.-C. iRNA-PseColl: Identifying the Occurrence Sites of Different RNA Modifications by Incorporating Collective Effects of Nucleotides into PseKNC. *Mol. Ther.-Nucleic Acids* **2017**, *7*, 155–163. [CrossRef]
18. Chen, W.; Lei, T.-Y.; Jin, D.-C.; Lin, H.; Chou, K.-C. PseKNC: A flexible web server for generating pseudo K-tuple nucleotide composition. *Anal. Biochem.* **2014**, *456*, 53–60. [CrossRef]
19. Chen, W.; Zhang, X.; Brooker, J.; Lin, H.; Zhang, L.; Chou, K.-C. PseKNC-General: A cross-platform package for generating various modes of pseudo nucleotide compositions. *Bioinformatics* **2014**, *31*, 119–120. [CrossRef]
20. Chen, W.; Feng, P.-M.; Deng, E.-Z.; Lin, H.; Chou, K.-C. iTIS-PseTNC: A sequence-based predictor for identifying translation initiation site in human genes using pseudo trinucleotide composition. *Anal. Biochem.* **2014**, *462*, 76–83. [CrossRef]
21. Chen, W.; Feng, P.-M.; Lin, H.; Chou, K.-C. iSS-PseDNC: Identifying Splicing Sites Using Pseudo Dinucleotide Composition. *BioMed Res. Int.* **2014**. [CrossRef] [PubMed]
22. Lin, H.; Liang, Z.Y.; Tang, H.; Chen, W. Identifying sigma70 promoters with novel pseudo nucleotide composition. *IEEE/ACM Trans. Comput. Biol. Bioinform.* **2017**. [CrossRef]
23. Chen, W.; Lin, H.; Chou, K.-C. Pseudo nucleotide composition or PseKNC: An effective formulation for analyzing genomic sequences. *Mol. BioSyst.* **2015**, *11*, 2620–2634. [CrossRef]
24. Chen, W.; Feng, P.; Ding, H.; Lin, H.; Chou, K.-C. iRNA-Methyl: Identifying N-6-methyladenosine sites using pseudo nucleotide composition. *Anal. Biochem.* **2015**, *490*, 26–33. [CrossRef]
25. Freier, S.M.; Kierzek, R.; Jaeger, J.A.; Sugimoto, N.; Caruthers, M.H.; Neilson, T.; Turner, D.H. Improved free-energy parameters for predictions of RNA duplex stability. *Proc. Natl. Acad. Sci. USA* **1986**, *83*, 9373–9377. [CrossRef] [PubMed]
26. Xia, T.; SantaLucia, J., Jr.; Burkard, M.E.; Kierzek, R.; Schroeder, S.J.; Jiao, X.; Cox, C.; Turner, D.H. Thermodynamic parameters for an expanded nearest-neighbor model for formation of RNA duplexes with Watson-Crick base pairs. *Biochemistry* **1998**, *37*, 14719–14735. [CrossRef] [PubMed]
27. Lu, X.J.; Olson, W.K.; Bussemaker, H.J. The RNA backbone plays a crucial role in mediating the intrinsic stability of the GpU dinucleotide platform and the GpUpA/GpA miniduplex. *Nucleic Acids Res.* **2010**, *38*, 4868–4876. [CrossRef]
28. Lorenz, R.; Bernhart, S.H.; Honer Zu Siederdissen, C.; Tafer, H.; Flamm, C.; Stadler, P.F.; Hofacker, I.L. ViennaRNA Package 2.0. *Algorithms Mol. Biol. AMB* **2011**, *6*, 26. [CrossRef]

29. Feng, C.Q.; Zhang, Z.Y.; Zhu, X.J.; Lin, Y.; Chen, W.; Tang, H.; Lin, H. iTerm-PseKNC: A sequence-based tool for predicting bacterial transcriptional terminators. *Bioinformatics* **2018**. [CrossRef]
30. Su, Z.D.; Huang, Y.; Zhang, Z.Y.; Zhao, Y.W.; Wang, D.; Chen, W.; Chou, K.C.; Lin, H. iLoc-lncRNA: Predict the subcellular location of lncRNAs by incorporating octamer composition into general PseKNC. *Bioinformatics* **2018**. [CrossRef]
31. Chen, W.; Yang, H.; Feng, P.; Ding, H.; Lin, H. iDNA4mC: Identifying DNA N4-methylcytosine sites based on nucleotide chemical properties. *Bioinformatics* **2017**, *33*, 3518–3523. [CrossRef] [PubMed]
32. Li, D.; Ju, Y.; Zou, Q. Protein Folds Prediction with Hierarchical Structured SVM. *Curr. Proteomics* **2016**, *13*, 79–85. [CrossRef]
33. Wang, S.P.; Zhang, Q.; Lu, J.; Cai, Y.D. Analysis and Prediction of Nitrated Tyrosine Sites with the mRMR Method and Support Vector Machine Algorithm. *Curr. Bioinform.* **2018**, *13*, 3–13. [CrossRef]
34. Yang, H.; Lv, H.; Ding, H.; Chen, W.; Lin, H. iRNA-2OM: A Sequence-Based Predictor for Identifying 2′-O-Methylation Sites in Homo sapiens. *J. Comput. Biol. J. Comput. Mol. Cell Biol.* **2018**, *25*, 1266–1277. [CrossRef] [PubMed]
35. Dao, F.Y.; Lv, H.; Wang, F.; Feng, C.Q.; Ding, H.; Chen, W.; Lin, H. Identify origin of replication in Saccharomyces cerevisiae using two-step feature selection technique. *Bioinformatics* **2018**. [CrossRef] [PubMed]
36. Feng, P.-M.; Chen, W.; Lin, H.; Chou, K.-C. iHSP-PseRAAAC: Identifying the heat shock protein families using pseudo reduced amino acid alphabet composition. *Anal. Biochem.* **2013**, *442*, 118–125. [CrossRef] [PubMed]
37. Song, J.; Wang, Y.; Li, F.; Akutsu, T.; Rawlings, N.D.; Webb, G.I.; Chou, K.C. iProt-Sub: A comprehensive package for accurately mapping and predicting protease-specific substrates and cleavage sites. *Briefings Bioinform.* **2018**. [CrossRef]
38. Zhu, X.J.; Feng, C.Q.; Lai, H.Y.; Chen, W.; Lin, H. Predicting protein structural classes for low-similarity sequences by evaluating different features. *Knowl.-Based Syst.* **2018**. [CrossRef]
39. Yang, H.; Qiu, W.R.; Liu, G.Q.; Guo, F.B.; Chen, W.; Chou, K.C.; Lin, H. iRSpot-Pse6NC: Identifying recombination spots in Saccharomyces cerevisiae by incorporating hexamer composition into general PseKNC. *Int. J. Biol. Sci.* **2018**, *14*, 883–891. [CrossRef]
40. Tang, H.; Zhao, Y.W.; Zou, P.; Zhang, C.M.; Chen, R.; Huang, P.; Lin, H. HBPred: A tool to identify growth hormone-binding proteins. *Int. J. Biol. Sci.* **2018**, *14*, 957–964. [CrossRef]
41. Chou, K.C. Some remarks on protein attribute prediction and pseudo amino acid composition. *J. Theor. Biol.* **2011**, *273*, 236–247. [CrossRef] [PubMed]
42. Feng, P.M.; Lin, H.; Chen, W. Identification of antioxidants from sequence information using naive Bayes. *Comput. Math. Methods Med.* **2013**, *2013*, 567529. [CrossRef] [PubMed]
43. Feng, P.M.; Ding, H.; Chen, W.; Lin, H. Naive Bayes classifier with feature selection to identify phage virion proteins. *Comput. Math. Methods Med.* **2013**, *2013*, 530696. [CrossRef] [PubMed]
44. Lai, H.Y.; Chen, X.X.; Chen, W.; Tang, H.; Lin, H. Sequence-based predictive modeling to identify cancerlectins. *Oncotarget* **2017**, *8*, 28169–28175. [CrossRef] [PubMed]
45. Yang, H.; Tang, H.; Chen, X.X.; Zhang, C.J.; Zhu, P.P.; Ding, H.; Chen, W.; Lin, H. Identification of Secretory Proteins in Mycobacterium tuberculosis Using Pseudo Amino Acid Composition. *BioMed Res. Int.* **2016**, *2016*, 5413903. [CrossRef] [PubMed]
46. Chen, X.X.; Tang, H.; Li, W.C.; Wu, H.; Chen, W.; Ding, H.; Lin, H. Identification of Bacterial Cell Wall Lyases via Pseudo Amino Acid Composition. *BioMed Res. Int.* **2016**, *2016*, 1654623. [CrossRef] [PubMed]

4

Adverse Drug Reaction Predictions using Stacking Deep Heterogeneous Information Network Embedding Approach

Baofang Hu [1,2,3], Hong Wang [1,3,*], Lutong Wang [1,3] and Weihua Yuan [1,3]

1 School of Information Science and Engineering, Shandong Normal University, Jinan 250014, China; hubaofang@sdwu.edu.cn (B.H.); wanglutong1002@163.com (L.W.); weihuayuan_qingdao@126.com (W.Y.)

2 School of Data and Computer Science, Shandong Women's University, Jinan 250014, China

3 Shandong Provincial Key Laboratory for Distributed Computer Software Novel Technology, Shandong Normal University, Jinan 250014, China

* Correspondence: wanghong106@163.com.

Academic Editor: Xiangxiang Zeng

Abstract: Inferring potential adverse drug reactions is an important and challenging task for the drug discovery and healthcare industry. Many previous studies in computational pharmacology have proposed utilizing multi-source drug information to predict drug side effects have and achieved initial success. However, most of the prediction methods mainly rely on direct similarities inferred from drug information and cannot fully utilize the drug information about the impact of protein–protein interactions (PPI) on potential drug targets. Moreover, most of the methods are designed for specific tasks. In this work, we propose a novel heterogeneous network embedding approach for learning drug representations called SDHINE, which integrates PPI information into drug embeddings and is generic for different adverse drug reaction (ADR) prediction tasks. To integrate heterogeneous drug information and learn drug representations, we first design different meta-path-based proximities to calculate drug similarities, especially target propagation meta-path-based proximity based on PPI network, and then construct a semi-supervised stacking deep neural network model that is jointly optimized by the defined meta-path proximities. Extensive experiments with three state-of-the-art network embedding methods on three ADR prediction tasks demonstrate the effectiveness of the SDHINE model. Furthermore, we compare the drug representations in terms of drug differentiation by mapping the representations into 2D space; the results show that the performance of our approach is superior to that of the comparison methods.

Keywords: adverse drug reaction prediction; heterogeneous information network embedding; stacking denoising auto-encoder; meta-path-based proximity

1. Introduction

Adverse drug reactions (ADRs) are side effects caused by the use of one or several drugs. Some ADRs may be part of the natural pharmacological action of a drug that cannot be avoided, but more often, they may be unpredictable at the development stage. ADRs have caused a global and substantial burden that accounts for considerable mortality and morbidity [1]. Before clinical application of a drug, it should go through two ADR detection stages, including preclinical in vitro safety profiling and clinical drug safety trials. However, since so many side effect types and drug combinations exist, many potential side effects cannot be detected during the early drug development stage [2].

Recently, with the development of data mining and computational prediction methods, researchers have collected extensive drug data from the literature, and reports and have utilized these data to predict unknown ADRs [3–6]. ADR predictions based on computational methods can point drug safety tests in the right direction and consequently shorten the time requirement and save financial costs during drug development. A large number of machine learning methods have been proposed to predict potential ADRs [7–9]. Vilar et al. [10] utilized known side effect information of drugs to construct an associated matrix of drugs and adverse effects and adopted the matrix completion method to predict unknown side effects. LaBrute et al. [11] processed multi-source drug target information to find association relationships between ADRs and drug targets. These prediction methods are based on single drug information, and these mined drug datasets usually contain much noise. For example, the SIDER dataset [5], which was extracted from the public ADR reports, may contain some fake or unconfirmed noise data. Researchers have established different drug databases that describe drug features from different aspects, including chemical, biological, phenotypic, and interaction relationships [4–6,12,13]. It is more logical to combine different drug information to reduce the prediction error. Integrating this useful complex drug information to obtain more accurate ADR predictions is more effective. Yamanishi et al. [7] used multi-source drug data from the SIDER, PubChem, DrugBank, and Matador databases to predict side effects. The prediction method they adopted was based on multiple kernel regression and canonical correlation analysis. Zhang et al. [14] integrated different drug information to calculate drug similarities and utilized the linear neighborhoods method to transform the similarities into the side effect space and predict side effects. These prediction methods are mainly for the side effects caused by a single drug. However, in real life, many patients, especially the elderly, are on multiple prescriptions to treat different diseases. Drug–drug interactions (DDIs) may change the effects of drugs and cause some potential ADRs. Therefore, predicting the potential side effects induced by DDIs is imperative. Segura-Bedmar et al. [15] utilized a text mining method to predict the occurrence of DDIs based on a shallow language learning model. Jin et al. [16] formulated the DDI type prediction problem as a multi-task dyadic regression problem and utilized the model to predict the side effect types induced by DDIs. Zhang et al. [17] collected a variety of drug data that might influence DDIs and adopted an ensemble learning method to predict the occurrence of DDIs. Motivated by the success of deep learning in many areas, Zitnik et al. [18] developed a new graph convolutional neural network for multi-relational link prediction in multimodal networks to predict the DDI types.

Although the above methods have achieved great success, the methods are mostly designed for specific ADR prediction tasks and lack generic abilities. With the development of the network embedding, learning combined characteristic embeddings of drugs has attracted great attention from researchers [19–21]. Every drug can be embedded into a low-dimensional feature vector, which integrates different drug information, including chemical, biological, phenotypic, and interaction relationships. The drug representations are more general and can be used for different ADR prediction tasks. Li et al. [22] proposed a matrix completion method to integrate multiple sources of drug data and predicted ADRs. Ma et al. [23] proposed a drug embedding method based on multi-view deep auto-encoders to predict ADRs. However, their works only considered immediately relevant information of side effects and neglected potential indirect information. There are some potential association relationships between different biological data. For example, drug targets propagate to another protein through the protein–protein interactions (PPI) network, because the biological function signal cascade propagates through different proteins via PPI [24,25]. When one drug acts on a known target protein, it may change another potential target protein through protein–protein interaction effects and consequently cause potential ADRs.

In this work, we propose a general drug embedding method to learn the representations of drugs and predict different types of ADRs. The flowchart for ADR prediction is shown in Figure 1. We firstly modeled different drug information in a drug heterogeneous information network (drug HIN) framework and then proposed a stacking deep heterogeneous information network embedding approach based on semantic meta-paths. The generated drug embedding integrates multi-source drug information and multi-relationship side effect information to improve the ADR prediction accuracy.

Especially, we utilized the target propagation strategy to recognize the potential drug targets and improve the prediction accuracy. At the target propagation stage, we need to search for the proteins that are more obviously affected by the known targets of the drug. Finding the nearest node of one node is challenging because tens of thousands of nodes exist in the PPI network. We propose using transition probability based on the random walk procedure [26,27], which is common in recommendation systems, to calculate the target propagation proximity and reconstruct the drug-target network based on the target propagation meta-path.

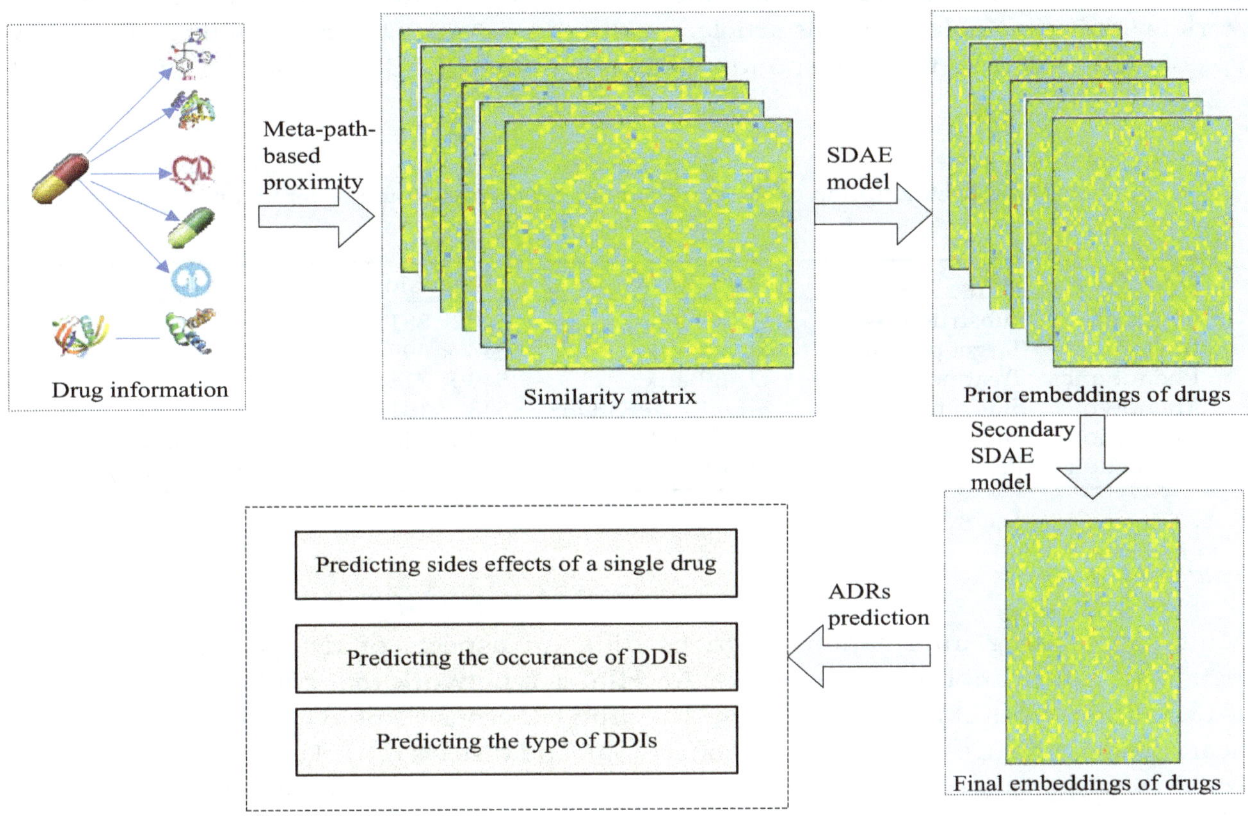

Figure 1. The flowchart for adverse drug reactions' (ADRs) prediction. SDAE, stacked denoising auto-encoder; DDIs, Drug-Drug interaction.

2. Datasets and Method

2.1. Datasets

We collected six types of drug data from seven public databases.

- Drug-Drug interaction information (DDI): Tatonetti [12] mined side effects induced by DDIs from the FDA Adverse Event Reporting System (FAERS, http://www.fda.gov/cder/aers/default.htm) and developed a database called "TWOSIDES". The database contains 645 drugs and ADRs caused by 63,473 combinations of different drugs.
- Protein-Protein interaction data (PPI): We downloaded the PPI network data from the Human Protein Reference Database (HPRD, http://www.hprd.org). The dataset contains 9519 proteins and 37,062 protein-protein interactions.
- Other drug information: We also obtained other drug information from four online drug information databases (DrugBank [4], the PubChem Compound database [6], the SIDER database [5], and the OFFSIDES database [12]). DrugBank is a widely-used public drug information database. From the DrugBank database, we collected drug target protein and disease treatment information. The PubChem system generates a binary substructure fingerprint

for chemical structures. From the PubChem database, we searched every drug's chemical substructure. We also extracted drug side effect information from the SIDER and OFFSIDES databases. These two databases include most associations between drugs and side effects, and we integrated the drug-side effect data obtained from the two databases.

We mapped drug ids in the TWOSIDES dataset to the other aforementioned datasets and finally constructed an integrated dataset that contained multi-source drug information, including DDIs, drug chemical substructures, drug targets, drug side effects, drug treatment, and PPI. The dataset used in this work is shown in Table 1. In this article, we did not consider the probabilities of ADR events. If a type of ADR occurred, the corresponding element in the DDI dataset or side effect dataset was labeled 1.

Table 1. Description of the drug data.PPI, protein–protein interaction.

Data Type	Data	Data Source	Dimension
Chemical	Substructures	PubChem	548 × 881
Biological	Target protein	DrugBank	548 × 695
Phenotypic	Treatment disease	DrugBank	548 × 718
Phenotypic	Side effect	SIDER, OFFSIDES	548 × 1318 (1318 ADR events)
Interaction	DDIs	TWOSIDES	548 × 548 × 1318 (1318 ADR events)
Interaction	PPI	HPRD	9519 × 9519 (37,062 interactions)

2.2. *Drug HIN*

Multi-source drug information describes different aspects of drugs and forms a typical heterogeneous information network (HIN). An HIN is a network that contains multiple types of objects or multiple types of relationships [28]. The drug HIN consists of five types of objects: drug (D), chemical substructure (C), protein (P), side effect (S_E), and disease (D_I). The five types of objects are connected through six types of links (as shown in Figure 2). A drug-drug link indicates a type of drug-drug interaction, whereas the link between a drug and its chemical substructure indicates that the drug consists of some type of chemical substructure. In Table 2, we present the semantics of the different link types in the drug HIN.

Table 2. Semantics of link types in the drug heterogeneous information network (HIN).

Link Types	Abbreviated Form	Semantics of Link Types
Drug-Drug	D-D	Drug-drug interactions
Drug-Chemical	D-C	The chemical substructure of a drug
Drug-Protein	D-P	The target protein of a drug
Protein-Protein	P-P	Protein-protein interactions
Drug-Disease	D-D_I	The therapeutic effect between a drug and a disease
Drug-Side Effect	D-S_E	The side effect between a drug and a disease

In HINs, two objects connect via different link types, which are called semantics meta-paths [29,30]. Given an HIN, a meta-path is a sequence of objects connected by different link types. Different types of meta-paths in the drug HIN are shown in Figure 3. Because our final goal is to learn drug representations, we only consider the meta-paths in which the starting objects are all drugs. The detailed meta-paths used in this study are summarized in Table 3.

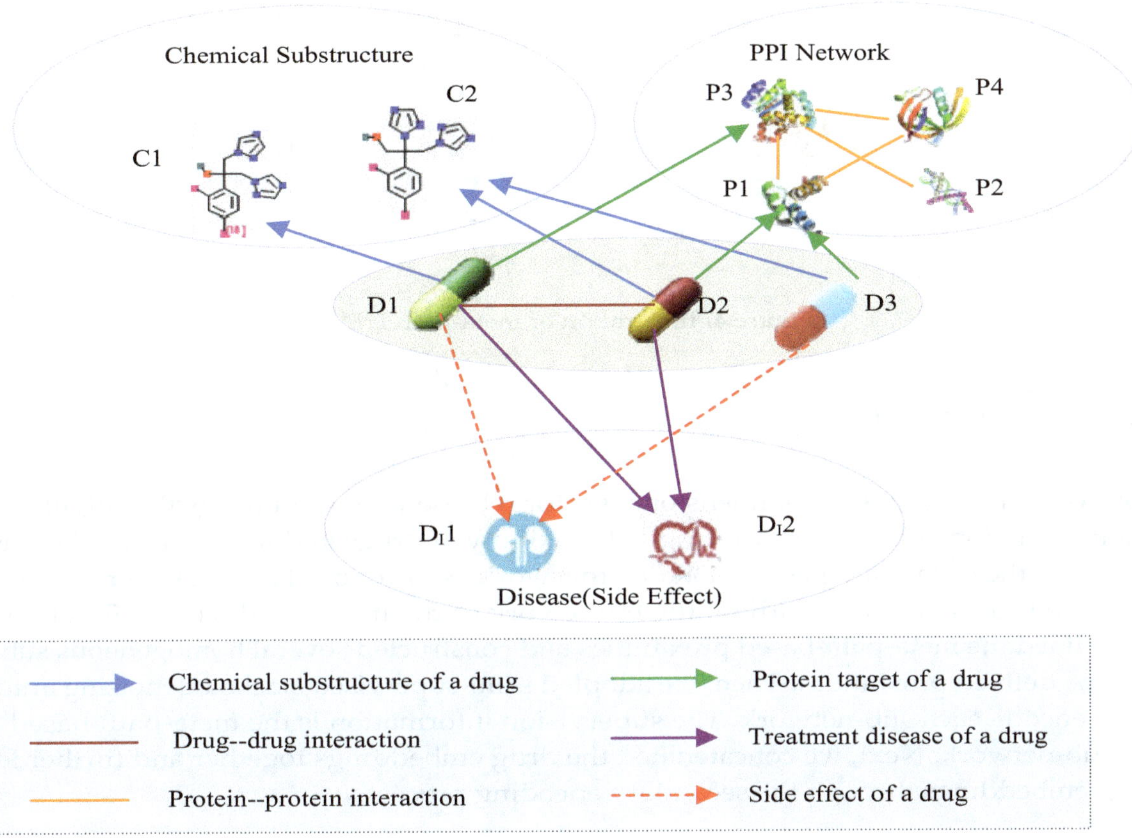

Figure 2. Heterogeneous drug information. PPI, protein-protein interaction.

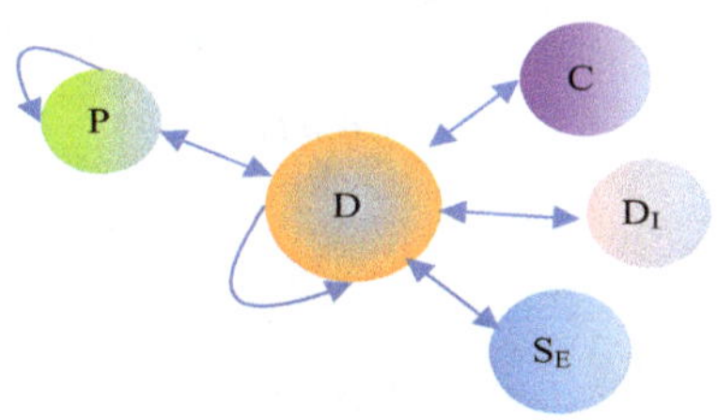

Figure 3. Meta-paths in drug HIN.

Table 3. Meta-paths in drug HIN.

Meta-Paths	Abbreviated Form	Semantics of Meta-Paths
Drug-Drug	DD	Drug-Drug interactions (at the drug embedding stage, interaction types are not considered).
Drug-Chemical-Drug	DCD	Two drugs have a similar chemical substructure.
Drug-Protein-Drug	DPD	Two drugs have the same target protein.
Drug-Protein-. . . -Protein-Drug	$DP^{(n)}D$ ($n \geq 2$)	There are protein-protein interactions between the targets of two drugs. For example, the path $D_1P_1P_2D_2$ in Figure 4 indicates that the targets of D_1 and D_2 are P_1 and P_2, respectively. Meanwhile, there is an interaction between P_1 and P_2 (in meta-path $DP^{(n)}D$, there are $n-1$ protein-protein interactions).
Drug-Disease-Drug	DD_ID	Two drugs have the same therapeutic effect.
Drug-Side Effect-Drug	DS_ED	Two drugs have the same side effect.

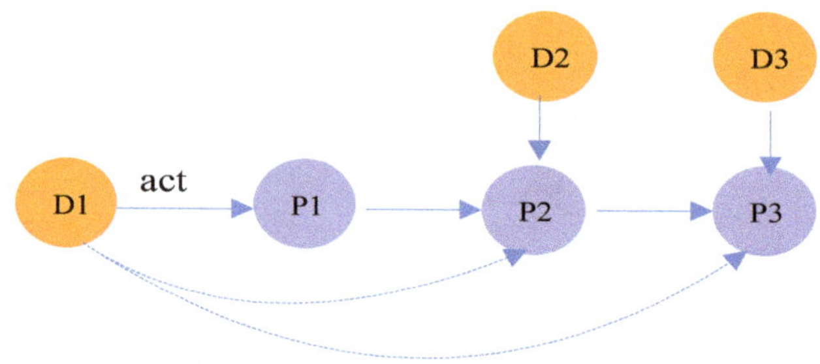

Figure 4. Illustration of meta-path $DP^{(n)}D$.

2.3. *Stacking Deep HIN Embedding*

Our goal is to learn the low-dimensional vector representations of drugs that highly summarize the drug information and preserve the original proximity of drugs in different drug relationships, and then to predict the different types of ADRs. In this work, we proposed a semi-supervised deep model SDHINE to perform HIN embedding; the framework of the model is shown in Figure 5. In detail, first, we defined the meta-path-based proximities and constructed several homogeneous sub-networks based on the defined proximities. Then, we adopted semi-supervised stacked denoising auto-encoders (SDAE) to encode each sub-network. The supervision information is the meta-path-based proximity in every sub-network. Next, we concatenated the drug embeddings together and further learned the final drug embeddings through the secondary encoding process.

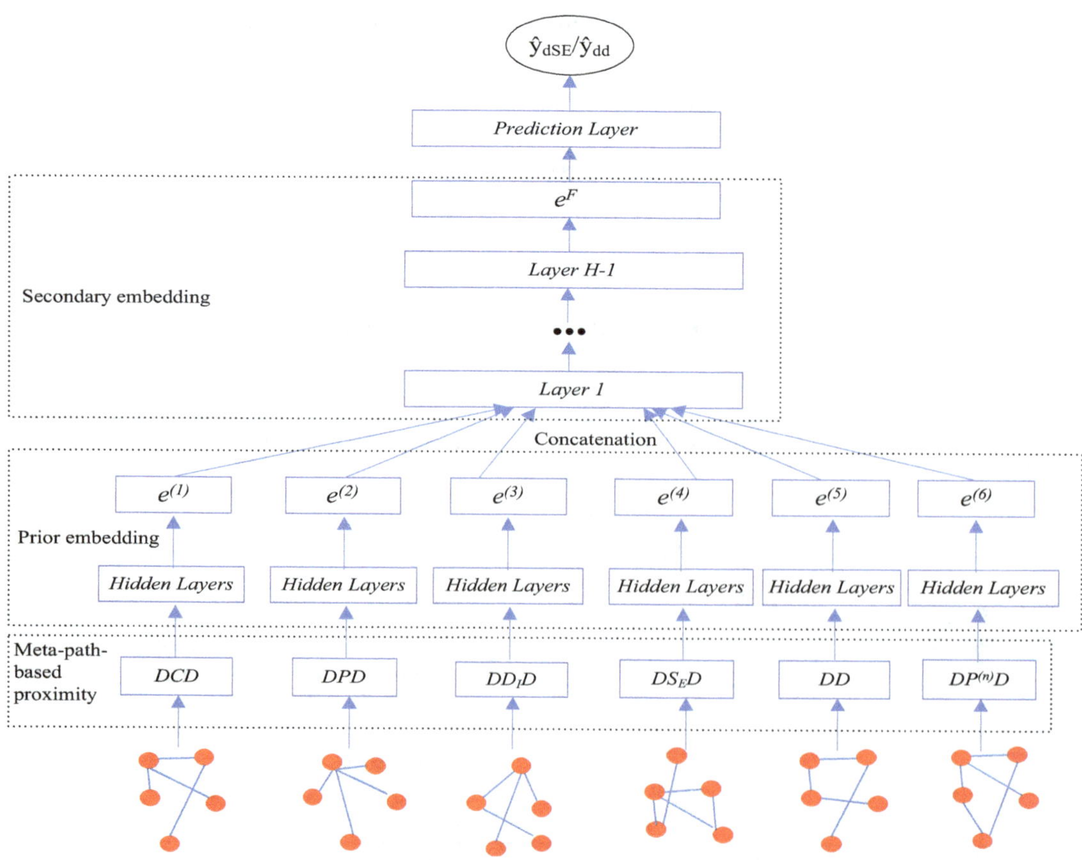

Figure 5. The framework of our proposed semi-supervised deep model SDHINE.

2.3.1. Meta-Path-Based Proximity

We defined three types of meta-path-based proximities and constructed corresponding sub-networks based on the defined proximities.

- Constructing drug-drug interaction sub-network.

For the Drug-Drug (DD) meta-path shown in Table 3, in which the nodes are all drug nodes, we utilized the Jaccard similarity as the edge weights to construct the Drug-Drug interaction sub-network. Notably, we do not consider the interaction types at the drug embedding stage. The proximity between drug i and drug j based on meta-path DD can be calculated as shown in Equation (1).

$$s(i,j) = \frac{|D_i \cap D_j|}{|D_i \cup D_j|} \tag{1}$$

where D_i is a column vector with 0 and 1 elements that represent the Drug-Drug interactions between the ith drug and other drugs.

- Constructing sub-networks using PathSim proximity.

The meta-paths Drug-Chemical-Drug (DCD), Drug-Protein-Drug (DPD), DD_ID, and DS_ED in drug HIN contain different types of nodes and path semantic information. For example, meta-path $D_1C_1D_2$ indicates that drug D_1 and drug D_2 have the same chemical substructure C_1 and there is a path from D_1 to D_2 via C_1. Therefore, we adopted PathSim [29] as the proximity measure in these meta-paths. The PathSim proximity $S(i,j)$ is defined in Equation (2).

$$s(i,j) = \frac{2 \times |\{p_{i\to j} : p_{i\to j} \in P\}|}{|\{p_{i\to i} : p_{i\to i} \in P\}| + |\{p_{j\to j} : p_{j\to j} \in P\}|} \tag{2}$$

where $p_{i\to j}$ is a path instance between i and j, $p_{i\to i}$ is a path instance between i and i, and $p_{j\to j}$ is a path instance between j and j.

The proximities of drugs under meta-paths DCD, DPD, DD_ID, and DS_ED (as shown in Table 3) are directly calculated using PathSim. Then, we constructed corresponding sub-networks using the proximities as the edge weights to form corresponding sub-networks.

- Reconstructing the drug-target sub-network using the target propagation method.

One innovation of our proposed approach is calculating target protein transition probabilities based on the PPI network and reconstructing the drug-target sub-network. As previously mentioned, potential association relationships exist between different biological data, especially for drug target information. When one target protein is activated by a drug, another potential protein may be activated by protein-protein interactions and consequently cause an unreported ADR. Therefore, we should reconstruct the drug-target sub-network using the target propagation strategy according to the meta-path $DP^{(n)}D(n \geq 2)$. The target propagation in the meta-path $DP^{(k)}D$ can be seen as a random walk procedure. A walker walks on the PPI network and achieves the destination protein via $k-1$ steps. Suppose node transition probabilities in the PPI network converge after n steps. The global proximity based on meta-paths $DP^{(n)}D(n \geq 2)$ is:

$$s(i,j) = \sum_{k=2}^{n} S^{(k)}(i,j) = \sum_{k=2}^{n} (DP^{(k)}D)_{ij} \tag{3}$$

In random walk theory, the k-step random walk transition probability is the kth power of the transition probability matrix P. For example, as shown in Figure 4, the probability that drug D_1 acts on protein P_3 is equal to the product of the transition probability from P_1 to P_2 and the probability from P_2 to P_3.

The proximity between drug i and drug j based on meta-path $DP^{(k)}D$ can be unfolded as follows:

$$s^{(k)}(i,j) = (DP^{(k)}D)_{ij} = (DP^kD)_{ij} = D_iPP^{k-2}(D_jP)^T = D_iP \times \underbrace{\text{P} \times \text{P} \times \cdots \times \text{P}}_{k-2} \times (D_jP)^T \tag{4}$$

where P is the transition probability matrix in the PPI network, D is the original drug-target matrix, and D_iP represents the target row vector of the ith drug.

$$s(i,j) = \sum_{k=2}^{n} S^{(k)}(i,j) = \sum_{k=2}^{n} (DP^kD)_{ij} = D_iP(\sum_{k=2}^{n} P^{k-2})(D_jP)^T \tag{5}$$

Given tens of thousands of nodes in the PPI network, calculating P^k is very difficult. However, P^k is very common in the random walk theory. During the random walk procedure, a walker starts from an initial node and moves to neighbors with probability μ and back to the initial node with probability $1-\mu$. Based on the Katz model [31], which is a method of computing similarities between nodes in a graph, taking into account not only the direct edges but also the indirect edges, Equation (5) can be rewritten as follows:

$$s(i,j) \approx D_iP[\sum_{k=2}^{n} (\mu P)^{k-2}](D_jP)^T = D_iP[(I-\mu P)^{-1} - I](D_jP)^T \tag{6}$$

where I is the identity matrix and the damping factor μ usually is 0.98. The inverse of the matrix in Equation (6) can be calculated using the SVD-based matrix factorization method.

2.3.2. Prior Drug Embedding

A stacking auto-encoder is a multi-layer deep neural network based on layer-wise training in which different multi-granularity data features are learned layer by layer and higher complex features are learned in higher layers. To enhance the robustness of sub-network embedding, we adopted stacked denoising auto-encoders (SDAE) in which the input neurons in every layer were randomly discarded by assigning some of the input neurons to 0 with a certain probability.

Traditional SDAE is an unsupervised model, which is composed of the encoder stage and decoder stage. At the encoder stage, the input data x_i are mapped into representation vector space, whereas at the decoder stage, the output data $\hat{x}_i$ are the reconstructed data from x_i. The optimizer objective function of the SDAE is to minimize the reconstruction error of the output and input. The loss function is shown as follows:

$$L_1 = \sum_{i=1}^{n} \|\hat{x}_i - x_i\|_2^2 \tag{7}$$

Here, to protect the meta-path-based proximity of every sub-network, we adopted a semi-supervised SDAE framework [32]. The different meta-path-based proximities are the supervision information that preserves the proximity of the representation of two nodes. The optimizer objective function for this goal is defined as follows:

$$L_2 = \sum_{i,j=1}^{n} S^{(p)}{}_{ij} \left\| e^{(p)}{}_i - e^{(p)}{}_j \right\|_2^2 \tag{8}$$

where $S^{(p)}{}_{ij}$ is the proximity of drug i and drug j based on meta-path p and $e^{(p)}{}_i$ is the embedding of x_i based on the corresponding meta-path.

The objective function of the semi-supervised SDAE model, which combines Equations (7) and (8), is as follows:

$$L = L_1 + \alpha L_2 + \beta L_{reg} \tag{9}$$

where L_{reg} (as shown in Equation (12)) is an $L2$-norm regularizer term to prevent overfitting and α and β are hyperparameters. $W^{(k)}$ and $\hat{W}^{(k)}$ are the kth layer weight matrices at the encoder and decoder stages, respectively.

$$L_{reg}=\sum_{k=1}^{K}(\left\|W^{(k)}\right\|_F^2+\left\|\hat{W}^{(k)}\right\|_F^2) \tag{10}$$

2.3.3. Secondary Drug Embedding

After obtaining the sub-network embeddings, we concatenated the embeddings together and used the secondary semi-supervised stacking denoising auto-encoder to obtain the final drug embedding. Given the embedding node $e^{(p)}{}_i$ of drug i in a different meta-path p, we concatenated them to obtain a new representation vector e_i. Then, we utilized auto-encoder layers to learn the final embedding $e^F{}_i$ of drug i (as shown in Equation (11)).

$$\begin{aligned} e^{F(1)}{}_i &= \sigma(W^{F(0)}e_i + b^{F(0)}) \\ e^{F(2)}{}_i &= \sigma(W^{F(1)}e^{F(1)}{}_i + b^{F(1)}) \\ &\cdots \\ e^F{}_i = e^{F(h)}{}_i &= \sigma(W^{F(h-1)}e^{F(h-1)}{}_i + b^{F(h-1)}) \end{aligned} \tag{11}$$

Here, we continued to adopt a semi-supervised SDAE framework to protect the original proximities of drugs in every sub-network. The supervision information in the optimizer objective function is defined as follows:

$$L_2=\sum_{i,j=1}^{n}\frac{\sum_{p=1}^{K}\alpha^{(p)}S^{(p)}{}_{ij}}{\sum_{p=1}^{K}\alpha^{(p)}}\left\|e^F{}_i-e^F{}_j\right\|_2^2 \tag{12}$$

where $S^{(p)}{}_{ij}$ is the proximity of drug i and drug j based on meta-path p and $\alpha^{(p)}$ is a hyperparameter, which is the weight coefficient of the meta-path p. At the experimental stage, the best hyperparameters $\alpha^{(p)}$ are learned using 10-fold cross-validation on 10% labeled data with a grid search over $\alpha^{(p)} \in \{0.1, 0.2, 0.3, 0.4, 0.5\}$.

The objective function of this part is shown in Equation (13).

$$L_G=\sum_{i=1}^{n}\|\hat{e}_i-e_i\|_2^2+\alpha L_2+\beta L_{reg} \tag{13}$$

2.4. Prediction Formulation

For prediction tasks, learning a classifier that can be generalized to unknown ADRs is desirable. We predict the labels on training data using a fully-connected layer $y = h(e^F{}_i) = \sigma(W^P e^F{}_i + b)$. The prediction loss is formulated by Equation (14).

$$L=\sum_{y\in Y_{tarin}}[-y\ln y'-(1-y)\ln(1-y')]+\lambda\left\|W^P\right\|_F^2 \tag{14}$$

3. Experiment

3.1. Implementation and Evaluation Strategy

Proposed model: We implemented the proposed model with TensorFlow 1.2 and trained the model using the adaptive learning rate optimizer Adam [33]. All neurons were activated by the sigmoid function. We optimized the hyperparameters in the model using validation data and then fixed them for all denoising auto-encoder layers.

Baseline: In addition, we implemented the following three network embedding baselines for comparison:

- Concatenate drug features: This method is a simple original HIN embedding method [28]. The approach constructs a feature vector for each drug by concatenating the PCA representation of each correlation matrix, which represents one aspect of the drug character.
- GraphCNN [34]: GraphCNN is a recently-proposed network embedding method based on spectral convolutional operation and achieves state-of-the-art performance on important prediction problems in recommender systems. Here, first, we linearly integrated similarity matrices based on all meta-paths except the target propagation meta-path $DP^{(n)}D$ and then learned the drug embeddings using the same GraphCNN structure described in [35].
- metapath2vec++ [36]: metapath2vec++ is a heterogeneous information network embedding method based on a meta-path-guided random walk strategy.

For further validation of the impact of target propagation on improving the quality of ADR predictions, we designed a network embedding algorithm without regard for the impact of protein–protein interactions and discarded the PPI dataset; this algorithm was named SDHINE-no-target propagation.

The subsequent ADR prediction methods after the network embedding stage were all based on the same loss function in every prediction task.

Evaluation: We evaluated and compared these algorithms using a 10-fold cross-validation methodology. We randomly selected a fixed percentage (10%) of drugs as the test set and moved all ADRs associated with these drugs from the dataset. The side effects and DDIs of these drugs were all set to 0. The other 90% of the drugs were further divided into the training set and validation set. The training set was formed with 95% of the remaining drugs and was used to train the model. The validation set was formed with the other 5% of the drugs and was used to test the model performance. The independent validation experiments were repeated 30 times with different random divisions of the data for the three sets.

The metrics used to evaluate the model performance were two common ranking metrics: mean average precision at K (MAP@K) and area under the receiver operating characteristic curve (ROC-AUC).

Average precision at K (AP@K) reflects the accuracy of the top-ranked ADRs by a model and can be computed as the mean of Precision@k for each drug or drug pair in the test set. The formula for computing AP@K is given as follows:

$$AP@K = \sum_{k=1}^{K} \Pr ecision(k) / \min(L, K) \quad (15)$$

where Precision(k) is the precision at cut-off k in the return list. L is the total number of true ADRs for the test drug or drug pairs.

3.2. Experimental Results

3.2.1. Visualization Results

First, we compared the performances of all network embedding approaches for the visualization task, which aimed to layout the drug HIN in a 2-dimensional vector space. We mapped the representation vectors of drugs obtained from all comparison approaches to a 2D vector space using the t-SNE [37]. Once a drug is successfully developed, the chemical substructure is fixed. The targets and side effects of a drug are all affected by the chemical substructures of the drug. Therefore, to compare the dimensional reduction performance of different network embedding approaches, the drugs are firstly clustered into different clusters based on their chemical substructures.

The results are shown in Figure 6, in which drugs belonging to the same cluster are represented by the same color. The concatenated drug features method and metapath2vec++ could not separate drugs from different groups. GraphCNN and SDHINE-no-target propagation basically separated drugs from different groups, but some dark green points were mixed with the other groups. The results obtained with SDHINE were the best among these methods, because it separated most of the drugs from the different groups. This result was consistent with the fact that deep integration of different characteristics can effectively eliminate noise from data and recover the original signal.

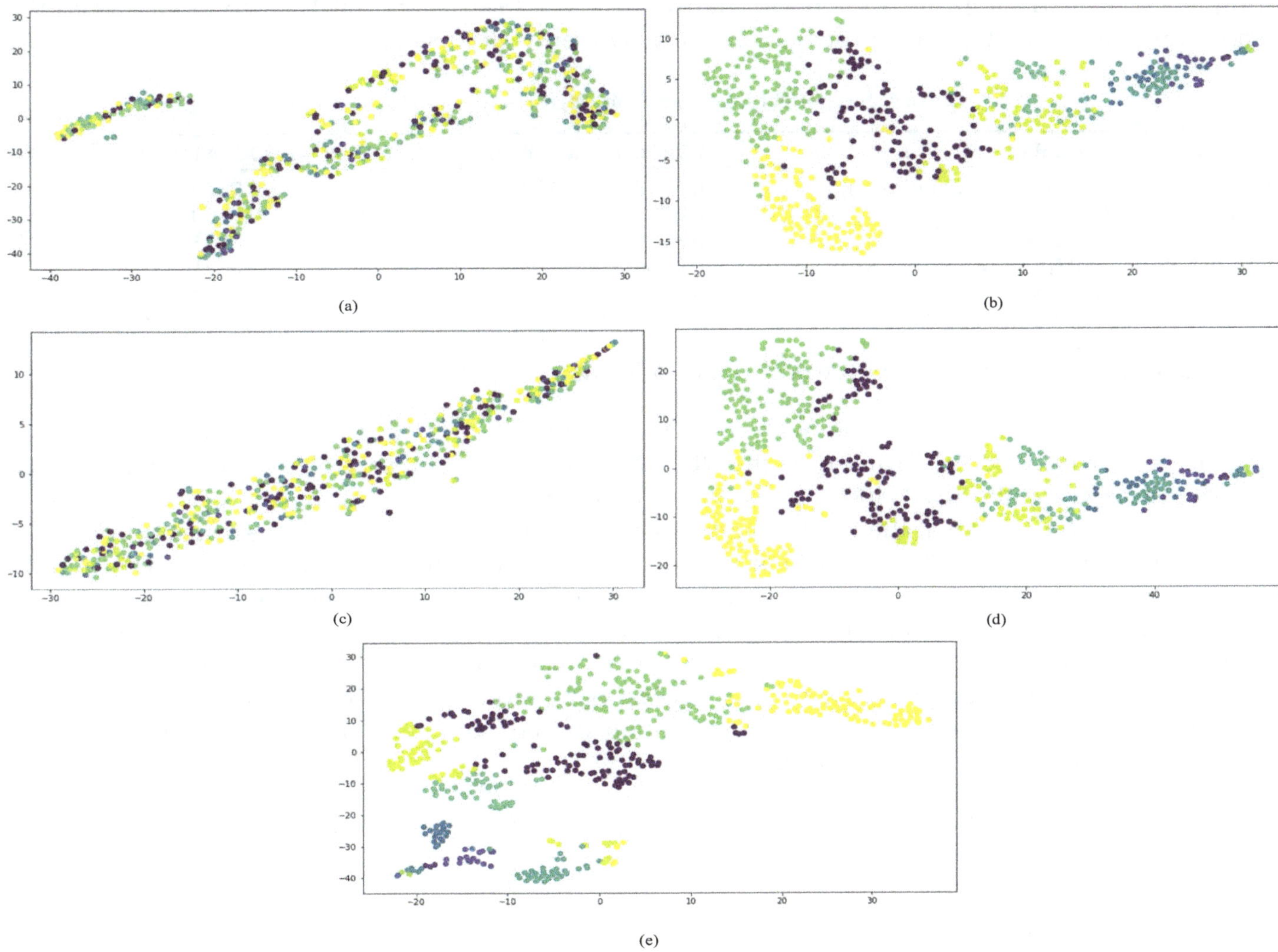

Figure 6. Visualization of the different representations: (**a**) concatenate drug features; (**b**) GraphCNN; (**c**) metapath2vec++ ; (**d**) SDHINE-no-target propagation; (**e**) SDHINE.

3.2.2. Prediction Results

Our experiments further evaluated the drug embeddings obtained through different network embedding methods on different tasks, including side effect predictions for a single drug, binary predictions of the occurrence of DDIs, and multi-label predictions of specific DDI types.

- Task 1: Predicting side effects of a single drug.

To demonstrate the side effect prediction performance based on our network embedding approach, we performed comparison experiments with the aforementioned three baselines and our two proposed models. Predicting the types of side effects caused by one single drug can be formulated into a multi-label classification problem. The output value y of the prediction formulation in Section 2.4 is a column vector with 1318 dimensions, and W^P is a weighted matrix. Each type of side effect was trained one by one. The negative sampling method was adopted to settle the sample unbalanced problem.

Detailed comparisons of the experimental results are shown in Table 4. Our model based on target propagation clearly performed best compared with other models without a target propagation process in terms of MAP@20 and MAP@100. It was also very close to the best result in terms of MAP@50. Analogously, our approach improved ROC-AUC by 5.87% (84.07% vs. 78.20%) compared to the worst result. From the perspective of the approaches based on deep architecture, GraphCNN and the two SDHINE models performed better than the other models. Meanwhile, the model based on target propagation clearly improved the performance by 3.86% (84.07% vs. 80.21%) compared with the similar model without a target propagation process in terms of ROC-AUC.

Table 4. Side effect identification performance comparison.

Models	MAP@20	MAP@50	MAP@100	ROC-AUC
Concatenate drug features	0.5590	0.5475	0.5310	0.7820
GraphCNN	0.6510	**0.6493**	0.6321	0.8190
metapath2vec++	0.5835	0.5760	0.5628	0.7845
SDHINE-no-target propagation	0.6508	0.6416	0.6356	0.8021
SDHINE	**0.6653**	0.6479	**0.6361**	**0.8407**

- Task 2: Binary prediction of the occurrence of DDIs.

When one drug is administered with another drug, the effect of the drug may be changed, and an unknown side effect may be caused by the DDI. Detecting the occurrence of DDIs is preparation for further research on the ADRs induced by DDIs. When predicting the occurrence of a DDI without regard for the type of DDI, the prediction task can be modeled as a binary classification problem. In this situation, a probability value can be the output layer of the prediction formulation in Section 2.4. W^P can be written as a weighted vector. The input layer is formed by the embedding vectors of the two drugs. Table 5 shows a detailed comparison of the experimental results obtained from the binary prediction task of the occurrence of DDIs. The model based on target propagation performed better compared with the similar model without a target propagation process in terms of the mean average precision at k and ROC-AUC. The target propagation strategy and deep architecture were still useful for improving the prediction of DDI occurrence.

Table 5. DDI occurrence identification performance comparison.

Models	MAP@20	MAP@50	MAP@100	ROC-AUC
Concatenate drug features	0.6122	0.5624	0.5432	0.7409
GraphCNN	0.6874	0.6715	0.6219	0.7918
metapath2vec++	0.6542	0.6326	0.5986	0.7332
SDHINE-no-target propagation	0.6813	0.6718	0.6211	0.7814
SDHINE	**0.7015**	**0.6854**	**0.6328**	**0.8124**

- Task 3: Multi-label prediction of specific adverse DDI types.

Compared with the prediction of DDI occurrence, most often, we need to address which types of side effects are caused by the DDI. This issue is a multi-label classification problem in which the output layer y in the prediction formulation is a column vector with 1318 DDI events. The input layer is concatenated by two drug representation vectors, and W^P is a weighted matrix.

Detailed comparison experimental results for specific adverse DDI type identification tasks are shown in Table 6. The model based on target propagation and deep architecture was superior to the models without a target propagation process or deep architecture in terms of not only the mean average precision at k, but also the ROC-AUC value.

Table 6. DDI type identification performance comparison.

Models	MAP@20	MAP@50	MAP@100	ROC-AUC
Concatenate drug features	0.6596	0.6144	0.5045	0.74322
GraphCNN	0.6823	0.6681	**0.6137**	0.7851
metapath2vec++	0.6766	0.6567	0.5118	0.7543
SDHINE-no-target propagation	0.6804	0.6622	0.6119	0.7996
SDHINE	**0.6881**	**0.6745**	0.6126	**0.8175**

Based on the results of the three prediction tasks, the network embedding approach with target propagation performance was superior to the approaches without target propagation processing. Moreover, approaches based on deep architecture performed better than the other linear network embedding methods and the combination methods. This result indicates the feasibility of predicting ADRs based on target propagation and proves that the deep learning process is effective at heterogeneous information network embedding.

3.3. Performance Comparison of Different Embedding Dimensions

To examine the impact of embedding size on prediction performance, we compared SDHINE and SDHINE-no-propagation with different dimensions of drug embeddings for three prediction tasks in terms of ROC-AUC. The results are shown in Figure 7. The prediction performances gradually increased with the increase of embedding dimension and reached the top when embedding dimensions were 64. The prediction performances at 256 dimensions were worse than that at 64 dimensions. This is because the higher dimensional embedding reduced the drug's differentiability. From the results, we also can find that SDHINE performed better than the same model without the target propagation process at the same embedding dimension on all three prediction tasks. It further verified our assumption that target propagation based on PPI can improve ADRs' prediction performance.

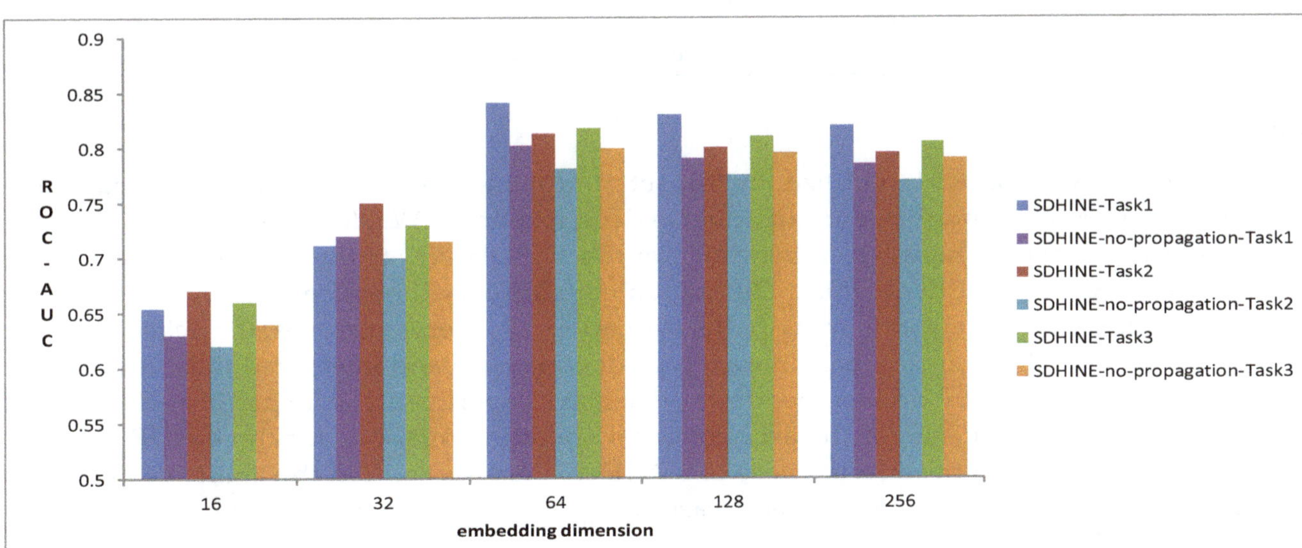

Figure 7. Performance comparison of different embedding dimensions.

3.4. Case Studies

We examined how the proposed network embedding method predicted potential unknown side effects based on learned drug embeddings. In this article, we only can query the drugs that are in the selected datasets by inputting their id. We took triamcinolone, which is an intermediate-acting synthetic glucocorticoid given orally, by muscular or intra-articular injection, or as a topical ointment or cream and is used to treat various medical conditions (e.g., eczema and ulcerative colitis), as an example. It has been confirmed that it may cause many kinds of side effects, such as cough, headache, influenza, and so on. In the SIDER database, triamcinolone may cause 147 kinds of side effects, and 37 of them have been confirmed in preclinical in vitro safety profiling and clinical drug safety trials.

Table 7 shows the top 10 side effects of triamcinolone based on our model. We found from the results that most of the returned side effects were confirmed and that only two side effects were not confirmed. We analyzed the reasons for this result by taking eye redness as an example. First, we analyzed the target protein of triamcinolone from the DrugBank database and found that triamcinolone activated the target protein NR3C1. Then, we calculated the similarity based on the target propagation meta-path and found that NR3C1 might walk to protein EGFR with a high transition probability in the PPI network. EGFR is the only target protein of the drug gefitinib. We searched the side effects of gefitinib in the SIDER and OFFSIDES databases and found that gefitinib had an eye redness side effect. Thus, we found that eye redness is a potential side effect of triamcinolone based on the above logical inference. The works in [38–40] reported that peribulbar injections of triamcinolone may cause intraocular pressure (IOP) elevation, keratitis, and cataract. It is reasonable to believe that eye redness could be one of the ADRs of triamcinolone.

Table 7. Prediction of the top 10 side effects for triamcinolone based on SDHINE.

Top K	Side Effect	Confirmation
K = 1	headache	yes
K = 2	cough	yes
K = 3	fever	yes
K = 4	eye redness	no
K = 5	sneezing	yes
K = 6	nausea	yes
K = 7	rash	yes
K = 8	fatigue	yes
K = 9	dry skin	no
K = 10	conjunctivitis	yes

4. Conclusions

In this work, we proposed to utilize the impact of protein–protein interactions on drug targets to improve the prediction performance of adverse drug reactions. We designed a meta-path-based heterogeneous information network embedding approach (SDHINE) to integrate multi-source drug information, especially the PPI network. Different meta-path-based proximity calculation methods are designed for different semantic meta-paths. We adopted a semi-supervised stacked denoising auto-encoder to learn drug embeddings in each type of meta-path and integrated them into a second auto-encoder to learn the final drug embeddings. Extensive experiments were performed to compare our algorithm with several state-of-the-art network embedding methods for three ADR prediction tasks, which demonstrated the effectiveness of SDHINE. We also verified the ability of SDHINE to distinguish side effect types and performed a case study by examining the impact of protein-protein interactions on side effects.

In this work, we only considered the meta-paths in which the start and end nodes are all drugs. In future work, we will investigate how to use meta-paths starting from other objects (e.g., side effect nodes) under the guarantee of rationality and interpretability. As a major issue in ADR prediction, we will also consider how to further enhance the interpretability of prediction methods and results based on the semantics of meta-paths.

Author Contributions: H.W. guided the research; B.H. designed the experiments and wrote the manuscript; W.Y. analyzed the data; L.W. performed the experiments.

References

1. Giacomini, K.M.; Krauss, R.M.; Dan, M.R.; Eichelbaum, M.; Hayden, M.R.; Nakamura, Y. When good drugs go bad. *Nature* **2007**, *446*, 975–977. [CrossRef] [PubMed]
2. Whitebread, S.; Hamon, J.; Bojanic, D.; Urban, L. Keynote review: In vitro safety pharmacology profiling: An essential tool for successful drug development. *Drug Discov. Today* **2005**, *10*, 1421–1433. [CrossRef]
3. Kanehisa, M.; Goto, S.; Furumichi, M.; Tanabe, M.; Hirakawa, M. KEGG for representation and analysis of molecular networks involving diseases and drugs. *Nucleic Acids Res.* **2010**, *38*, 355–360. [CrossRef] [PubMed]
4. Knox, C.; Law, V.; Jewison, T.; Liu, P.; Ly, S.; Frolkis, A.; Pon, A.; Banco, K.; Mak, C.; Neveu, V. DrugBank 3.0: A comprehensive resource for 'Omics' research on drugs. *Nucleic Acids Res.* **2011**, *39*, D1035. [CrossRef] [PubMed]
5. Kuhn, M.; Campillos, M.; Letunic, I.; Jensen, L.J.; Bork, P. A side effect resource to capture phenotypic effects of drugs. *Mol. Syst. Biol.* **2010**, *6*, 343. [CrossRef] [PubMed]
6. Li, Q.; Cheng, T.; Wang, Y.; Bryant, S.H. PubChem as a public resource for drug discovery. *Drug Discovery Today* **2010**, *15*, 1052–1057. [CrossRef] [PubMed]
7. Yamanishi, Y.; Pauwels, E.; Kotera, M. Drug side effect prediction based on the integration of chemical and biological spaces. *J. Chem. Inf. Model.* **2012**, *52*, 3284–3292. [CrossRef]
8. Li, J.; Zheng, S.; Chen, B.; Butte, A.J.; Swamidass, S.J.; Lu, Z. A survey of current trends in computational drug repositioning. *Brief. Bioinf.* **2016**, *17*, 2–12. [CrossRef]
9. Xu, B.; Shi, X.F.; Zhao, Z.H.; Zheng, W. Leveraging Biomedical Resources in Bi-LSTM for Drug Drug Interaction Extraction. *IEEE Access* **2018**, *17*, 33432–33439. [CrossRef]
10. Vilar, S.; Tatonetti, N.P.; Hripcsak, G. 3D Pharmacophoric Similarity improves Multi Adverse Drug Event Identification in Pharmacovigilance. *Sci. Rep.* **2015**, *5*, 8809. [CrossRef]
11. Labute, M.X.; Zhang, X.; Lenderman, J.; Bennion, B.J.; Wong, S.E.; Lightstone, F.C. Adverse drug reaction prediction using scores produced by large-scale drug-protein target docking on high-performance computing machines. *PLoS ONE* **2014**, *9*, e106298. [CrossRef] [PubMed]
12. Tatonetti, N.P.; Ye, P.P.; Daneshjou, R.; Altman, R.B. Data-Driven Prediction of Drug Effects and Interactions. *Sci. Transl. Med.* **2012**, *4*, 125–131. [CrossRef] [PubMed]
13. Ping, Z.; Fei, W.; Hu, J. Towards Drug Repositioning: A Unified Computational Framework for Integrating Multiple Aspects of Drug Similarity and Disease Similarity. *AMIA Annu. Symp. Proc.* **2014**, *2014*, 1258–1267.
14. Zhang, W.; Chen, Y.; Tu, S.; Liu, F.; Qu, Q. Drug side effect prediction through linear neighborhoods and multiple data source integration. In Proceedings of the IEEE International Conference on Bioinformatics and Biomedicine, Shenzhen, China, 15–18 December 2016; pp. 427–434.
15. Segura-Bedmar, I. Using a shallow linguistic kernel for drug–drug interaction extraction. *J. Biomed. Inf.* **2011**, *44*, 789–804. [CrossRef] [PubMed]
16. Jin, B.; Yang, H.; Xiao, C.; Zhang, P.; Wei, X.; Wang, F. Multitask Dyadic Prediction and Its Application in Prediction of Adverse Drug-Drug Interaction. In Proceedings of the Thirty-First AAAI Conference on Artificial Intelligence, San Francisco, CA, USA, 4–9 February 2017; pp. 1367–1373.
17. Zhang, W.; Chen, Y.; Liu, F.; Luo, F.; Tian, G.; Li, X. Predicting potential drug–drug interactions by integrating chemical, biological, phenotypic and network data. *BMC Bioinf.* **2017**, *18*, 18. [CrossRef] [PubMed]
18. Zitnik, M.; Agrawal, M.; Leskovec, J. Modeling polypharmacy side effects with graph convolutional networks. *Bioinformatics* **2018**, *34*, i457–i466. [CrossRef] [PubMed]
19. Yan, S.; Xu, D.; Zhang, B.; Zhang, H.J.; Yang, Q.; Lin, S. Graph Embedding and Extensions: A General Framework for Dimensionality Reduction. *IEEE Trans. Pattern Anal. Mach. Intell.* **2007**, *29*, 40–51. [CrossRef]
20. Cao, S.; Lu, W.; Xu, Q. Deep neural networks for learning graph representations. In Proceedings of the Thirtieth AAAI Conference on Artificial Intelligence, Phoenix, AZ, USA, 12–17 February 2016; pp. 1145–1152
21. Huang, Z.; Mamoulis, N. Heterogeneous Information Network Embedding for Meta Path based Proximity. Available online: https://arxiv.org/abs/1701.05291 (accessed on 19 Jan 2017).
22. Li, R.; Dong, Y.; Kuang, Q.; Wu, Y.; Li, Y.; Zhu, M.; Li, M. Inductive matrix completion for predicting adverse drug reactions (ADRs) integrating drug–target interactions. *Chemom. Intell. Lab. Syst.* **2015**, *144*, 71–79. [CrossRef]

23. Ma, T.; Xiao, C.; Zhou, J.; Wang, F. Drug Similarity Integration Through Attentive Multi-view Graph Auto-Encoders. Available online: https://arxiv.org/abs/1804.10850 (accessed on 28 Apr 2018).
24. Kelley, B.P.; Sharan, R.; Karp, R.M.; Sittler, T.; Root, D E.; Stockwell, B.R.; Ideker, T. Conserved pathways within bacteria and yeast as revealed by global protein network alignment. *Proc. Nat. Acad. Sci. USA* **2003**, *100*, 11394–11399. [CrossRef]
25. Yeh, C.Y.; Yeh, H.Y.; Arias, C.R.; Soo, V.W. Pathway Detection from Protein Interaction Networks and Gene Expression Data Using Color-Coding Methods and A* Search Algorithms. *Sci. World J.* **2012**, *2012*, 315797. [CrossRef]
26. Codling, E.A.; Plank, M.J.; Benhamou, S. Random walk models in biology. *J. R. Soc. Interface* **2008**, *5*, 813–834. [CrossRef] [PubMed]
27. Zou, Q.; Li, J.; Song, L.; Zeng, X.; Wang, G. Similarity computation strategies in the microRNA-disease network: A Survey. *Brief. Funct. Genom.* **2016**, *15*, 55–64. [CrossRef] [PubMed]
28. Shi, C.; Li, Y.; Zhang, J.; Sun, Y.; Yu, P.S. A Survey of Heterogeneous Information Network Analysis. *IEEE Trans. Knowl. Data. Eng.* **2016**, *29*, 17–37. [CrossRef]
29. Shakibian, H.; Charkari, N.M. Mutual information model for link prediction in heterogeneous complex networks. *Sci. Rep.* **2017**, *7*, 44981. [CrossRef] [PubMed]
30. Chang, S.; Han, W.; Tang, J.; Qi, G. J.; Aggarwal, C.C.; Huang, T.S. Heterogeneous Network Embedding via Deep Architectures. In Proceedings of the 21th ACM SIGKDD International Conference on Knowledge Discovery and Data Mining, Sydney, New South Wales, Australia, 10–13 August 2015; pp. 119–128.
31. Katz, L. A new status index derived from sociometric analysis. *Psychmetrika* **1953**, *18*, 39–43. [CrossRef]
32. Wang, D.; Cui, P.; Zhu, W. Structural Deep Network Embedding. In Proceedings of the 22nd ACM SIGKDD international conference on Knowledge discovery and data mining, San Francisco, CA, USA, 13–17 August 2016; pp. 1225–1234.
33. Kingma, D.P.; Ba, J. Adam: A Method for Stochastic Optimization. Available online: https://arxiv.org/abs/1412.6980 (accessed on 22 December 2014).
34. Kipf, T.N.; Welling, M. Semi-Supervised Classification with Graph Convolutional Networks. Availableonline: https://arxiv.org/abs/1609.02907 (accessed on 9 September 2016).
35. Kipf, T.N.; Welling, M. Variational Graph Auto-Encoders. Available online: https://arxiv.org/abs/1611.0730821 (accessed on 21 November 2016).
36. Dong, Y.; Chawla, N.V.; Swami, A. In metapath2vec: Scalable Representation Learning for Heterogeneous Networks. In Proceedings of the ACM SIGKDD International Conference on Knowledge Discovery and Data Mining, New York, NY, USA, 24–27 August 2017; pp. 135–144.
37. Maaten, L.V.D.; Hinton, G. Viualizing data using t-SNE. *J. Mach. Learn. Res.* **2008**, *9*, 2579–2605.
38. Hashizume, K.; Nabeshima, T.; Fujiwara, T.; Machida, S.; Kurosaka, D. A case of herpetic epithelial keratitis after triamcinolone acetonide subtenon injection. *Cornea* **2009**, *28*, 463–464. [CrossRef] [PubMed]
39. Suarez-Figueroa, M.; Contreras, I.; Noval, S. Side-effects of triamcinolone in young patients. *Arch. Soc. Esp. Oftalmol.* **2006**, *81*, 405–407. [PubMed]
40. Chew, E.Y.; Glassman, A.R.; Beck, R.W. Ocular side effects associated with peribulbar injections of triamcinolone acetonide for diabetic macular edema. *Retina* **2011**, *31*, 284. [CrossRef]

5

Scoring Amino Acid Mutations to Predict Avian-to-Human Transmission of Avian Influenza Viruses

Xiaoli Qiang [1], Zheng Kou [1,*], Gang Fang [1] and Yanfeng Wang [2]

[1] Institute of Computing Science and Technology, Guangzhou University, Guangzhou 510006, China; qiangxl@mail.scuec.edu.cn (X.Q.); yuxiangqd@163.com (G.F.)

[2] Henan Key Lab of Information-Based Electrical Appliances, College of Electrical and Electronic Engineering, Zhengzhou University of Light Industry, Zhengzhou 450002, China; wangyanfeng@zzuli.edu.cn

* Correspondence: kouzhengcn@foxmail.com.

Abstract: Avian influenza virus (AIV) can directly cross species barriers and infect humans with high fatality. Using machine learning methods, the present paper scores the amino acid mutations and predicts interspecies transmission. Initially, 183 signature positions in 11 viral proteins were screened by the scores of five amino acid factors and their random forest rankings. The most important amino acid factor (Factor 3) and the minimal range of signature positions (50 amino acid residues) were explored by a supporting vector machine (the highest-performing classifier among four tested classifiers). Based on these results, the avian-to-human transmission of AIVs was analyzed and a prediction model was constructed for virology applications. The distributions of human-origin AIVs suggested that three molecular patterns of interspecies transmission emerge in nature. The novel findings of this paper provide important clues for future epidemic surveillance.

Keywords: avian influenza virus; interspecies transmission; amino acid mutation; machine learning

1. Introduction

Wild birds are regarded as the natural reservoir of avian influenza virus (AIV) [1]. Interspecies transmission might have been enabled long ago, when wild birds were domesticated by humans. A highly pathogenic subtype of AIV, avian influenza H5N1, originated in Asia in 1996 [2]. Human-origin H5N1 virus was first isolated from clinical samples in 1997, confirming that the H5N1 virus can directly cross species barriers and fatally infect the respiratory system [3,4]. Human infection by H5N1 has been continuously reported since 2003, attracting the attention of both researchers and wider society [5–8]. Moreover, viral subtypes other than H5N1 can infect humans by direct interspecies transmission. Two infectious cases of H9N2 virus have been reported; one in 1999, the other in 2003 [9,10]. H7N7 virus infected farmers in the Netherlands in 2003 [11], and H7N9 has continuously infected China's population since 2013 [12,13].

Interspecies transmission of AIV from its natural reservoir occurs in three steps: (1) the residence of AIVs in their wild animal hosts; (2) AIV contact with humans and direct infection with low probability; and (3) adaptation of AIVs to their new host and efficient human-to-human transmission thereafter. Thus far, AIV has not progressed beyond step 2, which represents initial adaption to the new host and low efficiency of transmission among the new host. The subtype viruses that can cross the species barrier and cause epidemics should be identified. Approximately twenty years has passed since human-originated AIV was first isolated from human samples in 1997. During this period, vast amounts of genomic data have accumulated in public databases. Therefore, after screening the important amino acid sites in the 11 viral proteins, the AIV risk can be predicted by machine learning methods and other mathematic models in the field of bioinformatics [14–18].

AIV transmission relies on amino acid mutations [19–21]. In a previous study, five amino acid factors (AA factors) summarized from 491 highly redundant amino acid attributes were associated with specific physiochemical amino acid properties, namely, polarity, secondary structure, molecular volume, codon diversity, and electrostatic charge [22]. In this paper, we use five AA factors to transform viral proteins and use the random forest (RF) method to select features from high-dimensional protein data and score them by their contributions to the data category. After ranking and screening the positions containing important mutation information, the classifier can predict the interspecies transmission phenotypes.

Two prediction models of AIVs have been published in the literature [23–25]. However, both of these models lack the protein data of hemagglutinin (HA) and neuraminidase (NA), and the biological meanings of the features were not clarified. To construct a more robust and meaningful model, we revise these models and screen the signature amino acid positions in HA, NA, and nine other viral proteins. To this end, we first identify 183 signature mutation positions by RF scoring, then predict AIV occurrence by four popular machine learning methods. Using the most effective classifier, we seek the important amino acid factors and the minimal range of signature positions. The study results will benefit epidemic surveillance and future studies on interspecies AIV transmission.

2. Results

2.1. Dataset

The cleaned dataset contained 869 high-quality AIV strains: 440 avian-origin AIVs (negative samples; H1–H14, H16 subtypes) and 429 human-origin AIVs (positive samples; H5N1, H5N6, H7N3, H7N7, H7N9, and H9N2 subtypes). The information related to these strains is summarized in Table S1.

2.2. Signature Amino Acid Residues

The importance score at each position in the 11 viral proteins was computed by RF. As shown in Figure 1a, the slope of the curve suddenly changes at an importance score of 9. Therefore, 9 was selected as the cutoff score, providing 183 signature positions for further machine learning.

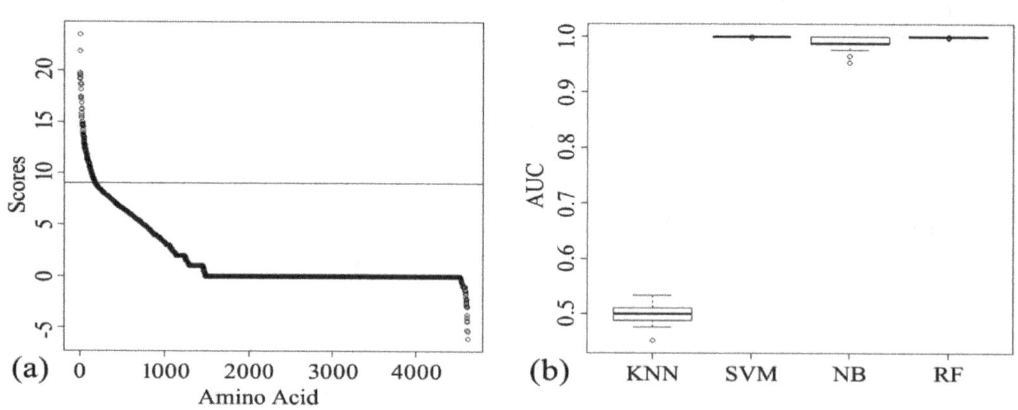

Figure 1. Importance score curve and the performances of k-nearest neighbor (KNN), support vector machine (SVM), naïve Bayes (NB), and random forest (RF) classifiers. (**a**) The ranked scores were calculated from five AA factors using the random forest method. The x and y coordinates denote the total length of the 11 protein alignments and the importance scores, respectively. The cutoff value (9) is indicated by the thin horizontal line. (**b**) Performances of the four classifiers were evaluated from 100 repeats of 10-fold cross-validation. The area under the curve (AUC) ranges from 0 to 1.

As shown in Table 1, the HA protein contained the largest number of signature positions (65 amino acid residues), suggesting that HA is very important for interspecies transmission of AIVs. HA is mainly involved in receptor-binding and fusion activities. Positions HA102–HA290 (Table 1) locate in or close to the region of host receptor binding [26,27], and H163 is reportedly related to the specificity of receptor binding [28]. HA91, HA96, HA328, HA377, and HA397 locate at or near the fusion peptide [29], which triggers fusion activity in acidic environments and favors transmission to humans. The four HA327 positions located in the cleavage site are important virulence sites [30].

NA protein contains 44 signature positions (Table 1). The three NA52s located in the stalk deletion region are related to the virulence and pathogenesis of H5N1 influenza A virus [31]. NA19–NA37 located in the N-terminal are associated with structural stability and enzyme activity [32]. The PB2 627 position has been implicated in increased replication or virulence of AIVs in mammals [33]. PB1 14, located in the binding region of polymerase, is related to viral genome replication [34]. M2 97, which is affiliated with viral particle ensembles [35], was also screened. NEP 14, NP 373, and NP 377 are reportedly involved in intracellular transport of viral proteins [36,37].

Table 1. Scores for the 183 signature amino acids of avian influenza viruses (AIVs).

Num	Pro [1]	Pos [2]	Score	Num	Pro	Pos	Score	Num	Pro	Pos	Score
1	PB2	389	11.95	62	HA	176	13.61	123	NA	65	10.98
2	PB2	478	9.81	63	HA	179	10.08	124	NA	66	9.93
3	PB2	598	17.36	64	HA	185	14.73	125	NA	72	10.96
4	PB2	627	9.83	65	HA	189	14.55	126	NA	79	11.38
5	PB2	648	15.55	66	HA	207	9.49	127	NA	85	9.57
6	PB2	676	9.94	67	HA	211	11.15	128	NA	88	10.13
7	PB1	14	19.16	68	HA	213	11.40	129	NA	100	11.34
8	PB1	48	18.13	69	HA	216	12.17	130	NA	187	10.48
9	PB1	113	18.58	70	HA	221	10.57	131	NA	205	9.62
10	PB1	149	11.09	71	HA	222	9.02	132	NA	233	10.13
11	PB1	257	13.74	72	HA	240	17.36	133	NA	249	9.05
12	PB1	383	12.14	73	HA	251	16.26	134	NA	257	17.24
13	PB1	384	9.34	74	HA	266	10.96	135	NA	265	9.29
14	PB1	387	11.50	75	HA	273	12.53	136	NA	285	10.46
15	PB1	525	9.95	76	HA	274	9.23	137	NA	287	10.65
16	PB1	573	13.38	77	HA	275	9.38	138	NA	288	10.28
17	PB1	628	9.59	78	HA	289	10.36	139	NA	333	10.07
18	PB1-F2	4	9.38	79	HA	290	11.74	140	NA	338	9.02
19	PB1-F2	26	9.24	80	HA	297	10.48	141	NA	347	9.82
20	PB1-F2	48	13.50	81	HA	315	11.98	142	NA	359	10.08
21	PB1-F2	50	11.81	82	HA	323	13.04	143	NA	368	11.05
22	PB1-F2	57	16.85	83	HA	327	12.84	144	NA	369	10.82
23	PB1-F2	77	11.29	84	HA	327	16.23	145	NA	399	11.71
24	PA	37	18.74	85	HA	327	19.25	146	NA	415	9.43
25	PA	61	12.34	86	HA	327	10.41	147	NA	416	13.74
26	PA	63	9.70	87	HA	328	16.24	148	NA	418	9.09
27	PA	129	9.34	88	HA	377	13.91	149	NA	445	12.13
28	PA	337	11.25	89	HA	397	16.18	150	NA	468	9.66
29	PA	356	12.77	90	HA	407	9.49	151	M1	15	9.79
30	PA	367	14.56	91	HA	431	13.52	152	M1	27	12.16
31	PA	405	10.01	92	HA	492	9.49	153	M1	37	14.66
32	PA	554	14.67	93	HA	495	11.15	154	M1	46	14.96
33	PA	607	11.97	94	HA	496	10.62	155	M1	101	13.28
34	PA	684	12.20	95	HA	500	11.88	156	M1	140	12.40
35	PA	712	9.25	96	HA	503	12.76	157	M1	142	11.31

Table 1. *Cont.*

Num	Pro [1]	Pos [2]	Score	Num	Pro	Pos	Score	Num	Pro	Pos	Score
36	HA	40	9.42	97	HA	526	11.91	158	M1	166	17.35
37	HA	42	9.21	98	HA	530	11.26	159	M1	205	11.09
38	HA	45	11.92	99	HA	531	11.67	160	M1	219	13.18
39	HA	46	16.27	100	HA	534	12.77	161	M1	224	23.52
40	HA	53	9.87	101	NP	34	17.45	162	M1	232	14.80
41	HA	57	9.42	102	NP	77	12.39	163	M1	242	19.59
42	HA	65	10.99	103	NP	105	10.61	164	M1	248	11.25
43	HA	66	11.13	104	NP	373	14.73	165	M2	13	13.66
44	HA	79	12.71	105	NP	377	21.88	166	M2	21	10.53
45	HA	81	12.03	106	NP	482	19.71	167	M2	97	15.79
46	HA	84	10.27	107	NA	19	9.20	168	NS1	77	10.59
47	HA	91	17.33	108	NA	23	11.02	169	NS1	80	12.48
48	HA	96	14.98	109	NA	37	9.57	170	NS1	81	12.55
49	HA	102	9.04	110	NA	41	11.30	171	NS1	82	12.01
50	HA	112	12.67	111	NA	42	9.33	172	NS1	83	14.52
51	HA	114	19.46	112	NA	47	10.12	173	NS1	84	10.21
52	HA	115	9.66	113	NA	48	11.23	174	NS1	172	14.21
53	HA	121	10.42	114	NA	49	10.85	175	NS1	179	11.18
54	HA	124	10.28	115	NA	50	9.14	176	NS1	197	9.32
55	HA	131	12.31	116	NA	52	12.38	177	NS1	212	14.19
56	HA	142	12.01	117	NA	52	10.34	178	NEP	14	13.01
57	HA	163	10.07	118	NA	52	9.75	179	NEP	22	15.38
58	HA	164	9.03	119	NA	53	9.03	180	NEP	40	10.28
59	HA	167	14.22	120	NA	58	11.05	181	NEP	60	9.17
60	HA	173	12.81	121	NA	60	9.34	182	NEP	100	10.58
61	HA	174	10.16	122	NA	63	9.44	183	NEP	115	11.10

[1] Viral protein; [2] Position of amino acid residue as H3 subtype numbering.

The AA factors and RF method screened 183 signature positions, some of which are reported to be associated with the mechanism of interspecies transmission. All of the residues were useful, not only for constructing the prediction model but also for further investigating the molecular mechanisms underlying the interspecies transmission of AIVs.

2.3. Performance of the Prediction Model

The performances of the four classifiers are presented as boxplots in Figure 1b. The results were obtained from 100 repeats of 10-fold cross-validation. The area under the curve (AUC) medians in the support vector machine (SVM) and RF classifiers were almost 1. The AUC was clearly lower in the *k*-nearest neighbor (KNN) classifier, possibly because of the nonlinear prediction rules. Although the naïve Bayes (NB) classifier achieved a similar AUC score to the SVM classifier, its performance was poorer and less stable than those of the SVM and RF classifiers. Considering the complexity of the computation, the SVM classifier was selected as the optimal machine learning model for predicting avian-to-human transmission of AIVs.

2.4. Contributions of the AA Factors

The AIV strains were characterized by five AA factors. To understand the mechanism of interspecies transmission, the performance of the SVM classifier was calculated for all combination patterns of these AA factors. The result reveals the importance of the five AA factors. Most of the stable performances of the SVM classifier were contributed by AA Factor 3 or AA Factor 4 (Figure 2a). Notably, the median AUC values were almost 1 and remained stable under AA Factor 3 or AA Factor 4 alone. The SVM classifiers were unstable under AA Factor 1, AA Factor 2, and AA Factor 5 alone. Moreover, AA Factor 3 yielded a slightly better result than AA Factor 4. These results indicate an important role for AA Factor 3 in the avian-to-human transmission of AIVs. Therefore, AA Factor 3 was employed in further analysis.

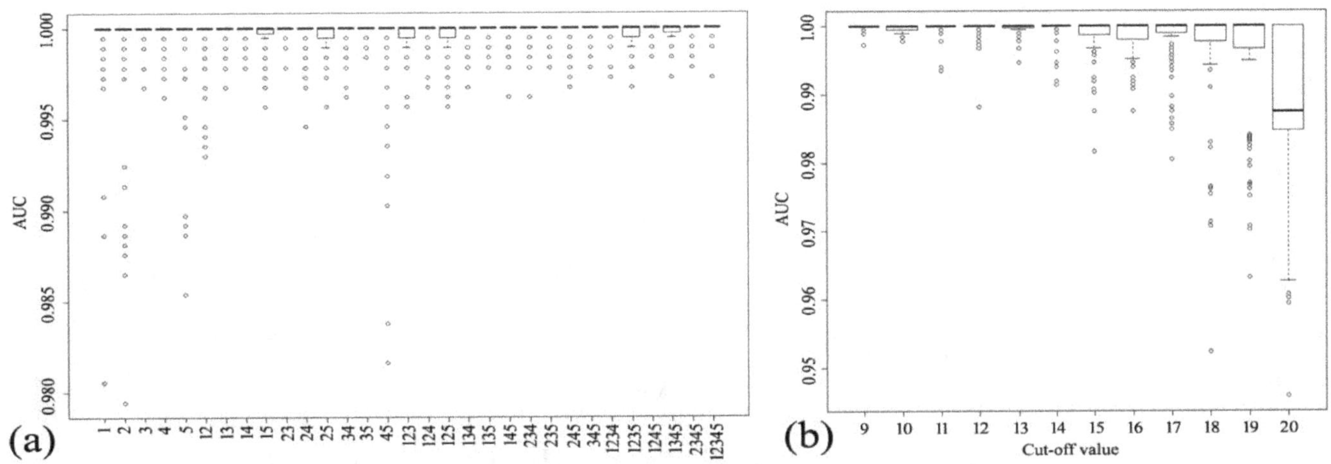

Figure 2. Contributions of AA factors and different mutation sets. (**a**) Performance of SVM classifier for different combinations of the five AA factors. The x and y coordinates denote the 31 combination patterns and the AUC values (from 0 to 1), respectively. Along the x axis, '13' denotes that the set of 183 amino acid residues was transformed using AA Factor 1 and AA Factor 3 together, for example. (**b**) Contributions of mutation positions for different cutoff values (range 9–20). The y coordinate shows the AUC values.

2.5. Contributions of Mutation Positions at Different Cutoff Values

As mentioned above, 183 mutation sites survived a cutoff value of 9. To further explore the mechanism of interspecies transmission, we should reduce the range of crucial positions. To this end, the cutoff value was incremented in steps of 1 (thereby decreasing the number of mutation sites), and the performance of the SVM classifier was calculated with the five AA factors. As shown in Figure 2b, the SVM classifier achieved stable and high performance at cutoffs up to 14. The SVM classifier destabilized at higher cutoffs.

Considering the results under AA factor combinations and cutoff values, the performance of the SVM classifier with AA Factor 3 alone was assessed for different cutoffs. In this situation, the SVM classifier performed stably and well up to a cutoff of 13 (Figure 3a). The analysis results confirm that 13 is the extreme cutoff, giving 50 signature positions (Figure 3b). This set was regarded as the minimal mutation position set for predicting AIVs. We transformed these 50 signature residues using AA Factor 3 alone, and obtained the patterns of the human-origin AIVs (positive samples) by the multidimensional scaling method (Table S2). The resulting clusters are shown in Figure 3c. Cluster 1 comprises three H9N2 viruses (A/Hong Kong/1073/1999; A/Korea/KBNP-0028/2000; A/Bangladesh/0994/2011), two H7N3 viruses (A/Canada/rv504/2004; A/Mexico/InDRE7218/2012), two H7N7 viruses (A/Netherlands/219/2003; A/Italy/3/2013), and one H5N1 virus (A/Hong Kong/482/1997). Cluster 2 includes only H5N1 viruses isolated from 2003 to 2015. Cluster 3 is composed of H7N9 viruses, two H5N6 viruses (A/Yunnan/14563/2015; A/Yunnan/0127/2015), and two H9N2 viruses (A/Hong Kong/308/2014; A/Hunan/44558/2015). The distribution of the human-origin AIVs suggests that three molecular patterns of avian-to-human interspecies transmission emerge in nature.

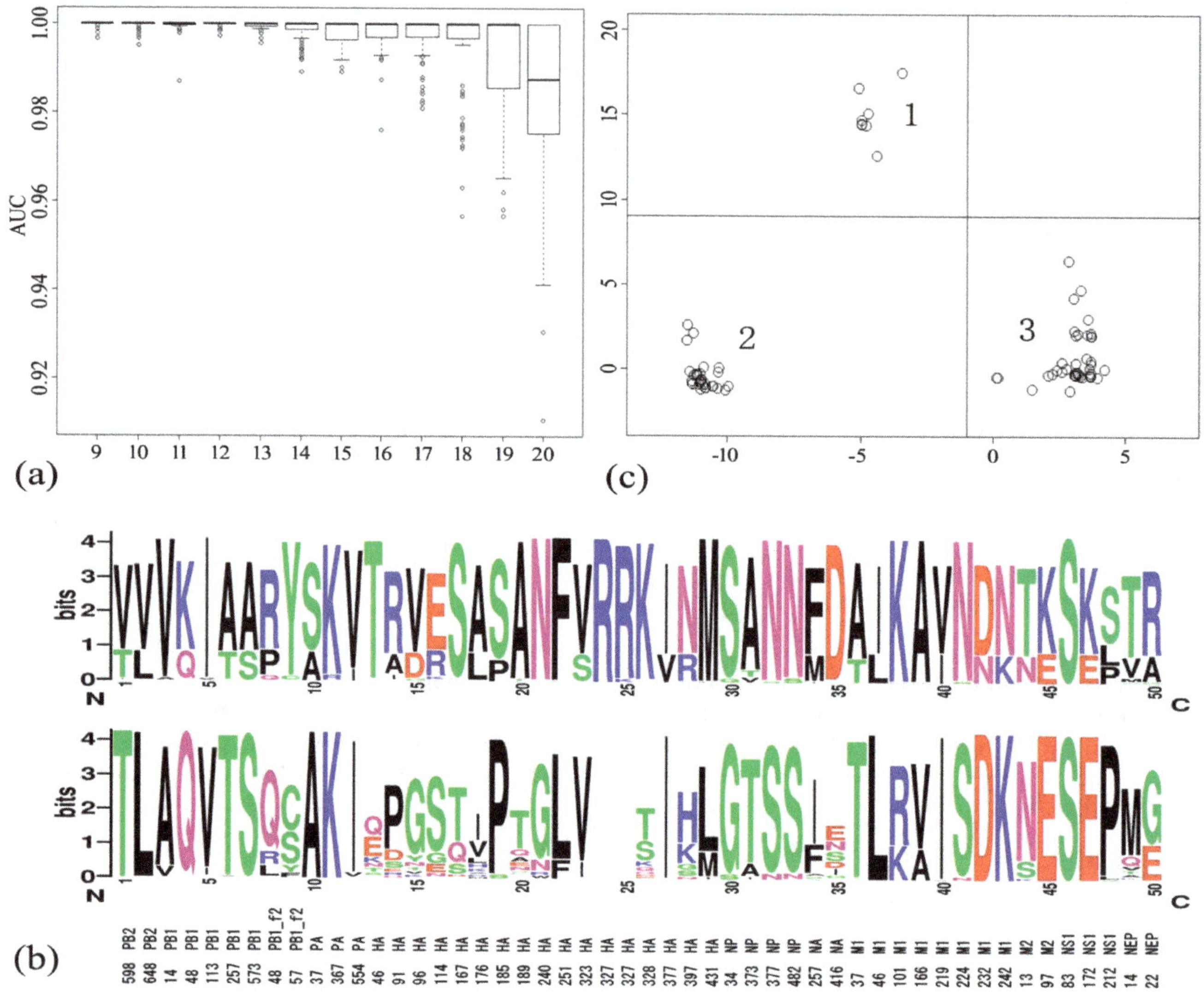

Figure 3. Minimal amino acid set for predicting AIVs. (**a**) Contributions of reduced mutation position sets. The x and y coordinates denote the cutoff (range 9–20) and the AUC values (range 0–1), respectively. (**b**) Profiles of 50 signature positions from human-origin (top) and avian-origin (bottom) AIVs. (**c**) Three patterns of human-origin AIVs clustered by the multidimensional scaling (MDS) method.

3. Discussion

Avian influenza viruses can cross the species barrier, potentially causing a human pandemic. In this paper, human AIV transmission was predicted by a machine learning model with excellent performance (namely, SVM). We firstly screened 183 mutation positions in 11 viral proteins after ranking them by random forest (RF). Most of the screened amino acid positions locate in the important functional regions of receptor binding, fusion peptides, intracellular transport, protein active sites, or virus assembly [26–37]. Some of the residues at these positions have been related to interspecies transmission in earlier reports, such as HA102–HA290 [26,27], H163 [28], HA91, HA96, HA328, HA377 and HA397 [29], HA327 [30], NA52 [31], and PB2 627 [33]. The signature positions guarantee the accuracy of the classifier and are biologically meaningful, which will benefit epidemic surveillance and further studies on interspecies AIV transmission. The proposed method provides important clues for future surveillance and is a useful pre-screening tool for phenotype screening in high-level biological safety laboratories.

The signature positions related with the phenotype of interspecies transmission were screened by the method of random forest. Some yielded a modest score (PB2 627, for example). PB2 E627K was firstly identified in a mouse model [33] and found in the protein of other human-origin avian influenza viruses [12]. In part of the PB2 protein of the human seasonal influenza virus from the public database, PB2 E627 still existed. It is possible that the mutation PB2 E627K is not a strong marker for interspecies transmission, which is consistent with our results. In the future, we need to update the model with new molecular evidence in the field of virology and with more powerful technology in the field of machine learning.

Amino acid mutations in the HA protein are essential for AIV transmission in mammals [21], but mutations in other viral proteins are also necessary [19,20]. Mutations of different proteins introduce synergy and nonlinearity in interspecies transmission. This concept was supported by the present study. Specifically, the linear classifier (the KNN model) showed poor predictive performance on the initial set of 183 signature positions. Moreover, the minimal signature position set was 50 amino acid long and distributed among different viral proteins. This synergistic effect should be notable in further study.

The molecular characteristics of AA Factor 3 are related to molecular size or volume with high factor coefficients of bulkiness, residue volume, average volume of buried residues, side chain volume, and molecular weight [22]. Molecular size or volume is strong related with the binding of biology molecules, such as viral surface protein, host receptor, enzyme, and substrate. In this paper, the AA Factor 3 makes an important contribution to the prediction in terms of high accuracy, which agrees with previous results concerning the receptor binding of viral surface protein [26–28], enzyme activity of viral neuraminidase [32], and RNA binding of viral polymerase [34]. The slightly poor performance of other factors may suggest that host receptor binding, virus partial release triggered by viral neuraminidase, and viral polymerase activity play key roles for the interspecies transmission of avian influenza virus.

The patterns of human-origin AIVs were clarified by the MDS method. Cluster 1 was composed of one H5N1 virus from 1997; three H9N2 viruses from 1999, 2000, and 2011; two H7N3 viruses from 2004 and 2012; and two H7N7 viruses from 2003 and 2013. Cluster 2 contained only H5N1 viruses isolated from 2003 to 2015. Cluster 3 contained H7N9 viruses, two H5N6 viruses from 2015, and two H9N2 viruses from 2014 and 2015. The distribution of human-origin AIVs implies that three molecular patterns of avian-to-human interspecies transmission have emerged. Further investigations on the appearance of novel patterns should be undertaken in future.

The proposed method is applicable to other infectious pathogens that can cross species barriers. As deep learning technology develops, powerful methods that omit feature selection and complex computations might emerge. To better understand the interspecies transmission mechanism of AIVs, the prediction model could be supplemented with information on the host's genetic background [38].

4. Materials and Methods

4.1. Dataset

The avian- and human-origin AIVs were collected from the EpiFlu public database (http://platform.gisaid.org/epi3/frontend) and processed using multiple public bioinformatics tools and algorithms (Figure 4). The details of each procedure are described below.

Step 1: In total, 6305 avian-origin and 644 human-origin AIV strains were obtained from the public influenza virus database. The strains were isolated between January 1996 and February 2016. GISAID deposits high-quality genomic sequences along with their clinical information.

Step 2: Our prediction classifiers were based on eleven viral proteins (PB2, PB1, PBI-F2, PA, HA, NP, NA, M1, M2, NS1, and NEP) with reported roles in interspecies transmission. AIV strains lacking any of these 11 protein sequences in the GISAID database, and strains without subtype information, were excluded in this step.

Step 3: The amino acid residues in the 11 proteins were numbered by the multiple sequence alignment tool MUSCLE [39], using the seasonal human H3 subtype virus as the reference. This step eliminated strains lacking more than 3 amino acids at any protein terminal. The missing residues were replaced by the corresponding residues in the protein sequence with highest identity.

Step 4: To reduce redundancy in the dataset, the AIV strains should differ by at least one amino acid. The amino acid sequences were compared using the CD-Hit tool [40].

Step 5: If the genome sequences of the avian-origin and human-origin AIV strains share high identity, the interspecies transmission capabilities of the avian-origin strains are ambiguous. Therefore, this step eliminated avian-origin strains in which any nucleotide sequence of the eight genome segments shared > 97% identity with that of the human-origin strains. The elimination was performed by the BLAST + tool [41].

Step 6: Ambiguous amino acid residues such as 'X' and 'B' were replaced by the corresponding residues in the protein sequence with highest identity.

The final dataset for predicting AIV interspecies transmission contained 429 positive samples (human-origin AIVs) and 440 negative samples (avian-origin AIVs). All of these strains are listed in Table S1.

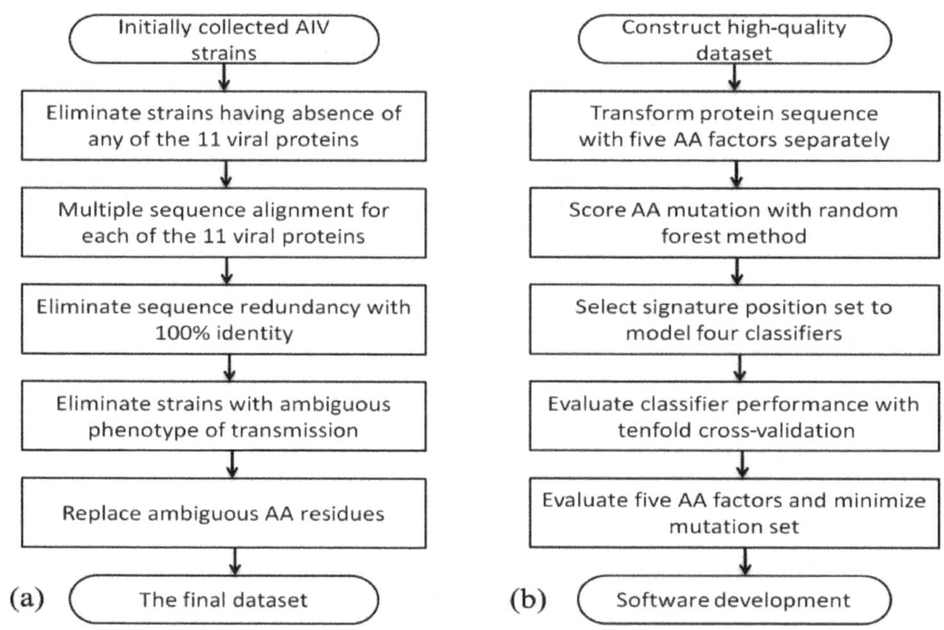

Figure 4. Flowchart of methods used in this paper. (**a**) High-quality dataset construction; (**b**) Machine learning algorism.

4.2. Recognition of Signature Positions

The random forest method is very popularly used for feature selection of prediction problems and can rank the importance of the features in a large scale to discriminate the different categories. The signature positions in the 11 viral proteins were recognized by the RF method (RF, https://cran.r-project.org/web/packages/randomForest/index.html). In each strain, the 11 proteins were concentrated in the following order: PB2 > PB1 > PB1-F2 > PA > HA > NP > NA > M1 > M2 > NS1 > NEP. The proteins with the length of 4620 amino acids were then transformed into numerical sequences of the amino acid factor. Any deletions or insertions in the protein were replaced by zeros. The strains were processed sequentially and accumulated into the total dataset, which was input to the RF. The positive samples (human-origin AIVs) and negative samples (avian-origin AIVs) were classified by their importance scores at each amino acid position. As the classification was based on five factors, the final importance score at each position was the sum of five calculations. Therefore, highly scoring positions were important for distinguishing positive and negative samples. These high scorers were regarded as important amino acid mutations in the interspecies transmission of AIVs. Breiman's random forest algorithm was used as default.

4.3. Constructing the Classifier Model

The machine learning method can solve the classification problem and the numeric features of the positive and negative samples are essential for classification. After screening the signature positions as mentioned above, each strain was represented by an amino acid residue set. These amino acid sets were again transformed into numerical sequences of the five AA factors. Each strain was represented as a numeric vector of length 5N, where N is the number of amino acids in an amino residue set. The interspecies AIV transmission was then predicted by four popular machine learning models that are widely used in bioinformatics and computational biology: (1) support vector machine (SVM, https://cran.r-project.org/web/packages/e1071/index.html), (2) random forest (RF, https://cran.r-project.org/web/packages/randomForest/index.html), (3) naïve Bayes (NB, https://cran.r-project.org/web/packages/e1071/index.html), and (4) *k*-nearest neighbor (KNN, https://cran.r-project.org/web/packages/class/index.html). The present prediction task is a two-class classification problem (in which human-origin and avian-origin AIVs are classified as positive and negative, respectively). The four classifiers were implemented in the R environment and related packages.

The SVM classifier performs the classification in a high-dimensional feature space, which was transferred from the input feature vector with the kernel function. If the samples from two categories were partly overlapped in the original feature space, the SVM will have good performance. In this paper, the optimal hyperplane is determined with the regularization parameter C (C = 1) and the radial basis function (RBF) as default. The RF classifier is an ensemble of many decision trees. Each tree is fully grown using part of the samples in the training dataset selected with the bootstrap technique. The NB is constructed based on the Bayes theorem. Both RF and NB were implemented with the default parameter in the package. The KNN classifier is a nonparametric method to determine a sample category by a majority vote of its neighbors; the number of neighbors in this paper was set to be 3 ($k = 3$).

4.4. Evaluating the Performance of Different Classifiers

The four classifiers were trained on 387 positive samples and 396 negative samples randomly selected from the AIV dataset. The remaining 10% of samples (42 positive and 44 negative samples) were reserved as an independent test dataset for assessing the performances of the classifiers. The classifier performances were evaluated by 10-fold cross-validation and the receiver operating characteristic (ROC) curve. The area under the ROC curve (AUC) reveals the optimal parameters in the four classifiers. To compare the classifier performances, we repeated the evaluation process 100 times and plotted the distributions of the resulting AUC values. The ROC curve relates the true and false positive rates, where both rates range from 0 to 1. The AUC was calculated by the 'ROCR' package in R (https://cran.r-project.org/web/packages/ROCR/index.html). As both rates range from 0 to 1, AUC also ranges from 0 to 1. A higher AUC value denotes a higher performance of the classifier. The human-origin AIVs were shown by the multidimensional scaling method in R (MDS, https://cran.r-project.org/web/packages/MASS/index.html) and the amino acid profile was drawn by the WebLogo server (http://weblogo.berkeley.edu/logo.cgi).

4.5. Prediction Software

By integrating the features at the signature positions with the best-performing classifier, we constructed a software program for predicting avian-to-human transmission of AIVs (delivery by request).

Author Contributions: Conceptualization, X.Q. and Z.K.; Methodology, X.Q. and G.F.; Software, Y.W.; Writing-review and editing, Z.K.

Acknowledgments: We would like to acknowledge the originating and submitting laboratories of the viral sequences from GISAID's EpiFlu Database. We thank Leonie Pipe for manuscript editing.

References

1. Webster, R.G.; Bean, W.J.; Gorman, O.T.; Chambers, T.M.; Kawaoka, Y. Evolution and ecology of influenza A viruses. *Microbiol. Rev.* **1992**, *56*, 152–179. [CrossRef] [PubMed]
2. Xu, X.; Subbarao, K.; Cox, N.J.; Guo, Y. Genetic characterization of the pathogenic influenza A/Goose/Guangdong/1/96 (H5N1) virus: Similarity of its hemagglutinin gene to those of H5N1 viruses from the 1997 outbreaks in Hong Kong. *Virology* **1999**, *261*, 15–19. [CrossRef] [PubMed]
3. Claas, E.C.J.; Osterhaus, A.D.; van Beek, R.; De Jong, J.C.; Rimmelzwaan, G.F.; Senne, D.A.; Krauss, S.; Shortridge, K.F.; Webster, R.G. Human influenza A H5N1 virus related to a highly pathogenic avian influenza virus. *Lancet* **1998**, *351*, 472–477. [CrossRef]
4. Subbarao, K.; Klimov, A.; Katz, J.; Regnery, H.; Lim, W.; Hall, H.; Perdue, M.; Swayne, D.; Bender, C.; Huang, J.; et al. Characterization of an avian influenza A (H5N1) virus isolated from a child with a fatal respiratory illness. *Science* **1998**, *279*, 393–396. [CrossRef] [PubMed]
5. Chen, H.; Smith, G.J.; Li, K.S.; Wang, J.; Fan, X.H.; Rayner, J.M.; Vijaykrishna, D.; Zhang, J.X.; Zhang, L.J.; Guo, G.T.; et al. Establishment of multiple sublineages of H5N1 influenza virus in Asia: Implications for pandemic control. *Proc. Natl. Acad. Sci. USA* **2006**, *103*, 2845–2850. [CrossRef] [PubMed]
6. Li, K.S.; Guan, Y.; Wang, J.; Smith, G.J.; Xu, K.M.; Duan, L.; Rahardjo, A.P.; Puthavathana, P.; Buranathai, C.; Nguyen, T.D.; et al. Genesis of a highly pathogenic and potentially pandemic H5N1 influenza virus in eastern Asia. *Nature* **2004**, *430*, 209–213. [CrossRef] [PubMed]
7. Zhu, Q.Y.; Qin, E.D.; Wang, W.; Yu, J.; Liu, B.H.; Hu, Y.; Hu, J.F.; Cao, W.C. Fatal infection with influenza A (H5N1) virus in China. *N. Engl. J. Med.* **2006**, *354*, 2731–2732. [CrossRef] [PubMed]
8. Shu, Y.L.; Yu, H.J.; Li, D.X. Lethal avian influenza A (H5N1) infection in a pregnant woman in Anhui province, China. *N. Engl. J. Med.* **2006**, *354*, 1421–1422. [CrossRef] [PubMed]
9. Peiris, M.; Yuen, K.Y.; Leung, C.W.; Chan, K.H.; Ip, P.L.; Lai, R.W.; Orr, W.K.; Shortridge, K.F. Human infection with influenza H9N2. *Lancet* **1999**, *354*, 916–917. [CrossRef]
10. Butt, K.M.; Smith, G.J.; Chen, H.; Zhang, L.J.; Leung, Y.H.; Xu, K.M.; Lim, W.; Webster, R.G.; Yuen, K.Y.; Malik Peiris, J.S.; et al. Human infection with an avian H9N2 influenza A virus in Hong Kong in 2003. *J. Clin. Microbiol.* **2005**, *43*, 5760–5767. [CrossRef] [PubMed]
11. Fouchier, R.A.; Schneeberger, P.M.; Rozendaal, F.W.; Broekman, J.M.; Kemink, S.A.; Munster, V.; Kuiken, T.; Rimmelzwaan, G.F.; Schutten, M.; van Doornum, G.J.J.; et al. Avian influenza A virus (H7N7) associated with human conjunctivitis and a fatal case of acute respiratory distress syndrome. *Proc. Natl. Acad. Sci. USA* **2004**, *101*, 1356–1361. [CrossRef] [PubMed]
12. Gao, R.; Cao, B.; Hu, Y.; Feng, Z.; Wang, D.; Hu, W.; Chen, J.; Jie, Z.; Qiu, H.; Xu, K.; et al. Human infection with a novel avian-origin influenza A (H7N9) virus. *N. Engl. J. Med.* **2013**, *368*, 1888–1897. [CrossRef] [PubMed]
13. Cao, H.F.; Liang, Z.H.; Feng, Y.; Zhang, Z.N.; Xu, J.; He, H. A confirmed severe case of human infection with avian-origin influenza H7N9: A case report. *Exp. Ther. Med.* **2015**, *9*, 693–696. [CrossRef] [PubMed]
14. Zeng, X.; Liu, L.; Lv, L.; Zou, Q. Prediction of potential disease-associated microRNAs using structural perturbation method. *Bioinformatics* **2018**. [CrossRef] [PubMed]
15. Zeng, X.; Zhang, X.; Zou, Q. Integrative approaches for predicting microRNA function and prioritizing disease-related microRNA using biological interaction networks. *Brief. Bioinform.* **2016**, *17*, 193–203. [CrossRef] [PubMed]

16. Gang, F.; Zhang, S.; Dong, Y. Optimizing DNA assembly based on statistical language modelling. *Nucleic Acids Res.* **2017**, *45*, e182. [CrossRef]

17. Zeng, X.; Liao, Y.; Liu, Y.; Zou, Q. Prediction and validation of disease genes using HeteSim Scores. *IEEE/ACM Trans. Comput. Biol. Bioinform.* **2017**, *14*, 687–695. [CrossRef] [PubMed]

18. Zeng, X.; Lin, W.; Guo, M.; Zou, Q. A comprehensive overview and evaluation of circular RNA detection tools. *PLoS Comput. Biol.* **2017**, *13*, e1005420. [CrossRef] [PubMed]

19. Herfst, S.; Schrauwen, E.J.; Linster, M.; Chutinimitkul, S.; de Wit, E.; Munster, V.J.; Sorrell, E.M.; Bestebroer, T.M.; Burke, D.F.; Smith, D.J.; et al. Airborne transmission of influenza A/H5N1 virus between ferrets. *Science* **2012**, *336*, 1534–1541. [CrossRef] [PubMed]

20. Imai, M.; Watanabe, T.; Hatta, M.; Das, S.C.; Ozawa, M.; Shinya, K.; Zhong, G.; Hanson, A.; Katsura, J.; Watanabe, S.; et al. Experimental adaptation of an influenza H5 HA confers respiratory droplet transmission to a reassortant H5 HA/H1N1 virus in ferrets. *Nature* **2012**, *486*, 420–428. [CrossRef] [PubMed]

21. Glaser, L.; Stevens, J.; Zamarin, D.; Wilson, I.A.; García-Sastre, A.; Tumpey, T.M.; Basler, C.F.; Taubenberger, J.K.; Palese, P. A single amino acid substitution in 1918 influenza virus hemagglutinin changes receptor binding specificity. *J. Virol.* **2005**, *79*, 11533–11536. [CrossRef] [PubMed]

22. Atchley, W.R.; Zhao, J.; Fernandes, A.D.; Drüke, T. Solving the protein sequence metric problem. *Proc. Natl. Acad. Sci. USA* **2005**, *102*, 6395–6400. [CrossRef] [PubMed]

23. Kou, Z.; Lei, F.; Wang, S.; Zhou, Y.; Li, T. Molecular patterns of avian influenza A viruses. *Chin. Sci. Bull.* **2008**, *53*, 2002–2007. [CrossRef]

24. Qiang, X.; Kou, Z. Prediction of interspecies transmission for avian influenza A virus based on a back-propagation neural network. *Math. Comput. Model.* **2010**, *52*, 2060–2065. [CrossRef]

25. Wang, J.; Kou, Z.; Duan, M.; Ma, C.; Zhou, Y. Using amino acid factor scores to predict avian-to-human transmission of avian influenza viruses: A machine learning study. *Protein Pept. Lett.* **2013**, *20*, 1115–1121. [CrossRef] [PubMed]

26. Stevens, J.; Corper, A.L.; Basler, C.F.; Taubenberger, J.K.; Palese, P.; Wilson, I.A. Structure of the uncleaved human H1 hemagglutinin from the extinct 1918 influenza virus. *Science* **2004**, *303*, 1866–1870. [CrossRef] [PubMed]

27. Hulse, D.J.; Webster, R.G.; Russell, R.J.; Perez, D.R. Molecular determinants within the surface proteins involved in the pathogenicity of H5N1 influenza viruses in chickens. *J. Virol.* **2004**, *78*, 9954–9964. [CrossRef] [PubMed]

28. Mishin, V.P.; Novikov, D.; Hayden, F.G.; Gubareva, L.V. Effect of hemagglutinin glycosylation on influenza virus susceptibility to neuraminidase inhibitors. *J. Virol.* **2005**, *79*, 12416–12424. [CrossRef] [PubMed]

29. Chen, J.; Skehel, J.J.; Wiley, D.C. N- and C-terminal residues combine in the fusion-pH influenza hemagglutinin HA (2) subunit to form an N cap that terminates the triple-stranded coiled coil. *Proc. Natl. Acad. Sci. USA* **1999**, *96*, 8967–8972. [CrossRef] [PubMed]

30. Schrauwen, E.J.A.; de Graaf, M.; Herfst, S.; Rimmelzwaan, G.F.; Osterhaus, A.D.M.E.; Fouchier, R.A.M. Determinants of virulence of influenza A virus. *Eur. J. Clin. Microbiol. Infect. Dis.* **2014**, *33*, 479–490. [CrossRef] [PubMed]

31. Zhou, H.; Yu, Z.; Hu, Y.; Tu, J.; Zou, W.; Peng, Y.; Zhu, J.; Li, Y.; Zhang, A.; Yu, Z.; et al. The special neuraminidase stalk-motif responsible for increased virulence and pathogenesis of H5N1 influenza A virus. *PLoS ONE* **2009**, *4*, e6277. [CrossRef] [PubMed]

32. Barman, S.; Adhikary, L.; Chakrabarti, A.K.; Bernas, C.; Kawaoka, Y.; Nayak, D.P. Role of transmembrane domain and cytoplasmic tail amino acid sequences of influenza a virus neuraminidase in raft association and virus budding. *J. Virol.* **2004**, *78*, 5258–5269. [CrossRef] [PubMed]

33. Hatta, M.; Gao, P.; Halfmann, P.; Kawaoka, Y. Molecular basis for high virulence of Hong Kong H5N1 influenza A viruses. *Science* **2001**, *293*, 1840–1842. [CrossRef] [PubMed]

34. Perez, D.R.; Donis, R.O. Functional analysis of PA binding by influenza a virus PB1: Effects on polymerase activity and viral infectivity. *J. Virol.* **2001**, *75*, 8127–8136. [CrossRef] [PubMed]

35. Iwatsuki-Horimoto, K.; Horimoto, T.; Noda, T.; Kiso, M.; Maeda, J.; Watanabe, S.; Muramoto, Y.; Fujii, K.; Kawaoka, Y. The cytoplasmic tail of the influenza A virus M2 protein plays a role in viral assembly. *J. Virol.* **2006**, *80*, 5233–5240. [CrossRef] [PubMed]

36. Bullido, R.; Gomez-Puertas, P.; Albo, C.; Portela, A. Several protein regions contribute to determine the nuclear and cytoplasmic localization of the influenza A virus nucleoprotein. *J. Gen. Virol.* **2000**, *81*, 135–142. [CrossRef] [PubMed]
37. Iwatsuki-Horimoto, K.; Horimoto, T.; Fujii, Y.; Kawaoka, Y. Generation of influenza A virus NS2 (NEP) mutants with an altered nuclear export signal sequence. *J. Virol.* **2004**, *78*, 10149–10155. [CrossRef] [PubMed]
38. Srivastava, B.; Błazejewska, P.; Heßmann, M.; Bruder, D.; Geffers, R.; Susanne, M.; Gruber, A.D.; Schughart, K. Host genetic background strongly influences the response to influenza A virus infections. *PLoS ONE* **2009**, *4*, e4857. [CrossRef] [PubMed]
39. Edgar, R.C. MUSCLE: multiple sequence alignment with high accuracy and high throughput. *Nucleic Acids Res.* **2004**, *32*, 1792–1797. [CrossRef] [PubMed]
40. Li, W.; Godzik, A. Cd-hit: A fast program for clustering and comparing large sets of protein or nucleotide sequences. *Bioinformatics* **2006**, *22*, 1658–1659. [CrossRef] [PubMed]
41. Altschul, S.; Gish, W.; Miller, W.; Myers, E.; Lipman, D. Basic local alignment search tool. *J. Mol. Biol.* **1990**, *215*, 403–410. [CrossRef]

6

Causal Discovery Combining K2 with Brain Storm Optimization Algorithm

Yinghan Hong [1,2], Zhifeng Hao [1,3], Guizhen Mai [1,*], Han Huang [4] and Arun Kumar Sangaiah [5]

1 School of Computer Science and Technology, Guangdong University of Technology, Guangzhou 510006, China; honyinghan@163.com (Y.H.); zfhao@gdut.edu.cn (Z.H.)
2 School of Physics and Electronic Engineering, Hanshan Normal University, Chaozhou 521041, China
3 School of Mathematics and Big Data, Foshan University, Foshan 528000, China
4 School of Software Engineering, South China University of Technology, Guangzhou 510006, China; hhan@scut.edu.cn
5 School of Computing Science and Engineering, Vellore Institute of Technology, Vellore-632014, Tamil Nadu, India; sarunkumar@vit.ac.in or arunkumarsangaiah@gmail.com
* Correspondence: mgz0323@126.com.

Abstract: Exploring and detecting the causal relations among variables have shown huge practical values in recent years, with numerous opportunities for scientific discovery, and have been commonly seen as the core of data science. Among all possible causal discovery methods, causal discovery based on a constraint approach could recover the causal structures from passive observational data in general cases, and had shown extensive prospects in numerous real world applications. However, when the graph was sufficiently large, it did not work well. To alleviate this problem, an improved causal structure learning algorithm named brain storm optimization (BSO), is presented in this paper, combining K2 with brain storm optimization (K2-BSO). Here BSO is used to search optimal topological order of nodes instead of graph space. This paper assumes that dataset is generated by conforming to a causal diagram in which each variable is generated from its parent based on a causal mechanism. We designed an elaborate distance function for clustering step in BSO according to the mechanism of K2. The graph space therefore was reduced to a smaller topological order space and the order space can be further reduced by an efficient clustering method. The experimental results on various real-world datasets showed our methods outperformed the traditional search and score methods and the state-of-the-art genetic algorithm-based methods.

Keywords: Bayesian causal model; causal direction learning; K2; brain storm optimization

---

1. Introduction

In recent years, the application of causal inference in bioinformatics has become more extensive, and plays a very important role in the development of this field. For instance, it is used for the discovery of the causal relationships between genes and the development of symptoms [1], and how to analyze the phenomenon of synthetic lethality [2,3] in biomedicine, which arises when a combination of mutations in two or more genes leads to cell death. Causal inference is different from the traditional statistical learning methods. The causal inference is the internal generative mechanism of the research data and the traditional statistical learning is the joint distribution of observation variables. The most significant difference between causality and relevance is whether or not to reflect the intrinsic relationship between the data. In scientific research, understanding the causal relationship of objects is crucial to predicting the laws of objects. Causal inference has already been applied in many fields, such as gene therapy, economic prediction, etc.

The problem of causal discovery or causal inference is generally formulated by a probabilistic graphical model where causal directions are represented by the directed edges [4]. In the causal inference algorithm, the techniques commonly used in local causality are conditional independent test (CI) method [5] and score & search method [4].

For example, Peter-Clack algorithms (PC algorithms) [5] determine causal relationships by finding out all the CIs in the given dataset, and the K2 algorithm [1] obtains the maximum score by searching for an optimal structure to discover causal relationships.

In general, a CI test method is used to detect the V-structure, and we can even infer the directed acyclic graph (DAG) [6] by the extension of the partial directed acyclic graph (PDAG). The accuracy of the above methods in causal inference is highly impacted by the number of the detected V-structures. In special cases, for example, without detecting the V-structure, the effect is poor. Therefore, the method cannot completely determine all edges and cannot distinguish the Markov equivalence classes, therefore often fails to uncover the true causal relationships contained in the given dataset if the number of equivalence classes is sufficiently large.

To distinguish causal direction in a non-experimental setting, some researchers recently resorted to using asymmetric relationships among variables under various hypothetical conditions. The additive noise model (ANM) proposed by Shimizu [7,8] is proved to be effective if the given data is generated by following linear non-Gaussian structural equation model. This method was later extended to nonlinear cases for continuous cases [9,10] as well as discrete cases [11,12].

Concretely, the existing ANM-based algorithms can be formulated as follows: assume there are two variables x and y satisfying a causal functional model $y = f(x) + \varepsilon$, where $f(*)$ is an arbitrary square-integrable function and ε is an independent noise of x. If the joint distribution $P(x,y)$ allows an ANM for one (forward) direction rather than the other one (backward), i.e., x cannot be obtained by a function of y plus an independent noise term, then the forward causal direction $x \rightarrow y$ is accepted as the true causal direction. The Post-Nonlinear (PNL) model [13] further extends ANM by making an additional function on the function $f(*)$ such that $y = g(f(x) + \varepsilon)$ with a bijective function $g: R \rightarrow R$. More recently, some researchers have aimed to detect the asymmetry from an information-geometric perspective [14–16]. We can see that these methods assume that reversible and deterministic mappings can get the random variables independently. According to the previous works, these methods are used to examine the asymmetry causality by different techniques, and effect in the low dimension is very good, but poor in the nonlinear high dimensional causal inference between variables.

There are also some hybrid algorithms such as the hybrid algorithm (HYA) [1] and three phases causal discovery method (TPCDM) algorithm [17], to some extent, are able to find the causal relationships of multidimensional networks. The additive noise method (ANM) differentiates the parent nodes and the child nodes in the HYA algorithm and also detects the relationship between the neighbor sets and the sink nodes in the TPCDM algorithm. However, the experimental results show that the effect of the methods above are not very accurate, because it is difficult to detect a one-to-many network structure by ANM methods [10,18–27].

We can see that all these methods for learning causal structure are unreliable, or the time complexity is so very high that we often cannot get the result in an acceptable time. In this situation, we resort to optimization algorithms.

Then, we study the optimization algorithms. Problems existing in many real worlds can be classified as optimization problems. The traditional optimization algorithm is based on a single point, such as gradient descent algorithm, which is a point that moves in the opposite direction of the gradient function. The traditional optimization algorithm mainly solves the problem of a single peak; it is easy to obtain the local optimal solution in the case of complex multiple modes and nonlinear problems.

In recent years, the swarm intelligence (SI) algorithm has been a topical research topic in solving the problem of multiple peaks. Swarm intelligence algorithms are used to solve problems by learning some life phenomena or natural phenomena in nature, which includes the characteristics of self-organization, self-learning and adaptability of natural life phenomena. Especially in 2011, a new

SI algorithm [28] called "Brain Storm Optimization" (also known as Brainstorm optimization, BSO) was proposed, which was inspired by human brainstorming activities. The paper demonstrates the ability of BSO to solve optimization problems by testing two basic functions. Based on the idea of human creative problem-solving, a new swarm intelligence algorithm, Shi's [9] optimization algorithm, was proposed. Unlike traditional swarm intelligence algorithms, such as ant colony optimization (ACO) and artificial bee colony (ABC), the BSO algorithm is the first one to solve the problem based on human creative thinking. Humans are the smartest animals in the world, and the BSO algorithm, which is inspired by their social behavior, is considered a promising method [9]. Shi [9,10] elaborated the thought and implementation process of BSO algorithm, and used the classical test function to simulate the BSO algorithm, and the results showed the effectiveness of BSO algorithm. However, there is still a problem of precocious maturity, and it is necessary to further study to optimize the algorithm itself, improve the effect of BSO algorithm [11].

In this study, we design an efficient method to support causal discovery by combining K2 with Brain Storm Optimization Algorithm (K2-BSO). We use the score returned by the K2 algorithm as the fitness function, and design an elaborate distance function for the clustering step in the BSO according to the mechanisms of K2. The graph space therefore was reduced to a smaller topological order space and the order space can be further reduced by an efficient clustering method. After a optimal causal order is returned by BSO, we run K2 to search for the optimal causal structure, and output the causal skeleton. In the case of high dimensions, the following methods are first used to process the skeleton. We split the causal skeleton into n (the number of variables in the skeleton) smaller sub-skeleton, and employ ANM to detect the causal directions between the target variables and all its parents from each causal skeleton. Consequently, we obtain a partial DAG (PDAG) w.r.t. each sub-skeleton. By merging all the PDAGs, the whole structure corresponding to the high dimensional causal network w.r.t. the given dataset is finally reconstructed. K2-BSO is designed for a certain problem, and the most different thing from other BSO methods should be the clustering procedure, since in the our design, we need to measure the distance between two node sequences in term of the corresponding orders instead of two sequences perset, therefore the existing clustering methodologies used in other BSO methods like those mentioned in [29–31] are not applicable for our case.

The rest of this paper is organized as follows: Section 2 briefly summarizes these definitions. Then we focus on the introduction to the basic concepts, algorithm flow and advantages and disadvantages of K2 and BSO algorithms in Section 3. The details of Causal Discovery combining K2 with Brain Storm Optimization Algorithm are discussed in Section 4. The correctness and performance characteristics of three algorithms are shown in the Section 5. Section 6 gives detailed experimental results. Finally, the conclusions are drawn in Section 7.

2. Definitions

In this section, we will introduce several basic definitions applied in our method. The concepts of the D-separation, V-structure and Additive noise model, which is described as follows:

A causal network can be expressed as a directed acyclic graph (DAG) $G_N = \{V_N, E_N\}$, in which $E_N = \{e_1, e_2, \ldots, e_n\}$ and $V_N = \{x_1, x_2, \ldots, x_n\}$ denote the sets of edges and nodes in G_N.

A. D-separation

Definition 1. *(d-Separation). Assume L is a path from x_i to x_j, and is blocked by a set of variables Z if one of the following conditions holds:*

(1) L either contains a chain, $x_i \leftarrow x_k \leftarrow x_j$, and $x_k \in Z$,

(2) or a fork, $x_i \leftarrow x_k \rightarrow x_j$, and $x_k \in Z$,

(3) or a collider, $x_i \rightarrow x_k \leftarrow x_j$, and $x_k \notin Z$, and no descendent of x_k is contained in Z.

We say a set Z separates two disjointed sets X_i and X_j (X_i, $X_j \subseteq V_D$) if Z blocks each path between X_i and X_j.

B. V-structure

Definition 2. *(V-Structure). Given three variables x, y, and z. If x and z are the parent nodes of y, and no other edge is existing between x and z, then x, z and y together form a V-structure. As shown in Figure 1.*

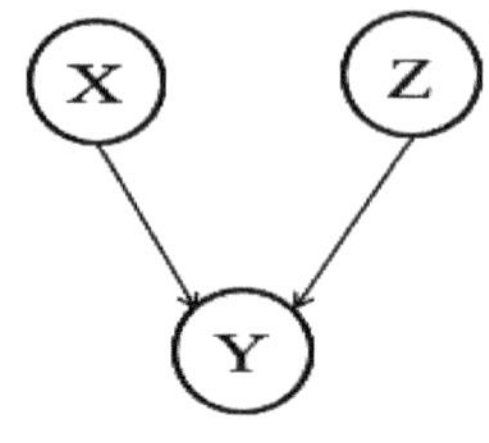

Figure 1. Illustration of a V-structure.

C. Additive noise model

Definition 3. *(Additive noise model (ANM for short)) ANM is represented by a collection of n equations* $S = \{S_1, S_2, \ldots S_n\}$: $S_i{:}x_i = f_i(x_{pa_{(i)}}) + \varepsilon_i$, $i = \{1,2, \ldots ,n\}$, *where* $x_{pa_{(i)}}$ *is the direct parent set of* x_i, *the noise terms* ε_i *are jointly independent, and are independent from* x_i.

It can be seen that the data-generating processes of X can be expressed as:

$$S_i : x_i = \varepsilon_i, i = \{1,\ 2,\ \ldots,\ k\}\ \text{(the root nodes)}$$

$$S_j : x_j = f_j\left(x_{pa_{(j)}}\right) + \varepsilon_j,\ j = \{1,\ 2,\ \ldots,\ n-k\}\ \text{(the other nodes)}$$

As shown aforementioned ANM provides a way for finding casualties by using the assumption of additional noise data generation process rather than satisfying Markov conditions.

3. The K2 and Brain Storm Optimization

In this section, we first introduce the K2 algorithm. Then, the basic concepts, algorithm flow and advantages and disadvantages of BSO algorithm are introduced in detail. All in all, the whole process of the K2 and Brain Storm Optimization can be described as follows.

3.1. The K2 Algorithm

K2 Algorithm, developed by Cooper and Herskovits in 1992, is a Bayesian Network Structure learning algorithm based on the score search method. It is a classical algorithm in the Bayesian Network Structure field with excellent learning performance [32].

As we all know, Bayesian Network Structure Learning aims to find the Bayesian Network Structure B_S which best connects with D through the analysis of data set D. That is the Bayesian Network Structure B_S with maximum posterior probability $P(B_S \mid D)$. Because $P(B_S \mid D) = P(B_S \mid D)/P(D)$ in which $P(D)$ is a constant, what we find is the network structure B_S that maximizes the probability $P(B_S \mid D)$, that is:

$$\max[P(B_S, D)] = c\prod_{i=1}^{n} \max\left[\prod_{j=1}^{q_1} \frac{(r_i - 1)!}{(N_{ij} + r_i - 1)!} \prod_{k=1}^{r_i} N_{ijk}!\right], \tag{1}$$

where c is the a priori probability $P(B_S \mid D)$ of each network structure, which is meant to be a constant because in the algorithm of K2, it is assumed that every network structure B_S has the same probability; n is the number of nodes; r_i is the number of values of node X_i; π_i is parent nodes set of node X_i; q_i is the number of configurations of π_i; N_{ijk} is the sample number of node X_i, which takes the value of k, and its parent set is the jth configuration in data set D; $N_{ij} = \sum_{k=1}^{r_i} N_{ijk}$.

As is showing above, K2 Algorithm uses Equation (1) as a score function to learn the Bayesian Network Structure. From Equation (1), the score function can be decomposed, that is, it can be seen as products of n local structures, which is made up of each node X_i, $i = 1, 2, \ldots, n$ and its corresponding parent nodes set. Then the following equation is derived:

$$g(X_i, \pi_i) = \sum_{j=1}^{q_i} \frac{(r_i - 1)!}{(N_{ij} + r_i - 1)!} \prod_{k=1}^{r_i} N_{ijk}! \tag{2}$$

$$G(B_S, D) = \sum_{i=1}^{n} g(X_i, \pi_i) \tag{3}$$

so we can maximize $G(B_s,D)$ if we maximize every local structure's scores $g(X_i,\pi_i)$, inevitably also maximizing the scores of the whole Bayesian Network Structure (Equation (1)). According to this idea, given nodes order ρ and the upper limits μ of each node's parent nodes, the K2 algorithm can use Greedy Searching to find each node's parent nodes in turn so as to finally construct a whole complete Bayesian Network. The concrete method is as follows: firstly, for every node X_i, i = 1,2, ... ,n, constantly choose the former nodes in former nodes' set from nodes order into parent set π_i of node X_i, making the score function $g(X_i,\pi_i)$ of π_i and X_i continuously increase. The above process cannot stop until function $g(X_i,\pi_i)$ does not increase any more when adding nodes. In that process, we need to limit that the parent node's number should be under μ.

As is known to all that the K2 algorithm has two prerequisites, given nodes order ρ and the upper limits μ of each node's parent nodes. With these two prerequisites, it can obtain a very good learning performance, but in most situations, we can't always meet the above prerequisites, causing difficulties in the application of the K2 algorithm.

3.2. *Brain Storm Optimization*

3.2.1. Brainstorming Algorithm Principle

Inspired by human behavior patterns, in 2011, a human brainstorming process was proposed for the first time by Shi et al., called Brainstorming Optimization Algorithm (BSO). Shi's article expounds the thought and realization process of BSO in detail, and simulates the BSO algorithm with classical test function, and the experimental results show that the BSO algorithm is effective. However, there are some deficiencies in the new algorithm, such as easily falling into local optima, resulting in premature convergence. Therefore, it is necessary to improve the BSO algorithm and optimize the algorithm so as to improve its effect [33–38].

The concept and theory of the basic BSO algorithm is derived from the simulation of the human brainstorming process. A brainstorming meeting needs a moderator, a number of owners to solve problems, and a group of parliamentarians with different backgrounds. Since parliamentarians have different backgrounds, different experiences and different ways of thinking, one problem will get different solutions. The moderator, in accordance with the four Rules of the Conference (see Table 1), presides over the meeting and gets solutions from as many as possible [38–43]. The algorithm needs a skilled host, with no or almost no problem-solving knowledge, so as not to lead host into bias, and also the host cannot engage in new ideas until all ideas are proposed. The host can divide it into K classes, and for each class, people can diversify their thinking and propose better solutions until they get the best solution. The BSO algorithm gets its inspiration from this model and then simulates the process. In the BSO algorithm, the feasible solution of each optimization problem is a quantity of information in the search space, all the information has an adaptive value which is determined by the function of optimization, and then the optimal information is iterated by clustering and learning all kinds of excellent information.

Table 1. Osborn's Original Rules for Idea Generation in a Brainstorming Process.

Rule 1	No bad ideas, every thought is good
Rule 2	Every thought has to be shared and recorded
Rule 3	Most ideas are based on existing ideas, and some ideas can and should be raised to generate new ideas
Rule 4	Try to produce more ideas

The brainstorming session procedure is as follows:

(A) Assemble as many parliamentarians with different backgrounds as possible;
(B) Get the solutions based on the brainstorming rules in Table 1;
(C) Choose a scheme as the best solution for the current problem from each of the problem-solving owners;
(D) Generate new schemes from the schemes selected in C according to the rules in Table 1
(E) Choose a solution from the idea of each problem-solving owner in D as the best solution for the current problem
(F) Randomly select a scheme as a clue to generate new schemes in the case of meeting the Rules in Table 1;
(G) Each problem-solving owner chooses a scheme from F to be the best solution for the current problem;
(H) Get the best solution that is desired by considering merging these programs.

3.2.2. BSO Algorithm Steps

The brainstorming algorithm is mainly composed of two modules: a clustering module and a learning module. In the clustering module, the algorithm uses the clustering method to gather the information into K classes, and the cluster center in each class is the optimal value. The algorithm is optimized by learning, also the information in each class is in parallel. Similarly the local search is promoted, and the algorithm jumps out of the local optimization through the cooperation between classes and the mutation operation, which promotes the global search. The convergence of the algorithm is ensured by the optimization process of cluster center, and the process of optimizing the information variation in the class ensures the diversity of the algorithm population. Each individual in the BSO algorithm represents a potential problem solution that is updated by the individual's evolution and fusion, a process similar to that of the human brainstorming process [44–46].

The implementation of BSO algorithm is simpler:

(1) Obtain the solution of n potential problems, then divide n individuals into M class by K-means method, the individual in each class is sorted by evaluating these n individuals, and the optimal individual is selected as the central point of the class;

(2) Randomly select the central individual of a class and determine whether it is replaced by a randomly generated individual according to the probability;

(3) to update the individual, the way is updated by the following four ways: (a) randomly select a class (the probability of selection is proportional to the number of individuals within the class), the random perturbation is added to the class center to produce a new individual; (b) randomly select a class (the probability of selection is proportional to the number of individuals within the class) and randomly select an individual in the selected class, plus a random perturbation to produce a new individual; (c) randomly selected two classes, the fusion of the class center and the random perturbation to produce a new individual; (d) randomly select two classes, randomly select an individual in each class, and then add a random perturbation to create a new individual. By adjusting the parameters to control the proportion of the above four ways to produce new individuals. After the new individual generation, compared with the original individual, the final

selection of the best one to retain to a new generation of individuals, repeat the above operation, the n individual to update each one, produced a new generation of n individuals.

This loops until the upper limit of the preset individual update algebra is reached. In the third step, the update of the individual has four ways to produce a new individual process; the selected amount of information plus a Gaussian random is worth the new amount of information, such as the following Equation (4):

$$X^d_{new} = X^d_{selected} + \varepsilon \times n(\mu, \sigma), \tag{4}$$

where X^d_{new} is the d dimension of the new information, $X^d_{selected}$ is the d dimension of the selected information, $n(\mu, \sigma)$ is the Gaussian function whose mean value is μ and variance is σ; ε is a weight coefficient which is described by Equation (5):

$$\xi = \log sig((0.5 \times \max\_iteration - current\_iteration)/k) \times rand(), \tag{5}$$

where log*sig*() is a s-type logarithmic transfer function, and *max_iteration* is the maximum number of iterations, while *current_iteration* means the current number of iterations; k can change the slope of the function log*sig*(), *rand*() is the random value between (0,1).

4. The K2-BSO Method

In this section, the details of the K2-BSO method are given, we show that this method is able to discover causation combining K2 with the Brain Storm Optimization algorithm. All in all, the whole process of causality is deduced, which is described as follows:

4.1. Skeleton Learning Phase Based on K2-BSO

The Additive Noise Model (ANM) could find out the causal relationships correctly between variables in sparse causal networks, but this model would encounter multiple challenges when applied to high-dimensional complex network structures [12]. First of all, high-dimensional causal networks contain a large number of variables, and the causal relationships between them are very complex, so the algorithm requires the ability to quickly search. Causal relationship references based on the traversal method will face all possible network structures, which leads directly to the insufferable computational complexity, the storage space overflow and other problems. The K2 algorithm needs to satisfy two prerequisites, given nodes order and the upper limits of each node's parent nodes. However, it is difficult to make it in fact. What's more, the K2 algorithm is easy to fall into the local optimal solutions while the BSO algorithm could get rid of local optimizations. Therefore, the combination of the algorithm K2 and BSO can effectively solve the structural learning problem of causal network structure. As discussed in the previous section, there are three points we need to note:

(1) What needs to be optimized is the causal order that will highly affect the accuracy of K2. Generally, an input order approaching the actual topological order of the underling causal network will return the highest score and most similar causal structure.
(2) The fitness function is easy to be chosen, that is the score return by K2.
(2) The clustering method of BSO should be redesigned; all the distance function likes [46] cannot be directly applied to this case, as what we consider is the topological order. We design a new distance function like this:

Step I. Given two orders R_1 and R_2, for each variable in R_1, we find the same variable in R_2, assume it is v_1.

Step II. Consider n variables in front of v_1 in R_1, and m variables in front of v_1 in R_2, we calculate the number of the repeated variables in $n + m$ variables.

Step III. By literately sum up the repeated variables w.r.t. every variable in R_1 (or R_2), we get a number, and let this number as the distance between R_1 and R_2.

We note that, the clustering step is crucial to the BSO, as shown before, our distance function is designed based on the mechanism of K2, which will highly improve the clustering performance in BSO.

4.2. Direction Learning Phase

Algorithm 1 can obtain the skeleton of network returned by K2-BSO. Because the K2 can only examine a set of Markov equivalence classes rather than the realistic causal structure, we aim to detect the remaining directions of the output skeleton for distinguishing this equivalence in this section. Because of the existence of Markov equivalence classes, the structural learning methods are generally difficult to infer all causal direction. On the other hand, the ANM provides an effective way to learn causal direction in low-dimensional cases. Note that, we get the causal skeleton, then we can separate the causal skeleton S into n sub-skeletons ($S_i, \ldots, S_n$) which contain a target node X_i and all its neighbor nodes N_i according to S. In general, these sub-skeletons are generally low-dimensional and therefore can be solved by using ANM. The way to orient the edges of a skeleton in ANM method is described as follows:

Algorithm 1. Skeleton learning based on K2-BSO.

Input: dataset X, population size $|V|$.
Output: the skeleton w.r.t. X.
Randomly generate n potential causal order $R = R_1 \sim R_n$;
Cluster R into m clusters $C = C_1 \sim C_m$;
For each R_i
$\text{Score}_\text{i} = \text{K2}(R_i)$;
End For
$\text{Score} = \text{Score}_1 \sim \text{Score}_\text{n}$;
$\text{R}_\text{optimal} = \text{BSO}\ (X, R, \text{Score}, C)$;
$\text{G}_\text{optimal} = \text{K2}(\text{R}_\text{optimal})$;
$\text{X} = \text{G}_\text{optimal}$;
return the causal skeleton X.

Firstly, consider a given dataset $X = \{X_1, X_2, \ldots, X_n\}$ with index $V = \{1, 2, \ldots, n\}$. X corresponds to an n-dimensional DAG $G = \{V, E\}$, where E represents the edges of V. Assume that X is generated by the following way: each variable $X_i \in X$ corresponds to one node $i \in V$ in G, and is determined by a causal function $X_i = f_i(x_{pa_{(i)}}) + \varepsilon_i$ in which f_i is nonlinear, $x_{pa_{(i)}}$ is the parent of x_i. The noise terms ε_i have a non-Gaussian distribution and are jointly independent.

In the issue of seeking out the causal direction, we aim to seek out all the parent nodes (contained in N_i) amount to each target X_i from S_i. On the basis of the mechanism of ANM, we denote the homologous remains between X_i and each candidate parent set C_{ik} as $X_i = f(C_{ik}) + \varepsilon_i$ by using GPR, and we test whether C_{ik} and ε are statistically independent. If they are independent we accept the model $C_{ik} \rightarrow X_i$; if not, we deny it. In this phase, we measure the independence by using the kernel-based conditional independence (KCI) test. The details of causal directions inference from a output causal skeleton is presented in Algorithm 2.

Algorithm 2. Learning causal direction from a sub-skeleton.

Input: sub-skeleton S_i and the corresponding target node X_i with all its neighbors N_i.
Output: the direction between X_i and (partial) N_i.
For each candidate parent set C_{ik};
fit X_i and C_{ik} to ANM;
if ε is independent of C_{ik} **then**
  accept $C_{ik} \rightarrow X_i$;
end if
end for

4.3. K2-BSO Framework (Algorithm 3)

We first present the details of the K2-BSO method:

Step 1. Learning the causal skeleton S by algorithm 1.
Step 2. Split S into n sub-skeleton $S_1, \ldots, S_n$ according to each node X_i contained in S.
Step 3. Perform Algorithm 2 for each sub-skeleton S_i.
Step 4. Merge all the partial results and output the final causal structure.

Algorithm 3. K2-BSO framework.

Input: Dataset X, threshold k
Output: Causal structure G.
1. Set Dimension X to n;
2. **if** $(n < k)$ **then**
3. $S = Algorithm\ 1(X);\ G = Algorithm\ 2(X,\ S);$
4. **else**
5. Split S into n sub-skeleton $S_1, \ldots, S_n$ according to each node X_i contained in S;
6. **For** each S_i in S
7. $S_i = Algorithm\ 1(X_i);\ PDAG_i = Algorithm\ 2(X_i,\ S_i);$
8. Merge all $PDAG_i$ to G;
9. **End for**
10. **end if**
11. **return** the final causal structure G.

5. The Correctness and Performance of the Algorithms

In this part, we analyze theoretically about the respective characteristics of the correctness and performance with the three algorithms (K2-Random, K2-BSO, K2-GA).

First, we discuss the K2-Random algorithm. It is a traditional method, and there is not much optimization process. The main process is: first step, randomly obtain p data sort, then sort the score from the top to the bottom and select the highest score. The second step is to continue to randomly obtain p data sort, found the highest score Tscore, until 10 consecutive times are the same highest score, and end the program; this method is very easy to enter the local optimization state, but the experimental result is unstable.

Second, we discuss the K2-BSO algorithm, which is the method proposed in this paper. It is better to avoid local optimization problems. The main process is: first step, randomly obtain p data sort, then sort the score from the top to the bottom and obtained m data sort by clustering method. The second step is to obtain m new subclasses by random perturbation about m subclasses by the BSO algorithm. Then we reevaluate the score until the score converges.

Third, we discuss the K2-GA algorithm. The main process is: The main process is: first step, randomly obtain p data sort, Then sort the score from the top to the bottom and select the highest score until the score converges. The second step is to obtain p new data sort by means of Genetic Algorithm (GA) method with randomly perturbation the highest ranking data. Then sort the score from the top to the bottom to obtain the highest score Tscore.

In summary, the first algorithm in time complexity is the best, but the accuracy rate is the lowest and unstable; the second algorithm and the third algorithm's time complexity are the same, especially with the increase of network dimensions, second algorithms tend to advance convergence faster than the third algorithms, and the accuracy of the second algorithms is better than the third algorithm. Next, we'll use real data to validate three algorithms in the next chapter.

6. Experiments

In this section, we evaluate our proposal on eight real-world datasets that cover a variety of applications including Small Networks (Asia, Sachs), Medium Networks (Child, Alarm),

Large Networks (Barley, Win95pts), and Very Large Networks (Pigs, MINUN) that cover a variety of applications, including, medicine (ASIA, SACHS, CHILD and ALARM), agricultural industry (BARLEY), system troubleshooting (WIN95PTS) and bioinformatics (PIGS and MUMIN) are available at "http://archive.ics.uci.edu/ml/datasets.html". The structural statistics of the eight networks are summarized in Table 2.

Table 2. Statistics on the network.

Network	Nodes	Edges	Avg Degree	Maximum in-Degree
ASIA	8	8	2	2
SACHS	11	17	3.09	3
CHILD	20	25	1.25	2
ALARM	37	46	2.49	4
BARLEY	48	84	3.5	4
WIN95PTS	76	112	2.95	7
PIGS	441	592	2.68	2
MUMIN	1041	1397	2.68	3

In this group of experiments, our proposed method is compared with other two mainstream causal discovery methods—K2-Random (Causal Discovery combining K2 with Random) method and K2-GA (Causal Discovery combining K2 with Genetic Algorithm) method. We evaluate these methods by different sample size at 250, 500, 1000, 2000, respectively. We use three criteria, Recall, Precision, and F1 to evaluate these methods, which are defined as follows:

$$Recall = (Inferred\ directions \cap Actual\ directions) / (Actual\ directions), \tag{9}$$

$$Precision = (Inferred\ directions \cap Actual\ directions) / (Inferred\ directions), \tag{10}$$

$$F1 = (2 \times Recall \times Precision) / (Recall + Precision) \tag{11}$$

Obviously Precision is the actual fraction of inferred causality with respect to a true graph. Similarly, Recall is the part of actual causality found by the algorithm. F1 is the organic combination of Precision and Recall which can serve as the accuracy standard for our algorithms.

The experimental environment is as follows:

(1) CPU of the physical host: CPU E5-2640 v3, 2.60 GHz (2-way 8-core);
(2) Platform belongs to the cloud platform version from Bingo Cloud: v6.2.4.161205143;
(3) Memory is 24 G.

As shown in Table 3, we can see that the K2-Random runs much faster than the other two algorithms. However, as showed in Figure 2, the accuracy of K2-Random is lowest, this means K2-Random easily falls into a local optimum. One can imagine that if we use K2-Random to test all possible causal orders detailed we can obtain the best score, but we usually cannot get the final result in an acceptable time, because the time complexity of such an exhaustive algorithm reaches the upper limit.

On the other hand, we can see that in the small networks, K2-GA runs faster than K2-BSO. However, as the size of the networks grows, the running time of K2-BSO increases slower than that consumed by K2-BSO, and the running time of the two methods tend to be very close. We can see that in the case of WIN95PTS, K2-BSO runs much faster than K2-GA. What is the most different between K2-BSO and K2-GA in the task is that K2-BSO performs a clustering step, which can greatly reduce the convergence time. Recall that, the clustering step in K2-BSO also costs time. Therefore, when the causal network spends more time in clustering step, K2-BSO is probably slower than K2-GA, while for a network to spend less time in the clustering step, theoretically K2-BSO runs much faster than K2-GA. Accordingly, the specific structure of a certain causal network weighs heavily on total time.

Table 3. Comparisons between three algorithms on execution time.

Dataset	Sample	K2-Random			K2-BSO			K2-GA		
		Best	Mean	Worst	Best	Mean	Worst	Best	Mean	Worst
ASIA	250	1.2555	1.749	2.6157	3.0185	4.5585	5.8543	2.8367	4.2504	5.5427
ASIA	500	1.085	1.4656	1.9224	3.2043	4.3753	6.3392	2.1125	4.8925	6.4698
ASIA	1000	1.5104	1.7994	2.2099	4.7037	5.0472	5.3491	3.1055	5.1146	8.2191
ASIA	2000	2.2676	2.4993	2.6629	3.3815	3.9774	4.3823	4.1878	5.506	8.1496
SACHS	250	4.5248	5.025	5.3196	6.2266	6.4963	6.6793	3.02	4.6432	7.501
SACHS	500	2.4084	3.6746	5.2433	8.7819	11.2107	15.6879	5.5198	8.1696	8.6039
SACHS	1000	2.7062	3.5759	4.0197	8.3591	11.7962	15.3562	5.4356	8.6128	12.8977
SACHS	2000	3.4966	4.6789	6.4649	9.3677	10.7807	11.5173	6.1085	6.1516	6.1762
CHILD	250	7.7092	9.117	12.5631	28.0133	29.961	31.325	21.1455	25.986	30.974
CHILD	500	8.9414	11.8924	14.1062	17.7665	28.9292	41.265	30.2484	34.1696	38.9523
CHILD	1000	10.059	17.5069	28.2669	33.4957	38.4505	43.9597	23.6723	24.608	25.3404
CHILD	2000	10.709	13.8898	18.3508	30.1317	58.6901	78.3929	21.3343	38.465	47.0478
ALARM	250	46.647	71.9888	86.3291	217.3178	266.3269	355.1756	147.9969	283.8744	371.7365
ALARM	500	94.125	103.2041	119.5914	227.3739	293.7428	377.54	133.6973	388.0371	519.749
ALARM	1000	62.140	92.5241	125.1668	157.303	247.1288	294.8618	144.2063	316.8503	475.2459
ALARM	2000	95.002	159.3802	232.1197	323.0716	394.6686	496.8733	219.1187	275.1111	345.3221
BARLEY	250	66.378	91.2914	136.5331	198.4244	325.1975	400.4625	160.181	205.9434	233.3933
BARLEY	500	75.057	99.7028	138.1832	304.5911	364.1937	368.7941	194.9004	358.4717	567.8317
BARLEY	1000	86.012	100.4193	116.8377	326.8585	370.2588	404.4023	255.2328	396.0264	505.5334
BARLEY	2000	96.159	103.1696	116.3037	478.3594	525.7576	549.7203	368.8762	810.5905	1.48×10^3
WIN95PTS	250	649.75	1.10×10^3	1.52×10^3	1.81×10^3	4.44×10^3	7.16×10^3	1.60×10^4	2.06×10^4	2.31×10^4
WIN95PTS	500	555.26	727.4609	819.2716	2.33×10^3	4.76×10^3	6.67×10^3	6.23×10^3	1.83×10^4	2.48×10^4
WIN95PTS	1000	693.01	746.7864	827.4684	2.73×10^3	4.38×10^3	6.00×10^3	2.01×10^4	2.57×10^4	3.19×10^4
WIN95PTS	2000	715.72	1.43×10^3	1.85×10^3	2.25×10^3	7.04×10^3	1.50×10^4	2.13×10^4	3.44×10^4	4.98×10^4
PIGS	250	1.87×10^4	2.52×10^4	3.96×10^4	2.60×10^5	3.89×10^5	4.85×10^5	1.45×10^5	2.48×10^5	3.77×10^5
PIGS	500	5.35×10^4	6.84×10^4	8.53×10^4	3.09×10^5	4.03×10^5	5.24×10^5	1.58×10^5	2.56×10^5	3.03×10^5
PIGS	1000	7.12×10^4	8.64×10^4	9.71×10^4	2.92×10^5	4.14×10^5	4.59×10^5	1.85×10^5	2.70×10^5	3.57×10^5
PIGS	2000	6.17×10^4	9.00×10^4	1.09×10^5	4.26×10^5	5.26×10^5	7.05×10^5	1.98×10^5	2.74×10^5	3.33×10^5
MINUN	250	1.98×10^5	2.70×10^5	4.33×10^5	1.71×10^6	2.70×10^6	3.46×10^6	4.54×10^5	7.74×10^5	1.09×10^6
MINUN	500	3.10×10^5	4.05×10^5	4.98×10^5	2.52×10^6	3.24×10^6	4.17×10^6	5.45×10^5	9.00×10^5	1.07×10^6
MINUN	1000	3.39×10^5	4.14×10^5	4.72×10^5	2.43×10^6	3.41×10^6	3.88×10^6	8.19×10^5	1.17×10^6	1.57×10^6
MINUN	2000	2.93×10^5	4.23×10^5	5.06×10^5	2.93×10^6	3.67×10^6	4.99×10^6	9.81×10^5	1.35×10^6	1.65×10^6

As shown in Figure 2, K2-BSO achieves the better score in the majority of cases, which means that the clustering step can not only improve the convergence speed on the basis of the number of iterations, but also prevent K2-BSO from falling into local optima. Even the largest network PIGS shows that the F1 score is 2% better than K2-GA.

Figure 2 also shows the main trends of the indexes (Recall (R), Precision (P), and F1) of the three algorithms (K2-R, K2-BSO, K2-GA), with different samples [250,500,1000,2000] in eight datasets, including ASIA, ALARM, SACHS, BARLEY, CHILD, Win95pt, PIGS and MINUN. The blue line 'o:' represents the numerical trend of the Recall of K2-Random; the blue line 'o–' indicates the numerical trend of the Precision of K2- Random; the blue line '*—' indicates the numerical trend of the F1 of K2-Random; The red line 'o:' represents the numerical trend of the Recall of K2-BSO; the red line 'o–' indicates the numerical trend of Precision of K2-BSO; the red line '*—' indicates the numerical trend of F1 of K2-BSO; The green line 'o:' represents numerical trend of the Recall of K2-GA; the green line "o–" indicates the numerical trend of the Precision of K2-GA, and the blue line "*—" indicates the numerical trend of the F1 of K2-GA.

Figure 2a shows the curves of the three methods (K2-R, K2-BSO, K2-GA) with different samples in the data set ASIA. It can be seen that the red curve basically goes above the green one and the blue one, which means that K2-BSO's indexes R, P, F1 are higher than that of K2-R and K2-GA, thus proves that K2-BSO algorithm is better than K2-R algorithm and K2-GA algorithm.

Figure 2b–e show the curves of the three algorithms (K2-R, K2-BSO, K2-GA) with different samples in data sets SACHS, CHILD, ALARM and BARLEY. It can be observed that the results are similar to that in Figure 2a, that K2- BSO's indexes R,P,F1 are higher than that of K2-R and K2- GA, thus also proves that K2-BSO algorithm is better than K2-R algorithm and K2-GA algorithm. Figure 2f–g shows the curves of the three methods (K2-R, K2-BSO, K2-GA) with different samples in data set WIN95PTS, PIGS and MINUN. WIN95PTS is a 76-dimensional network, PIGS is a 441-dimensional network and MINUN is a 1041-dimensional network, so they belongs to the high dimensional networks. We can see from Figure 2f that the curve of the blue value is the lowest; with sample 500 and 2000, the value of the green curve is slightly higher than that of the red curve, while on the whole, the red curve goes above the green. However, when we refer to Table 3, it is obvious that the execution time of K2-BSO is much less than that of K2-GA, which means the K2-BSO is better than the other algorithms in this network. On the other hand, Figure 2g shows that the curves are slightly different from the form's results, the Recall of the three methods grows with the increase of sample size while the Precision reduces with the increase of sample size.

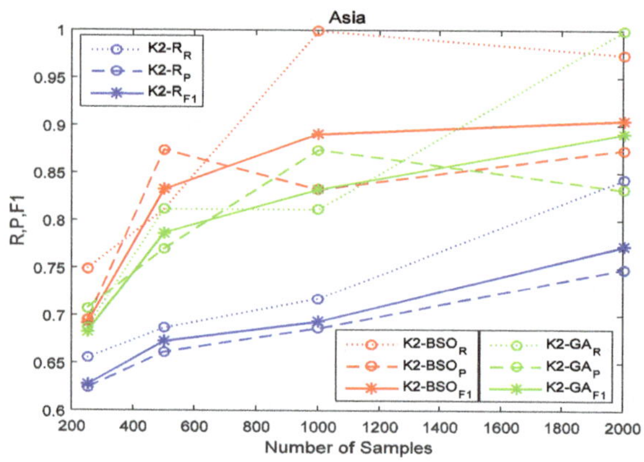

(**a**) ASIA dataset

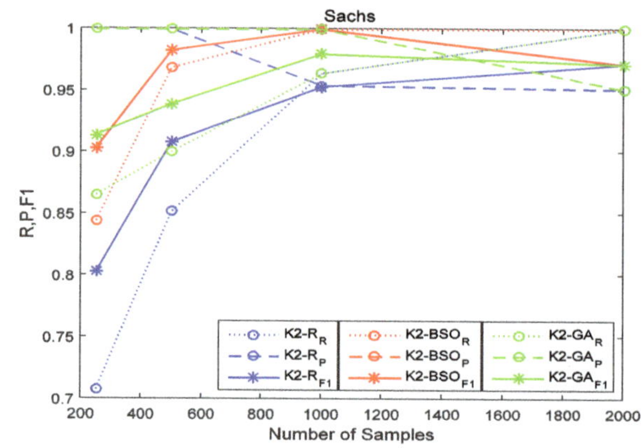

(**b**) SACHS dataset.

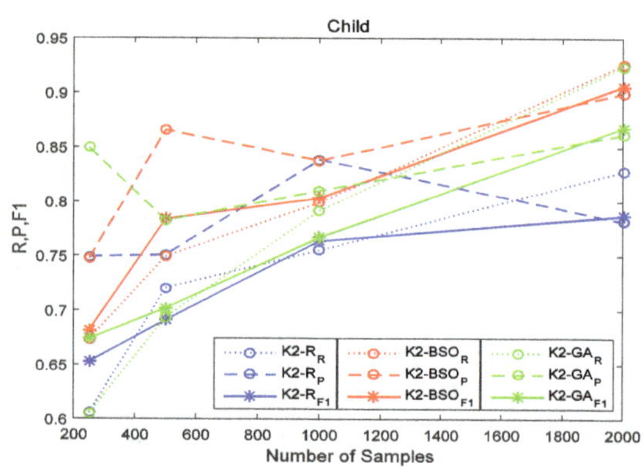

(**c**) CHILD dataset.

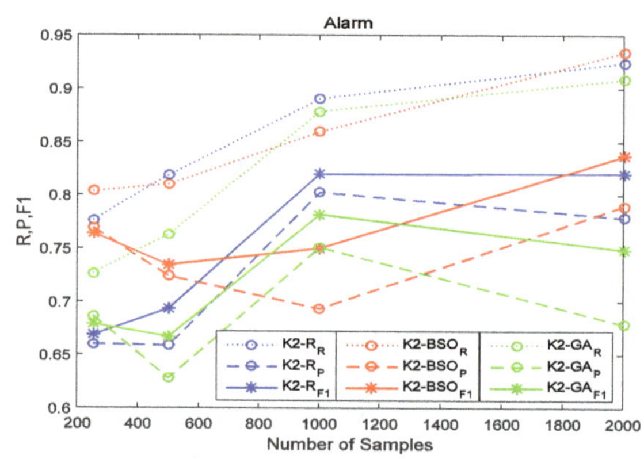

(**d**) ALARM dataset.

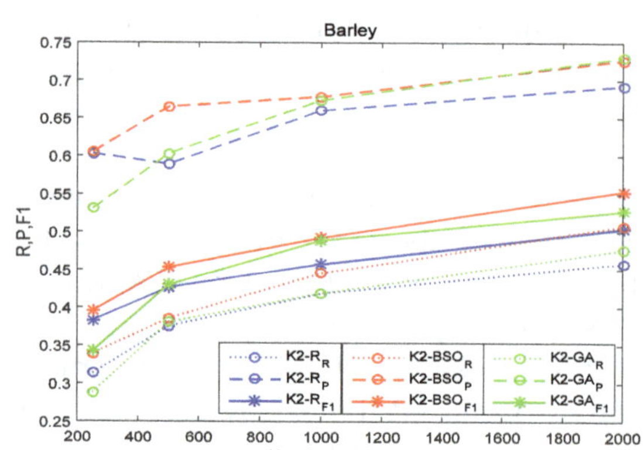

(**e**) BARLEY dataset.

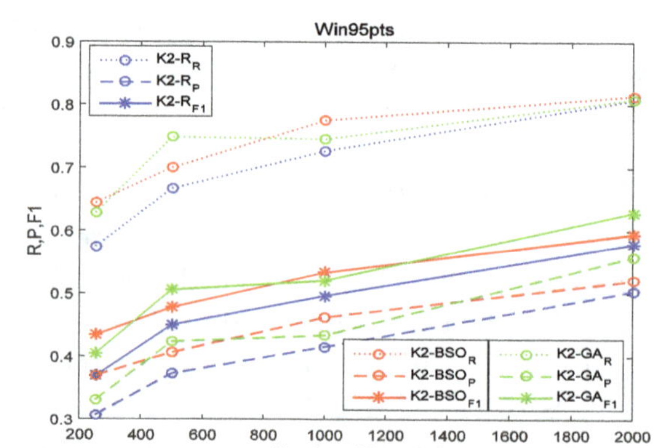

(**f**) WIN95PTS dataset.

Figure 2. *Cont.*

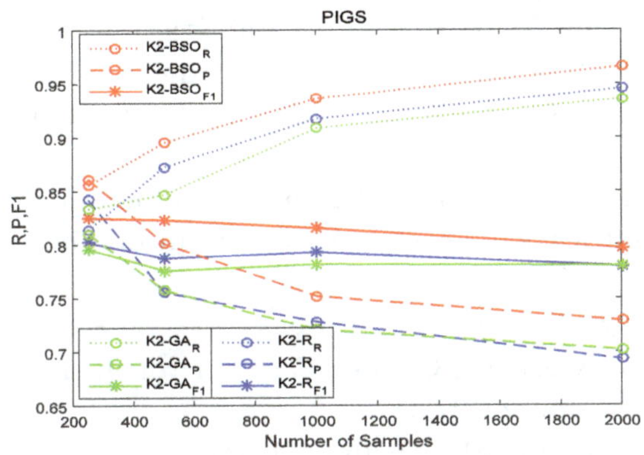

(g) PIGS dataset

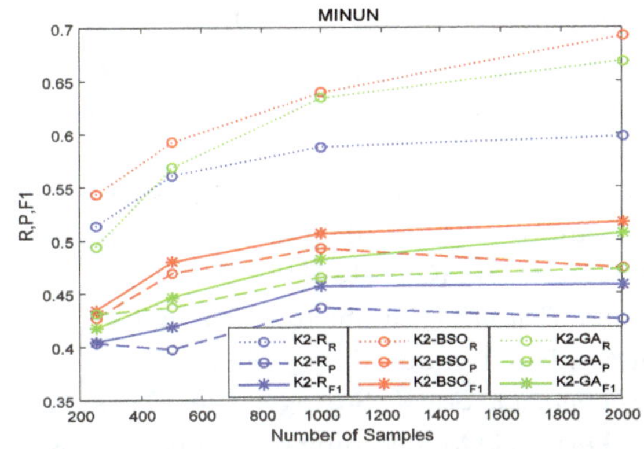

(h) MINUN dataset.

Figure 2. R, p & F1 of the K2-R, K2-BSO and K2-GA with eight dataset: (**a**) ASIA dataset; (**b**) SACHS dataset; (**c**) CHILD dataset; (**d**) ALARM dataset; (**e**) BARLEY dataset; (**f**) Win95pt dataset; (**g**) PIGS dataset; (**h**) MINUN dataset.

The reason for such a difference is that PIGS is a genetic network, that is, PIGS has a very complex structure where some nodes connect with many neighboring nodes, for example, the maximum degree is 41 (maximum in-degree is 2). Accordingly it is difficult for the subroutine K2 to remove these in-direct causal relationships. Even so, it can be seen that the F1 score of K2-BSO is still 2% better than those of K2-R and K2-GA. Figure 2h shows that even in the case of MINUN network of more than 1000 dimensionality, K2-BSO works much better than K2-GA and K2-R. These results demonstrate that our method K2-BSO is much reliable than the counterparts in more complexity and higher-dimensional cases, and also shows that K2-BSO is able to learn the causal structure from a dataset with hundreds of variables. In summary, K2-BSO performs better than K2-GA if the accuracy and execution time are combined, so in our future work, we will continue to perfect the K2-BSO algorithm, making it adapt to high-dimensional network accuracy problems at the cost of some appropriate execution time.

7. Conclusions

To reduce the search space of graphs is important in causal relationship discovery; however, the existing methods show inefficiency for large scale causal networks. In this work, an improved causal structure learning algorithm combining K2 with brain storm optimization (BSO) called K2-BSO is presented to alleviate this problem. In contrast to other evolutionary algorithms based on the search and score methods, K2-BSO has two significant advantages, (1) K2-BSO searches optimal topological order of nodes instead of graph space. The order space should be much smaller than the whole graph space. In this phase, an elaborate distance function is introduced for clustering nodes' orders in BSO based on the mechanism of K2. The graph space therefore is reduced to a smaller topological order space that can be further reduced by an efficient clustering method. (2) Our method is designed through the following split and merge strategy, the original dataset is split into a set of subdata sets in the first place. The BSO will run on these subdata sets to recover the corresponding substructures. Here we further use additive noise model approach to rectify the direction of the erroneous orientation or the side without direction. We eventually merge all these substructures and obtain the entire structure of the graph. The experimental results on various causal networks showed our method could outperform the traditional search and score method and the state-of-the-art genetic algorithm-based method.

Author Contributions: Y.H. and Z.H. designed the study. Y.H., Z.H., G.M., H.H. and A.K.S. performed the study. Y.H. and G.M. wrote the manuscript. G.M. was the principal investigator and corresponding author.

References

1. Hao, Z.; Huang, J.; Cai, R.; Wen, W. A hybrid approach for large scale causality discovery. In *Emerging Intelligent Computing Technology and Applications, Proceedings of the 8th International Conference, ICIC 2012,*

Huangshan, China, 25–29 July 2012; Springer: Berlin, Germany, 2013; Volume 375, pp. 1–6.

2. Li, X.J.; Mishra, S.K.; Wu, M.; Zhang, F.; Zheng, J. Syn-lethality: An integrative knowledge base of synthetic lethality towards discovery of selective anticancer therapies. *BioMed Res. Int.* **2014**, *2014*, 196034. [CrossRef] [PubMed]
3. Wu, M.; Li, X.; Zhang, F.; Li, X.; Kwoh, C.K.; Zheng, J. In silico prediction of synthetic lethality by meta-analysis of genetic interactions, functions, and pathways in yeast and human cancer. *Cancer Inform.* **2014**, *13*, 71–80. [CrossRef] [PubMed]
4. Pearl, J. *Causality*; Cambridge University Press: Cambridge, UK, 2009.
5. Spirtes, P.; Glymour, C.N.; Scheines, R.; Heckerman, D.; Meek, C.; Cooper, G.; Richardson, T. *Causation, Prediction, and Search*; Springer: New York, NY, USA, 1993; pp. 272–273.
6. Chickering, D.M. Learning equivalence classes of bayesian-network structures. *J. Mach. Learn. Res.* **2002**, *2*, 150–157.
7. Shimizu, S.; Hoyer, P.O.; Hyvärinen, A.; Kerminen, A. A linear non-gaussian acyclic model for causal discovery. *J. Mach. Learn. Res.* **2006**, *7*, 2003–2030.
8. Shimizu, S.; Inazumi, T.; Sogawa, Y.; Hyvärinen, A.; Kawahara, Y.; Washio, T.; Hoyer, P.O.; Bollen, K. Directlingam: A direct method for learning a linear non-gaussian structural equation model. *J. Mach. Learn. Res.* **2011**, *2*, 1225–1248.
9. Hoyer, P.O.; Janzing, D.; Mooij, J.M.; Peters, J.; Schölkopf, B. Nonlinear causal discovery with additive noise models. In Proceedings of the International Conference on Neural Information Processing Systems, Vancouver, BC, Canada, 8–10 December 2008; pp. 689–696.
10. Peters, J.; Mooij, J.M.; Janzing, D.; Schölkopf, B. Causal discovery with continuous additive noise models. *J. Mach. Learn. Res.* **2013**, *15*, 2009–2053.
11. Peters, J.; Janzing, D.; Scholkopf, B.; Teh, Y.W.; Titterington, M. Identifying cause and effect on discrete data using additive noise models. In Proceedings of the International Conference on Artificial Intelligence and Statistics, Sardinia, Italy, 13–15 May 2010; pp. 597–604.
12. Peters, J.; Janzing, D.; Scholkopf, B. Causal inference on discrete data using additive noise models. *IEEE Trans. Pattern Anal. Mach. Intell.* **2011**, *33*, 2436–2450. [CrossRef] [PubMed]
13. Zhang, K.; Hyvärinen, A. Distinguishing causes from effects using nonlinear acyclic causal models. In Proceedings of the 2008th International Conference on Causality: Objectives and Assessment, Vancouver, BC, Canada, 8–10 December 2008; Volume 6, pp. 157–164.
14. Daniusis, P.; Janzing, D.; Mooij, J.; Zscheischler, J.; Steudel, B.; Zhang, K.; Schölkopf, B. Inferring deterministic causal relations. In Proceedings of the Conference on UAI, Catalina Island, CA, USA, 8–11 July 2010; pp. 143–150.
15. Janzing, D.; Mooij, J.; Zhang, K.; Lemeire, J.; Zscheischler, J.; Daniušis, P.; Steudel, B.; Scholkopf, B. Information-geometric approach to inferring causal directions. *Artif. Intell.* **2012**, *182–183*, 1–31. [CrossRef]
16. Janzing, D.; Steudel, B.; Shajarisales, N.; Scholkopf, B. Justifying information-geometric causal inference. In *Measures of Complexity*; Springer: Cham, Switzerland, 2015; pp. 253–265.
17. Chen, W.; Hao, Z.; Cai, R.; Zhang, X.; Hu, Y.; Liu, M. Multiple-cause discovery combined with structure learning for high-dimensional discrete data and application to stock prediction. *Soft Comput.* **2016**, *20*, 4575–4588. [CrossRef]
18. Cai, R.; Zhang, Z.; Hao, Z. Causal gene identification using combinatorial v-structure search. *Neural Netw.* **2013**, *43*, 63–71. [CrossRef] [PubMed]
19. Cai, R.; Zhang, Z.; Hao, Z. SADA: A General Framework to Support Robust Causation Discovery. In Proceedings of the 30th International Conference on Machine Learning (ICML), Atlanta, GA, USA, 16–21 June 2013; Volume 28, pp. 208–216.
20. Cai, R.; Zhang, Z.; Hao, Z.; Winslett, M. Understanding Social Causalities Behind Human Action Sequences. *IEEE Trans. Neural Netw. Learn. Syst.* **2016**, *28*, 1801–1813. [CrossRef] [PubMed]
21. Cai, R.; Zhang, Z.; Hao, Z. BASSUM: A Bayesian semi-supervised method for classification feature selection. *Pattern Recognit.* **2011**, *44*, 811–820. [CrossRef]
22. Mooij, J.; Janzing, D.; Peters, J.; Scholkopf, B. Regression by dependence minimization and its application to causal inference in additive noise models. In Proceedings of the 26th Annual International Conference on Machine Learning, Montreal, QC, Canada, 14–18 June 2009; pp. 745–752.

23. Zhang, K.; Peters, J.; Janzing, D.; Scholkopf, B. Kernel-based conditional independence test and application in causal discovery. *Comput. Sci.* **2012**, *6*, 895–907.
24. Cheng, S.; Qin, Q.; Chen, J.; Shi, Y. Brain storm optimization algorithm: A review. *Artif. Intell. Rev.* **2016**, *46*, 445–458. [CrossRef]
25. Hong, Y.H.; Liu, Z.S.; Mai, G.Z. An efficient algorithm for large-scale causal discovery. *Soft Comput.* **2016**, *21*, 7381–7391. [CrossRef]
26. Hong, Y.H. Fast causal network skeleton learning algorithm. *J. Nanjing Univ. Sci. Technol.* **2016**, *40*, 315–321.
27. Hong, Y.H.; Mai, G.Z.; Liu, Z.S. Learning tree network based on mutual information. *Metall. Min. Ind.* **2015**, *12*, 146–151.
28. Duan, H.; Li, S.; Shi, Y. Predator–prey brain storm optimization for DC brushless motor. *IEEE Trans. Mag.* **2013**, *49*, 5336–5340. [CrossRef]
29. Shi, Y. Brain storm optimization algorithm. In Proceedings of the International Conference in Swarm Intelligence, Chongqing, China, 12–15 June 2011; Springer: Berlin/Heidelberg, Germany, 2011; pp. 303–309.
30. Zhan, Z.H.; Zhang, J.; Shi, Y.H.; Liu, H.L. A modified brain storm optimization. In Proceedings of the 2012 IEEE Congress on Evolutionary Computation, Brisbane, QLD, Australia, 10–15 June 2012; pp. 1–8.
31. Xue, J.; Wu, Y.; Shi, Y.; Cheng, S. Brain storm optimization algorithm for multi-objective optimization problems. In *Advances in Swarm Intelligence*; Springer: Berlin/Heidelberg, Germany, 2012; pp. 513–519.
32. Cooper, G.F.; Herskovits, E. A bayesian method for the induction of probabilistic networks from data. *Mach. Learn.* **1992**, *9*, 309–347. [CrossRef]
33. Shi, Y. An optimization algorithm based on brainstorming process. In *Emerging Research on Swarm Intelligence and Algorithm Optimization*; Information Science Reference: Hershey, Pennsylvania, 2015; pp. 1–35.
34. Zhou, D.; Shi, Y.; Cheng, S. Brain storm optimization algorithm with modified step-size and individual generation. *Adv. Swarm Intell.* **2012**, *7331*, 243–252.
35. Sun, C.; Duan, H.; Shi, Y. Optimal satellite formation reconfiguration based on closed-loop brain storm optimization. *IEEE Comput. Intell. Mag.* **2013**, *8*, 39–51. [CrossRef]
36. Jadhav, H.T.; Sharma, U.; Patel, J.; Roy, R. Brain storm optimization algorithm based economic dispatch considering wind power. In Proceedings of the 2012 IEEE International Conference on Power and Energy (PECon), Parit Raja, Malaysia, 2–5 December 2012; pp. 588–593.
37. Qiu, H.; Duan, H. Receding horizon control for multiple UAV formation flight based on modified brain storm optimization. *Nonlinear Dyn.* **2014**, *78*, 1973–1988. [CrossRef]
38. Shi, Y.; Xue, J.; Wu, Y. Multi-objective optimization based on brain storm optimization algorithm. *Int. J. Swarm Intell. Res.* **2013**, *4*, 1–21. [CrossRef]
39. Shi, Y. Brain storm optimization algorithm in objective space. In Proceedings of the 2015 IEEE Congress on Evolutionary Computation (CEC), Sendai, Japan, 25–28 May 2015; pp. 1227–1234.
40. Yang, Z.; Shi, Y. Brain storm optimization with chaotic operation. In Proceedings of the 2015 Seventh International Conference on Advanced Computational Intelligence (ICACI), Wuyi, China, 27–29 March 2015; pp. 111–115.
41. Yang, Y.; Shi, Y.; Xia, S. Advanced discussion mechanism-based brain storm optimization algorithm. *Soft Comput.* **2015**, *19*, 2997–3007. [CrossRef]
42. Jia, Z.; Duan, H.; Shi, Y. Hybrid brain storm optimisation and simulated annealing algorithm for continuous optimisation problems. *Int. J. Bio-Inspired Comput.* **2016**, *8*, 109–121. [CrossRef]
43. Cheng, S.; Shi, Y.; Qin, Q.; Gao, S. Solution clustering analysis in brain storm optimization algorithm. In Proceedings of the 2013 IEEE Symposium on Swarm Intelligence (SIS), Singapore, 16–19 April 2013; pp. 111–118.
44. Cheng, S.; Shi, Y.; Qin, Q.; Zhang, Q.; Bai, R. Population diversity maintenance in brain storm optimization algorithm. *J. Artif. Intell. Soft Comput. Res.* **2014**, *4*, 83–97. [CrossRef]
45. Cheng, S.; Shi, Y.; Qin, Q.; Ting, T.O.; Bai, R. Maintaining population diversity in brain storm optimization algorithm. In Proceedings of the 2014 IEEE Congress on Evolutionary Computation (CEC), Beijing, China, 6–11 July 2014; pp. 3230–3237.
46. Georgiou, D.N.; Karakasidis, T.E.; Nieto, J.J.; Torres, A. A study of entropy/clarity of genetic sequences using metric spaces and fuzzy sets. *J. Theor. Biol.* **2010**, *267*, 95–105. [CrossRef] [PubMed]

7

Correcting Errors in Image Encryption based on DNA Coding

Bin Wang [1,*], Yingjie Xie [2], Shihua Zhou [1], Xuedong Zheng [1] and Changjun Zhou [3,*]

[1] Key Laboratory of Advanced Design and Intelligent Computing, Dalian University, Ministry of Education, Dalian 116622, China; shihuajo@gmail.com (S.Z.); xuedongzheng@163.com (X.Z.)
[2] Applied Technology College, Dalian Ocean University, Dalian 116300, China; yingjieying@163.com
[3] College of Mathematics, Physics and Information Engineering, Zhejiang Normal University, Jinhua 321004, China
* Correspondence: wangbinpaper@gmail.com (B.W.); zhou-chang231@163.com (C.Z.);

Abstract: As a primary method, image encryption is widely used to protect the security of image information. In recent years, image encryption pays attention to the combination with DNA computing. In this work, we propose a novel method to correct errors in image encryption, which results from the uncertainty of DNA computing. DNA coding is the key step for DNA computing that could decrease the similarity of DNA sequences in DNA computing as well as correct errors from the process of image encryption and decryption. The experimental results show our method could be used to correct errors in image encryption based on DNA coding.

Keywords: image encryption; chaotic map; DNA coding; Hamming distance

---

1. Introduction

With wide usage of multimedia technologies and excessive spread of internet, the awareness of protecting information, especially image information, is heightened day by day. As we known, encrypting technology can usually be used to protect the security of image information. In image encryption, chaotic maps are usually employed to encrypt image, because they have the features of ergodicity, sensitivity to initial conditions, control parameters and so on [1–7]. Chen et al. proposed a novel 3D cat maps to design a real-time secure symmetric encryption scheme [1]. Lian et al. first analyzed the parameter sensitivity of standard map and proposed an improved standard map to encrypt image [3]. Wong et al. proposed a fast algorithm of image encryption, where the overall encryption time was reduced as fewer rounds were required [2]. Zhang et al. proposed a new image encryption algorithm based on the spatiotemporal chaos of the mixed linear-nonlinear coupled map lattices [7]. Wang et al. combined genetic recombination with hyper-chaotic system to design a novel image encryption and experiment results proved that the proposed algorithm was effective for image encryption [8]. Zhang et al. analyzed different kinds of permutation algorithms and proposed a new cryptosystem to address these drawbacks [5]. In the recent past, although these methods have made some progress, they lack the capability of parallel computing.

Inspired by the biological character of DNA sequences, such as parallel computing, low-energy and so on, DNA computing and DNA coding are widely used to encrypt image [6,9–17]. Zhang et al. combined DNA sequence addition operation with chaotic map to design a novel image encryption scheme [9]. The experimental results shown that the proposed scheme could achieve good encryption and resist some kind attacks. In Ref. [10], the authors transformed DNA sequences into its base pair for random times to confuse the pixels, generate the new keys according to the plain image and the common keys. Wei et al. further utilized DNA sequence addition operation and Chen's hyper-chaotic map to encrypt a color image [11]. Due to some disadvantages in One-Time-Pad (OTP) algorithm, the author used logistic chaotic map as an input of OTP algorithm and proposed an interesting encryption algorithm based on a chaotic selection between original message DNA strands and OTP DNA strands [12]. In Ref. [13], the authors used genetic algorithm to determine the best masks, which result from DNA and logistic map functions. Ozkaynak et al. broke a previous cryptosystem and proposed an improved image encryption algorithm [14]. Rehman et al. utilized whole set of DNA complementary rules dynamically and employed DNA addition operation to encrypt image [15]. Song and Qiao proposed a novel image encryption scheme based on DNA encoding and spatiotemporal chaos, which was of high key sensitivity and large key space [16]. In Ref. [17], DNA coding combined with an improved 1D chaotic systems to design image encryption. Kulsoom et al. employed an entire set of DNA complementary rules along with 1D chaotic maps to design an image encryption algorithm [6]. Wang et al. proposed a new chaotic image encryption scheme based on Josephus traversing and mixed chaotic map [18]. Parvaz and Zarebnia defined a combination chaotic system and studied its properties [19].

DNA computing was addressed to solve the seven-point Hamiltonian path problem by Adleman in 1994 [20]. Along with the development of research, there are a large number of applications about DNA computing, such as DNA logic gates [21], neural network [22], cryptography [4], data storage [23], image watermarking [24] and so on. Hybridization reaction is the key operation for DNA sequences and influences the reliability of DNA computing. However, the false hybridization is unavoidable because of the limit of biological technology, result from false positive and false negative. The lack of similarity between DNA sequences could result in false positive and generating hybridization reaction between two unmatched DNA sequences. The mistake in the biochemical operation result in false negative in which two matched DNA sequences did not hybridize each other [25]. Chai et al. encoded plain image by DNA matrix and permuted the image with a new wave-based permutation scheme [26]. In Ref. [27], DNA sequence operation combining with one-way coupled-map lattices was to structure a robust and lossless color image encryption algorithm and the three gray-level components of plain-image were converted into three DNA matrices and performed XOR operation twice. Designing DNA coding could obtain high quality DNA sequences which satisfy some constraints, such as Hamming distances, GC content and so on, to decrease the similarity between DNA sequences [28,29].

Inspired by Hybridization reaction is the kernel for DNA computing and influences the reliability of DNA computing. However, the false hybridization is unavoidable because of the limit of biological technology, result from false positive and false negative. The lack of similarity between DNA sequences could result in false positive and generating hybridization reaction between two unmatched DNA sequences. The mistake in the biochemical operation result in false negative in which two matched DNA sequences did not hybridize each other [25]. Designing DNA coding could obtain high quality DNA sequences which satisfy some constraints, such as Hamming distance, GC content and so on, to decrease the similarity between DNA sequences [28,29]. Inspired by communication theory, Hamming code can be used to correct errors. For example, d is the Hamming distance between two strings and then the bits of correcting errors are equal to $\left\lfloor \frac{d-1}{2} \right\rfloor$. So, in this paper, we introduce Hamming distance to decrease the similarity between DNA sequences as well as correct errors from hybridization reaction. Furthermore, to improve the accuracy of DNA computing, the constraints of DNA coding are used to decrease the generation of false positive. Finally, the experimental results show that the number of pixels change rate (NPCR) has achieved 99.57% and the unified average changing intensity (UACI) has achieved 32.38%. The proposed method could effectively correct the encrypted image contained 1000 errors and improve the accuracy of hybridization reaction.

2. Methods

2.1. DNA Coding

Hamming distance is widely used to design DNA coding. It is the number of positions at which the corresponding symbols are different when two strings have the equal length [30]. In the alphabet Σ = {A, C, G, T}, there exists a set S with length n and size of $|S| = 4^n$. A subset $C \subseteq S$ and let u, v any two codes in the C satisfy [31]:

$$\tau(u, v) \geq d \tag{1}$$

d is a positive integer, τ is the constraint criteria (or criterion) for designing DNA coding. In this paper, τ is denoted as the Hamming distance.

2.1.1. Sequences-Sequence Hamming Distance (SS)

Sequences-sequence Hamming Distance [31]: for the DNA sequences u, v with given length n (written from the $5'$ to the $3'$ end), the Hamming distance between u and v is denoted as H (u, v). The minimal H (u_i, v_j) in all DNA sequences is denoted as SS (u_i) and it should not be less than parameter d,

$$SS(u_i) = \min_{1 \leq j \leq n, j \neq i} \{H(u_i, v_j)\} \geq d \tag{2}$$

2.1.2. Sequences-Complementarity Hamming Distance (SC)

Sequences-complementarity Hamming Distance [31]: for the DNA sequences u, v with given length n (written from the $5'$ to the $3'$ end), H (u, v^C) denotes the Hamming distance between u and v^C, where v^C is the complementary sequence of v. For example, v = ACTG, then v^C = CAGT. The minimal H (u_i, v_j^C) in all DNA sequences is denoted as SS (u_i) and it should not be less than parameter d,

$$SC(u_i) = \min_{1 \leq j \leq n, j \neq i} \left\{H\left(u_i, v_j^C\right)\right\} \geq d \tag{3}$$

2.1.3. GC Content

In order to approximate the thermodynamic properties of DNA sequences, GC content constraint is used to combine with distance constraint, such as Hamming distance. The percentage of G or C bases within each DNA is denoted as GC content. In this paper, GC content is equal to 50%. The GC content is denoted as follows:

$$GC\_content = Num\_gc/n \times 100\% \tag{4}$$

2.1.4. DNA Coding Rule

Adenine (A), Cytosine (C), Guanine (G) and Thymine (T) are the four elements that make up the whole DNA sequence. When paring, the principle of complementary base pairing is observed, namely A with T and C with G [32]. There is a complementary relationship between 0 and 1 in the binary bit. Similarly, there is a complementary relationship between 01 and 10 as well as 00 and 11. In the previous works, the authors converted binary message to DNA sequences based on the DNA coding rule in Table 1 [6,9,13–15,17,23,33,34]. There are eight DNA coding methods to convert binary message to DNA sequences [9,24]. For example, the pixel 65 is firstly transformed into binary bit 01000001 and then 01000001 transformed into DNA sequence ACCA for the first rule.

Table 1. DNA coding list [9,24].

	1	**2**	**3**	**4**	**5**	**6**	**7**	**8**
A	01	01	00	00	10	10	11	11
T	10	10	11	11	01	01	00	00
C	00	11	01	10	00	11	01	10
G	11	00	10	01	11	00	10	01

From the Table 1, the information only is simple transformed between binary and DNA sequence. It does not consider the characters of DNA sequences, especially specific hybridization. So, in this paper, we designed DNA coding that satisfied three constraints above to encrypt image and use this DNA coding to correct the errors.

2.2. *New DNA Coding Rule for Correcting Errors*

As shown above, the DNA sequences used to encode pixels and chaotic orbits should satisfy these constraints to decrease the similarity between DNA sequences and correct the errors. In our previous work, we proposed a dynamic genetic algorithm to design DNA sequence sets which satisfy the combinational constraints [35]. DNA sequence set denotes that any pair of DNA sequences in this set satisfies the combinational constraints. In this paper, we use the DNA coding $A_4^{SS+SC+GC}(8,3) = 336$, namely the length is equal to 8 and Hamming distance is equal to 3. We randomly select 256 elements from this set to encode the pixels between 0 and 255. In this paper, we denote these 256 DNA sequences as DNA coding rule. Table 2 lists the first 50 DNA coding rule to encrypt image. The whole $A_4^{SS+SC+GC}(8,3)$ and DNA coding rule are shown in the Supplement.

Table 2. The first 50 DNA coding rule.

Pixel	**DNA Coding**	**Pixel**	**DNA Coding**	**Pixel**	**DNA Coding**	**Pixel**	**DNA Coding**	**Pixel**	**DNA Coding**
0	ATCATGCC	1	CTCGATCA	2	GCTCTTCT	3	AGTGGGAT	4	ACTCTCTG
5	AATCTGCG	6	ACTCACGT	7	CTTCCAAC	8	GCTTCTAG	9	TAGGAGGT
10	GATCGACT	11	TAACGCTG	12	TAAGCGGA	13	CTGTGATC	14	CCCTAATC
15	TGGAAGGA	16	TACTACCG	17	CTTATGGG	18	TCAGCAAG	19	CGACTTCT
20	AGTGTCGA	21	TGCGATTC	22	CAACGACA	23	GATCTGTC	24	GCCAACTA
25	ATGAGGGA	26	TAGAACGG	27	CCGTAACA	28	TAGACTGC	29	GCTGGATT
30	GTGAGTCA	31	TCATGGAC	32	ACCACTAC	33	TCCTAAGG	34	GGCTAAAG
35	CCAACTGA	36	TCGTCTTG	37	TTGGGAAC	38	AATAGCCC	39	CTGTCGAA
40	CCCCATAT	41	AACCTCTC	42	GGTTTACG	43	GCAGAAGA	44	TAGAGGAG
45	GAAAGGGA	46	ATCGACGA	47	GCAAGTAC	48	TCAGACAC	49	CTTGGTTG

2.3. *Process of Encrypting and Decrypting Image Based on DNA Coding*

2.3.1. Encrypting Image

Recently, there are some works on cryptanalysis of encrypting schemes based on chaotic map and DNA coding [36–38]. In this paper, in order to improving the security of our encrypting scheme, two logistic maps with different parameters and initial values are chosen to generate pseudorandom sequence. The different parameters and initial values for the Equation (5) are denoted as μ_1, μ_2, $x_1^1(0)$ and $x_1^2(0)$, respectively, where $\mu_1, \mu_2 \in [3.9, 4]$ and $x_1^1(0), x_1^2(0) \in (0, 1)$.

$$x_{i+1} = \mu x_i(1 - x_i) \tag{5}$$

The detailed of encrypting image is described as follows:

Step 1. The key with 16 elements is randomly generated as the initial key and the initial key is implemented XOR operation with every pixel value of the plain image. The result of XOR operation is regard as the relating key;

Step 2. According to initial condition of logistic maps, namely two parameters μ_1, μ_2 and two initial value $x_1^1(0)$, $x_1^2(0)$, the relating key is evenly dividing relating key into four parts. These logistic maps are to iterate for 100 times to get rid of the transient effect of chaotic systems;

Step 3. The logistic maps are continuingly iterated base on the number of pixels, namely one map for the half number and the pseudorandom sequence consists of the logistic chaotic orbits;

Step 4. In order to permute the plain image, the chaotic orbits are sorted in ascending order. This operation (permutation) only changes the location of pixels of plain image;

Step 5. The XOR operation is implemented between the pixels of the permuted image and the pseudorandom sequence from the logistic maps. This operation (diffusion) only changes the value of pixels of digital image;

Step 6. According to the new DNA coding rule, the encrypted image is encoded by DNA coding;

Step 7. Outputting the encrypted image.

The flowchart of encrypting image is illustrated in Figure 1.

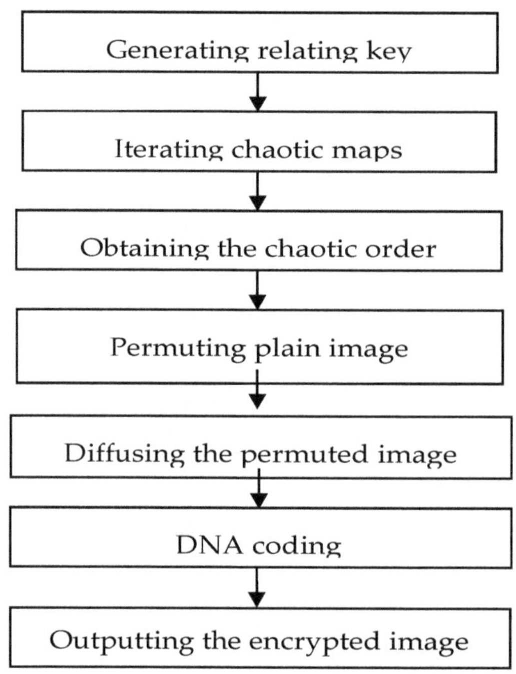

Figure 1. The flowchart of encrypting image.

2.3.2. Decrypting Image

The decryption process is similar to that of encryption procedure in the reversed order. It can be briefly stated as follows:

Step 1. According to the same relating key, the chaotic maps are to iterate for 100 times to get rid of the transient effect;

Step 2. The chaotic orbits are regenerated based on the same parameters and initial values as well as the encryption process;

Step 3. Decoding the cipher image based on the DNA coding rule;

Step 4. The XOR operation is implemented between the pixels of the cipher image and the pseudorandom sequence from the logistic maps and the permuted image is recovered;

Step 5. According to the order of chaotic sequences, the plain image is recovered from the permuted image;

Step 6. Outputting the plain image.

Note that the permutation–diffusion architecture is widely used into image encryption based on chaotic map and DNA coding. So, the whole architecture of the proposed method is the permutation–diffusion. However, the previous works are mainly to convert the pixel value into an 8-bit binary sequence and then perform a simple one-to-one correspondence between the binary (or ASCII codes) and the DNA sequence without the function of error correction. For example, the binary sequence of the pixel value 1 is 00000001 and the corresponding DNA sequence is AAAC (A for 00, C for 01, G for 10 and T for 11). In this paper, a DNA coding scheme with the function of error correction is proposed, where the pixel value of image is directly corresponded to a piece of DNA sequence with the function of error correction.

3. Experiment and Simulation

In order to resist the brute-force attack, the key space must be large enough for a secure image cryptosystem. 16 elements make up the key in our paper, $key = \{x_i\}, i = 1, 2, \ldots, 16, x_i \in [0, 255]$. It is sufficiently large to ensure the security of digital image when the key space reaches to $2^{128} \approx 3.4 \times 10^{38}$. All the following experiment have the same size for key space.

3.1. Key Sensitivity

The test of key sensitivity can be stated as follows:

Step 1. Generating the key 123456789012345 and using this key to encrypt the test images;

Step 2. Generating another key—123456789012346—with a slight difference and using this key to encrypt the same test image;

Step 3. Calculating the difference between different cipher images.

From the results, although the two different keys are only slightly different—by one bit—the cipher image with the key 123456789012345 is 99.63% different from the cipher image with the key 123456789012346. Figure 2 shows the results of test image Lena. For the same keys of Cameraman, there is 99.59% difference shown in Figure 3. There is 99.55% difference for Boat shown in Figure 4.

(a) (b)

Figure 2. *Cont.*

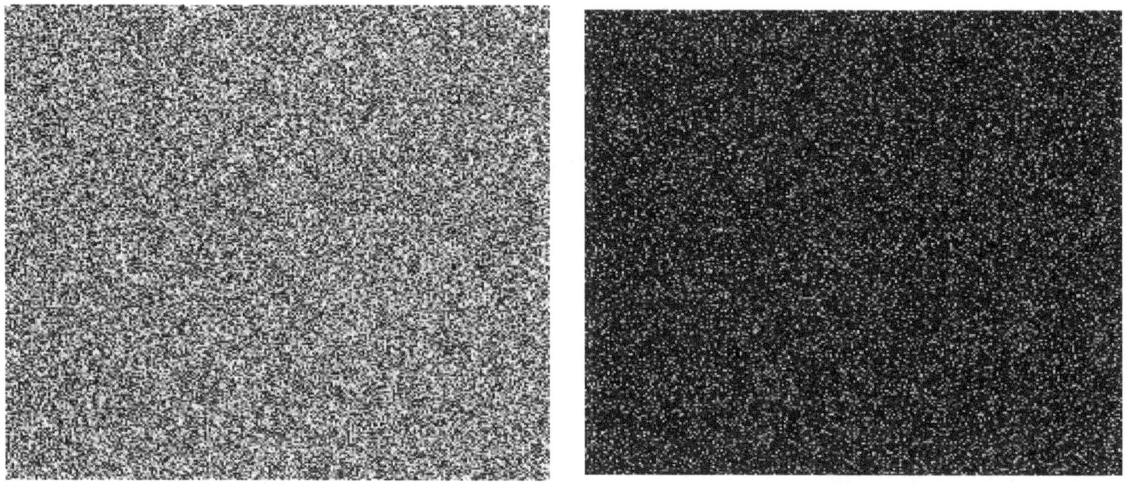

(c) (d)

Figure 2. Key sensitivity for Lena. (**a**) Plain-image of Lena; (**b**) Encrypted image by key: 123456789012345; (**c**) Encrypted image by key: 123456789012346; (**d**) Difference image.

(a) (b)

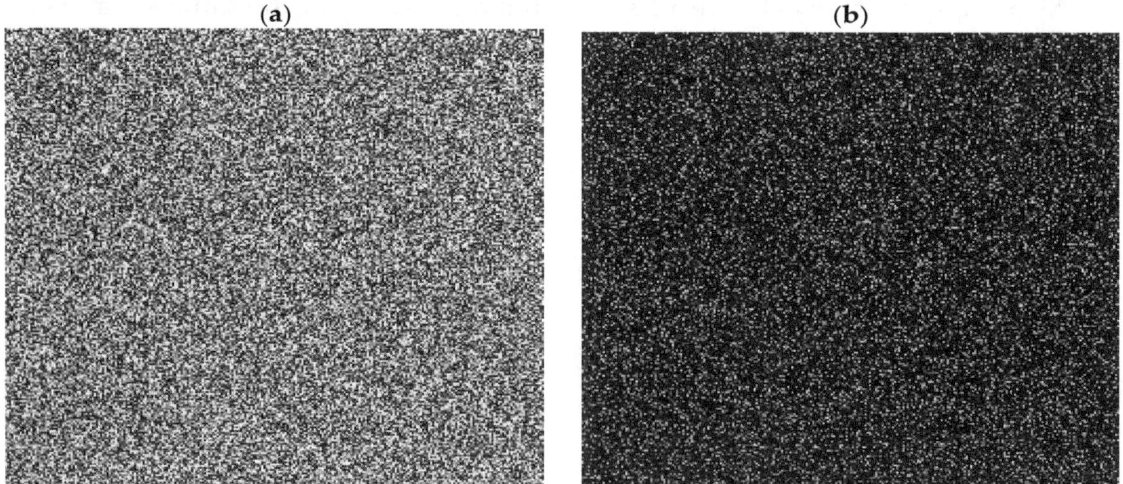

(c) (d)

Figure 3. Key sensitivity for Cameraman. (**a**) Plain-image of Cameraman; (**b**) Encrypted image by key: 123456789012345; (**c**) Encrypted image by key: 123456789012346; (**d**) Difference image.

Figure 4. Key sensitivity for Boat. (**a**) Plain-image of Boat; (**b**) Encrypted image by key: 123456789012345; (**c**) Encrypted image by key: 123456789012346; (**d**) Difference image.

3.2. Statistical Analysis

The statistical characteristics of digital image can be exploited to attack the encryption system. The correlation of two adjacent pixels, as one of statistical characteristics of digital image, is the main aspect of statistical attack. 1000 pairs of adjacent pixels are respectively selected from vertical pixels, horizontal pixels and diagonal pixels. The correlation coefficient of each pair is calculated by the following formulas [1]:

$$\mathrm{cov}(x,y) = E\{(x - E(x))(y - E(y))\} \tag{6}$$

$$r_{xy} = \frac{\mathrm{cov}(x,y)}{\sqrt{D(x)}\sqrt{D(y)}} \tag{7}$$

where x and y are grey-scale values of two adjacent pixels in the image. As digital image consists of discrete pixels, we adapt the following discrete formulas for calculating the correlation:

$$E(x) = \frac{1}{N}\sum_{i=1}^{N} x_i \tag{8}$$

$$D(x) = \frac{1}{N}\sum_{i=1}^{N} (x_i - E(x))^2 \tag{9}$$

$$\mathrm{cov}(x,y) = \frac{1}{N}\sum_{i=1}^{N}\{(x_i - E(x))(y_i - E(y))\} \tag{10}$$

Table 3 shows the results of horizontal, vertical and diagonal directions. The values outside the brackets indicate the correlation between two adjacent pixels for three different plaintext images and the correlation between the cipher text images is indicated in the brackets. From the experimental results, the proposed algorithm greatly reduces the correlation between pixel values of horizontally, vertically and diagonally adjacent images and improves the ability to resist statistical attacks.

Table 3. The correlation coefficient of adjacent pixels.

	Horizontal	Vertical	Diagonal
Lena	0.9727(0.0073)	0.9481(0.0058)	0.9250(−0.0091)
Cameraman	0.9561(−0.0053)	0.9213(−0.0062)	0.9145(−0.0059)
Boat	0.9334(0.0006)	0.9249(0.0009)	0.8891(−0.0002)

3.3. Differential Attack

Number of pixels change rate (NPCR) and Unified average changing intensity (UACI) are the common quantitative criteria for image cryptosystem to evaluate the property of resisting differential attack.

The NPCR and UACI are defined as follows [39–41]:

$$NPCR = \frac{\sum_{i,j} D(i,j)}{W \times H} \times 100\% \tag{11}$$

$$UACI = \frac{1}{W \times H}\left[\sum_{i,j}\frac{|C_1(i,j - C_2(i,j))|}{255}\right] \times 100\% \tag{12}$$

where C_1 and C_2 denotes two different cipher images. These cipher images only have one pixel difference. $C_1(i,j)$ and $C_2(i,j)$ respectively denote the pixel values at the same point (i,j) of C_1 and C_2; H and W are respectively the height and width of the image; $C_1(i,j)$ and $C_2(i,j)$ determine the value of $D(i,j)$, namely, if $C_1(i,j) = C_2(i,j)$ then $D(i,j) = 0$ otherwise, $D(i,j) = 1$.

The comparing results of NPCR and UACI list in the Table 4, where the image cryptosystem adapts the permutation—diffusion architecture with only one round.

The average of ten trials for our method is listed the table. According to the comparison, it shows that our method has higher security.

Table 4. The value of NPCR and UACI for Lena.

	NPCR	UACI
Proposed algorithm	99.57%	32.38%
Wang's work [42]	44.27%	14.874%
Gupta's work [43]	99.62%	17.30%

4. Correcting Errors

In this chapter, we simulated the process of correcting errors. First, we encode the cipher image to DNA sequences and randomly change 1000 bases. Each DNA sequence encoded pixel only change one base. Figure 5 shows the effect of correcting errors. Figure 5a shows the encrypted image contain 1000 errors. Figure 5b shows the image after correcting errors by Hamming code. Figure 5c shows the difference between Figure 5a and Figure 5b. Figure 5d shows the decrypted image after correcting errors. The experimental results express that the proposed method could effectively correct the errors and improve the accuracy of hybridization reaction.

Note that if the changed DNA sequence does not match the according to the DNA coding rule, we compulsively set this DNA sequence correspond to the pixel 255. For example, the pixel 16 match DNA sequence GCCTATCT according to DNA coding rule. If the third base is changed, namely GCGTATCT, there will be no pixel match this changed DNA sequence. So, we set GCGTATCT to correspond to pixel 255.

Figure 5. Correcting errors. (**a**) Containing errors image; (**b**) image after correcting errors; (**c**) Difference image; (**d**) Decrypted image after correcting errors.

5. Conclusions

In this paper, in order to improve the accuracy of DNA computing, we propose a novel method which could decrease the similarity of DNA sequences in DNA computing as well as correct errors from the process of image encryption and decryption. We first analyze the characteristic of DNA hybridization reaction and introduce the combinatorial constraints, namely Sequences-sequence Hamming Distance, Sequences-complementarity Hamming Distance and GC content, to design DNA coding. Then we use the chaotic map to generate pseudo-random sequences and encrypt the plain image by the permuting-diffusing architecture. Finally, we propose a novel DNA coding rule to encode the encrypted image. The experimental results show our method could be used to correct errors in image encryption based on DNA coding.

Bio-inspired computing models, such as membrane computing models [44–50], may provide intelligent methods for Image Encryption. As well, DNA coding strategies can provide biological ways in solving chemical information processing problems.

Author Contributions: Conceptualization, B.W.; Methodology, B.W.; Validation, B.W. and C.Z.; Formal Analysis, X.Z.; Investigation, Y.X.; Data Curation, S.Z.; Writing-Original Draft Preparation, B.W.; Writing-Review & Editing, B.W.; Project Administration, B.W.; Funding Acquisition, B.W.

References

1. Chen, G.; Mao, Y.; Chui, C.K. A symmetric image encryption scheme based on 3D chaotic cat maps. *Chaos Solitons Fractals* **2004**, *21*, 749–761. [CrossRef]
2. Wong, K.-W.; Kwok, B.S.-H.; Law, W.-S. A fast image encryption scheme based on chaotic standard map. *Phys. Lett. A* **2008**, *372*, 2645–2652. [CrossRef]
3. Lian, S.; Sun, J.; Wang, Z. A block cipher based on a suitable use of the chaotic standard map. *Chaos Solitons Fractals* **2005**, *26*, 117–129. [CrossRef]
4. Chang, W.L.; Guo, M.Y.; Ho, M.S.H. Fast parallel molecular algorithms for DNA-based computation: Factoring integers. *IEEE Trans. Nanobiosci.* **2005**, *4*, 149–163. [CrossRef]
5. Zhang, W.; Yu, H.; Zhao, Y.L.; Zhu, Z.L. Image encryption based on three-dimensional bit matrix permutation. *Signal Process.* **2016**, *118*, 36–50. [CrossRef]
6. Kulsoom, A.; Xiao, D.; Ur, R.A.; Abbas, S.A. An efficient and noise resistive selective image encryption scheme for gray images based on chaotic maps and DNA complementary rules. *Multimedia Tools Appl.* **2016**, *75*, 1–23. [CrossRef]
7. Zhang, Y.Q.; Wang, X.Y. A symmetric image encryption algorithm based on mixed linear-nonlinear coupled map lattice. *Inf. Sci.* **2014**, *273*, 329–351. [CrossRef]
8. Wang, X.Y.; Zhang, H.L. A novel image encryption algorithm based on genetic recombination and hyper-chaotic systems. *Nonlinear Dyn.* **2016**, *83*, 333–346. [CrossRef]
9. Zhang, Q.; Guo, L.; Wei, X.P. Image encryption using DNA addition combining with chaotic maps. *Math. Comput. Model.* **2010**, *52*, 2028–2035. [CrossRef]
10. Liu, H.J.; Wang, X.Y.; Kadir, A. Image encryption using DNA complementary rule and chaotic maps. *Appl. Soft Comput.* **2012**, *12*, 1457–1466. [CrossRef]
11. Wei, X.P.; Guo, L.; Zhang, Q.; Zhang, J.X.; Lian, S.G. A novel color image encryption algorithm based on DNA sequence operation and hyper-chaotic system. *J. Syst. Softw.* **2012**, *85*, 290–299. [CrossRef]
12. Babaei, M. A novel text and image encryption method based on chaos theory and DNA computing. *Nat. Comput.* **2013**, *12*, 101–107. [CrossRef]
13. Enayatifar, R.; Abdullah, A.H.; Isnin, I.F. Chaos-based image encryption using a hybrid genetic algorithm and a DNA sequence. *Opt. Lasers Eng.* **2014**, *56*, 83–93. [CrossRef]
14. Ozkaynak, F.; Yavuz, S. Analysis and improvement of a novel image fusion encryption algorithm based on DNA sequence operation and hyper-chaotic system. *Nonlinear Dyn.* **2014**, *78*, 1311–1320. [CrossRef]
15. Rehman, A.U.; Liao, X.F.; Kulsoom, A.; Abbas, S.A. Selective encryption for gray images based on chaos and DNA complementary rules. *Multimedia Tools Appl.* **2015**, *74*, 4655–4677. [CrossRef]
16. Song, C.Y.; Qiao, Y.L. A Novel Image Encryption Algorithm Based on DNA Encoding and Spatiotemporal Chaos. *Entropy* **2015**, *17*, 6954–6968. [CrossRef]
17. Wu, X.J.; Kan, H.B.; Kurths, J. A new color image encryption scheme based on DNA sequences and multiple improved 1D chaotic maps. *Appl. Soft Comput.* **2015**, *37*, 24–39. [CrossRef]
18. Wang, X.Y.; Zhu, X.Q.; Zhang, Y.Q. An Image Encryption Algorithm Based on Josephus Traversing and Mixed Chaotic Map. *IEEE Access* **2018**, *6*, 23733–23746. [CrossRef]
19. Parvaz, R.; Zarebnia, M. A combination chaotic system and application in color image encryption. *Opt. Laser Technol.* **2018**, *101*, 30–41. [CrossRef]
20. Adleman, L.M. Molecular computation of solutions to combinatorial problems. *Science* **1994**, *266*, 1021–1024. [CrossRef] [PubMed]
21. Li, T.; Wang, E.K.; Dong, S.J. Potassium-Lead-Switched G-Quadruplexes: A New Class of DNA Logic Gates. *J. Am. Chem. Soc.* **2009**, *131*, 15082–15083. [CrossRef] [PubMed]
22. Qian, L.; Winfree, E.; Bruck, J. Neural network computation with DNA strand displacement cascades. *Nature* **2011**, *475*, 368–372. [CrossRef] [PubMed]
23. Scudellari, M. Inner Workings: DNA for data storage and computing. *Proc. Natl. Acad. Sci. USA* **2015**, *112*, 15771–15772. [CrossRef] [PubMed]

24. Wang, B.; Zhou, S.; Zheng, X.; Zhou, C.; Dong, J.; Zhao, L. Image watermarking using chaotic map and DNA coding. *Opt.-Int. J. Light Electron Opt.* **2015**, *126*, 4846–4851. [CrossRef]
25. Zhang, Q.; Wang, B.; Wei, X.; Fang, X.; Zhou, C. DNA word set design based on minimum free energy. *IEEE Trans. NanoBiosci.* **2010**, *9*, 273–277. [CrossRef] [PubMed]
26. Chai, X.L.; Chen, Y.R.; Broyde, L. A novel chaos-based image encryption algorithm using DNA sequence operations. *Opt. Lasers Eng.* **2017**, *88*, 197–213. [CrossRef]
27. Wu, X.J.; Kurths, J.; Kan, H.B. A robust and lossless DNA encryption scheme for color images. *Multimedia Tools Appl.* **2018**, *77*, 12349–12376. [CrossRef]
28. Marathe, A.; Condon, A.E.; Corn, R.M. On combinatorial DNA word design. *J. Comput. Biol.* **2001**, *8*, 201–219. [CrossRef] [PubMed]
29. Shin, S.-Y.; Lee, I.-H.; Kim, D.; Zhang, B.-T. Multiobjective evolutionary optimization of DNA sequences for reliable DNA computing. *IEEE Trans. Evol. Comput.* **2005**, *9*, 143–158. [CrossRef]
30. Hamming, R.W. Error detecting and error correcting codes. *Bell Syst. Tech. J.* **1950**, *29*, 147–160. [CrossRef]
31. Zhang, Q.; Wang, B.; Wei, X.P.; Zhou, C.J. A Novel Constraint for Thermodynamically Designing DNA Sequences. *PLoS ONE* **2013**, *8*. [CrossRef] [PubMed]
32. Watson, J.D.; Crick, F.H.C. Molecular structure of nucleic acids—A structure for deoxyribose nucleic acid. *Nature* **1953**, *171*, 737–738. [CrossRef] [PubMed]
33. Zhang, Q.; Guo, L.; Wei, X.P. A novel image fusion encryption algorithm based on DNA sequence operation and hyper-chaotic system. *Optik* **2013**, *124*, 3596–3600. [CrossRef]
34. Kracht, D.; Schober, S. Insertion and deletion correcting DNA barcodes based on watermarks. *BMC Bioinform.* **2015**, *16*, 50. [CrossRef] [PubMed]
35. Wang, B.; Zhang, Q.; Zhang, R.; Xu, C.X. Improved Lower Bounds for DNA Coding. *J. Comput. Theor. Nanosci.* **2010**, *7*, 638–641. [CrossRef]
36. Su, X.; Li, W.; Hu, H. Cryptanalysis of a chaos-based image encryption scheme combining DNA coding and entropy. *Multimedia Tools Appl.* **2017**, *76*, 14021–14033. [CrossRef]
37. Zhang, Y. Breaking a RGB image encryption algorithm based on DNA encoding and chaos map. *Int. J. Inf. Secur.* **2014**, *1*, 22–28.
38. Liu, Y.; Tang, J.; Xie, T. Cryptanalyzing a RGB image encryption algorithm based on DNA encoding and chaos map. *Opt. Laser Technol.* **2014**, *60*, 111–115. [CrossRef]
39. Kwok, H.; Tang, W.K.S. A fast image encryption system based on chaotic maps with finite precision representation. *Chaos Solitons Fractals* **2007**, *32*, 1518–1529. [CrossRef]
40. Peng, J.; Zhang, D.; Liao, X. A digital image encryption algorithm based on hyper-chaotic cellular neural network. *Fundam. Inf.* **2009**, *90*, 269–282.
41. Wang, B.; Zheng, X.; Zhou, S.; Zhou, C.; Wei, X.; Zhang, Q.; Che, C. Encrypting the compressed image by chaotic map and arithmetic coding. *Opt.-Int. J. Light Electron Opt.* **2014**, *125*, 6117–6122. [CrossRef]
42. Wang, Y.; Wong, K.W.; Liao, X.; Chen, G. A new chaos-based fast image encryption algorithm. *Appl. Soft Comput.* **2011**, *11*, 514–522. [CrossRef]
43. Gupta, K.; Silakari, S. Novel Approach for fast Compressed Hybrid color image Cryptosystem. *Adv. Eng. Softw.* **2012**, *49*, 29–42. [CrossRef]
44. Yuan, S.; Deng, G.; Feng, Q.; Zheng, P.; Song, T. Multi-objective evolutionary algorithm based on decomposition for energy-aware scheduling in heterogeneous computing systems. *J. Univ. Comput. Sci.* **2017**, *23*, 636–651.
45. Wang, X.; Gong, F.; Zheng, P. On the computational power of spiking neural P systems with self-organization. *Sci. Rep.* **2016**. [CrossRef] [PubMed]
46. Song, T.; Wong, P.Z.D.M.; Wang, X. Design of logic gates using spiking neural p systems with homogeneous neurons and astrocytes-like control. *Inf. Sci.* **2016**, *372*, 380–391. [CrossRef]
47. Song, T.; Rodríguez-Patón, A.; Zheng, P.; Zeng, X. Spiking neural p systems with colored spikes. *IEEE Trans. Cognit. Dev. Syst.* **2018**. [CrossRef]
48. Wang, X.; Zheng, P.; Ma, T.; Song, T. Computing with bacteria conjugation: Small universal systems. *Moleculer* **2018**, *23*, 1307. [CrossRef] [PubMed]
49. Zhang, L.; Yuan, S.; Feng, L.; Guo, B.; Qin, J.; Xu, B.; Lollar, C.; Sun, D.; Zhou, H. Pore-Environment Engineering with Multiple Metal Sites in Rare-Earth Porphyrinic Metal-Organic Frameworks. *Angew. Chem.* **2018**. [CrossRef]

Application of Molecular Methods in the Identification of Ingredients in Chinese Herbal Medicines

Ke Han [1,*], Miao Wang [2], Lei Zhang [2] and Chunyu Wang [3]

1 School of Computer and Information Engineering, Harbin University of Commerce, Harbin 150028, China
2 Life sciences and Environmental Sciences Development Center, Harbin University of Commerce, Harbin 150010, China; w1993817m@163.com (M.W.); 13212921382@163.com (L.Z.)
3 School of Computer Science and Technology, Harbin Institute of Technology, Harbin 150001, China; chunyu@hit.edu.cn
* Correspondence: hanke@hrbcu.edu.cn.

Abstract: There are several kinds of Chinese herbal medicines originating from diverse sources. However, the rapid taxonomic identification of large quantities of Chinese herbal medicines is difficult using traditional methods, and the process of identification itself is prone to error. Therefore, the traditional methods of Chinese herbal medicine identification must meet higher standards of accuracy. With the rapid development of bioinformatics, methods relying on bioinformatics strategies offer advantages with respect to the speed and accuracy of the identification of Chinese herbal medicine ingredients. This article reviews the applicability and limitations of biochip and DNA barcoding technology in the identification of Chinese herbal medicines. Furthermore, the future development of the two technologies of interest is discussed.

Keywords: bioinformatics; identification of Chinese herbal medicines; biochip technology; DNA barcoding technology

---

1. Introduction

The traditional method of determining the authenticity of traditional Chinese medicine is to evaluate the color, shape, or nature of the medicinal materials using either physical/chemical methods or microscopy. These methods are still used, are relatively simple to carry out, and have led to substantial advancements in the screening of Chinese herbal medicines. However, these methods have some shortcomings and disadvantages [1]. It is assumed that there is a certain level of contamination of counterfeit and inferior medicinal ingredients in compound medications made by processing drugs of many ingredients, and it is difficult to identify the contaminating materials following the traditional method of identification. After thousands of years of development, traditional Chinese medicine has grown into a vast system, and its medicinal ingredients are numerous and complex [2]. Moreover, there are many disadvantages to the all too common practices of using the same name to describe many different herbs, or in having multiple names for a single medicinal material. This further increases the difficulty in the identification of Chinese herbal medicines [3,4]. In addition, other conventional Chinese herbal screening methods require the operators to have very rich work experience and expertise, or some subtle changes will lead to errors in the identification results [5]. Due to the error-prone nature of the identification process, it is often difficult to meet actual needs.

Bioinformatics is a discipline that integrates the most advanced knowledge of computer information technology and molecular biology [6–8]. This discipline uses the powerful data-processing ability of computers to analyze and compare proteomic, transcriptomic, genomic, and microorganism data, and so on, in order to identify and solve specific problems. At present, this technology has been applied to many fields, including Chinese herbal screening, as described in this review. It has also been applied to other medical fields, such as studying the effect of a drug on the prevention and treatment of diseases, and studying special drugs for cancer cell proliferation and death [9]. Therefore, it is of great significance and value for Chinese medicine modernization to further study the identification of Chinese herbal medicines by bioinformatics.

At present, bioinformatics assists in the identification and assessment of Chinese herbal medicines through the use of DNA barcoding technology and biochip technology. DNA barcoding seeks to identify biological material through sequencing a selected genetic marker, and then comparing that DNA sequence to sequences from the same genetic marker in other species [10,11]. Nowadays, this technology has been applied widely, and can be regarded as a mature tool for biological exploration and research. Not only can it be used to identify organisms that are difficult to distinguish by other methods, it can also be used to discover new organisms that have not been discovered by the biological research community to date [12]. If a specimen of a suspected new organism is collected, this technique can be used to extract its corresponding DNA barcode, and then compare and analyze it with DNA barcodes (nucleotide sequences) that are available in the public nucleotide sequence database, so that the relevant identity information of this organism can be determined step-by-step. On the other hand, biochip technology is a microarray that is composed of some biological components (nucleic acids, proteins, cells, etc.) wrapped on solid supports such as nylon film or a silicon wafer [13]. Through the use of a chip, this technology takes advantage of automation, speed, and big data. This review discusses the promotion and use of bioinformatics in the field of Chinese herbal medicine identification in recent years, and compares the advantages and disadvantages of various methods. With the broad applicability of these techniques, we hope that the identification of Chinese herbal medicines can be further developed.

2. Biochip Technology

Since the 1990s, due to continuous progress and development in science and technology, especially the rapid development of computer and network technology, a series of related disciplines emerged [14–18]. This also led to the development of biological chip technology for microanalysis research [19]. Biochip technology is a comprehensive technology discipline that combines chemical and physical technology. Biochip is the crystallization of DNA hybridization probe technology combined with semiconductor industry technology. The technique is to hybridize a large number of probe molecules with fluorescent-labeled DNA or other sample molecules (e.g., proteins, factors, or small molecules) after immobilizing them on the support. The number and sequence information of the sample molecules can be obtained by detecting the hybridization signal intensity of each probe molecule [20]. Fluorescence-labeled target molecules play a role in a variety of microbodies on the chip. A spectrophotometer is used for analysis of spectral/absorption characteristics, and the results will vary according to the intensity of the material. Next, a special instrument is used to collect and convert the data to be processed by a computer. In this way, the required biological information will be obtained [21,22]. Nowadays, biochips can be divided into tissue chips [23], protein chips, and gene chips. The latter two are widely used in Chinese herbal medicine identification.

2.1. Gene Chip

Gene chip technology is fast, highly efficient, automated, parallel, and economical. It has become an important technical method in the field of screening and the evaluation of inferior and counterfeit drugs of Chinese herbal medicine. The process involves obtaining the standard atlas of positive drugs and the atlas of prepared identification (query) drugs, and then analyzing and comparing the differences between the two. To obtain the two maps, one must first extract DNA from the corresponding samples, and then let them hybridize with DNA chips [24–27]. The detection steps of gene chip technology are shown in Figure 1.

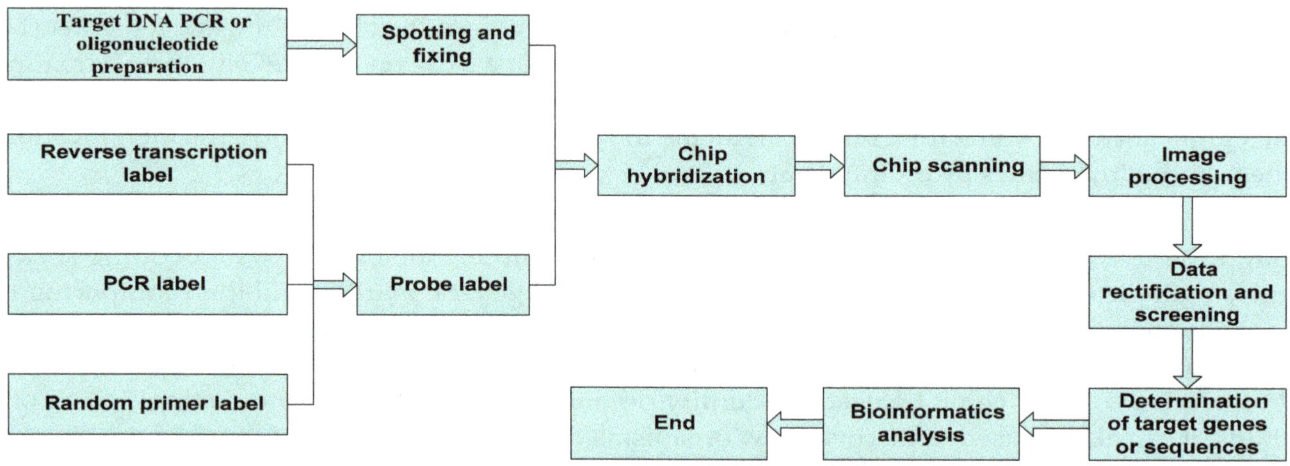

Figure 1. Steps of gene chip technology.

Zhang et al. [28] used gene chip technology to study the *Dendrobium nobile* (*Orchidaceae, Dendrobium Sw.*) species mixed in compound medication, and five species of *Dendrobium nobile* were successfully detected and loaded into China Pharmacopoeia (https://www.sinicave.com/pd_pharmacopoeia.cfm). It has been shown that this technique can be used in the field of Chinese herbal medicine taxonomy. Hao et al. [29] was the first to use AFLP (amplified fragment length polymorphism) technology in the study of the characteristics of Chinese herbal medicines, after which they drew the atlas of genetic diversity and AFLP fingerprints on the target taxa. This method can be used to easily and effectively differentiate between the characteristics of Chinese herbal medicines. The AFLP marker is then turned into a SCAR (sequence characterized amplified regions) marker, and the mounting glue can be recycled, or molecular cloning by PCR can be performed. The obtained PCR can be used to detect the characteristics of the target medicinal ingredient. Among expectorant and antitussive drugs, *Fritillaria cirrhosa* (*Liliaceae, Fritillaria*) has a positive effect, and is favored by many [30]. However, the high price of the highly sought-after drug encouraged sales of counterfeit or low-quality replicas at lower prices. In view of this, some scholars [31] began to sequence and study the 5S rRNA gene of *Fritillaria hupehensis*, *F. anhuiensis*, *F. thunbergii*, and *Fritillaria cirrhosa* by gene chip technology, and found that *Fritillaria cirrhosa* has a unique sequence of 5′-CTTTTGTCGATCA-3′, which is absent in other *Fritillaria* species. This sequence was used to make the gene chip. This technique is used to detect and extract the gene sequences of some tested products, and then compare them with the gene sequence of the positive *Fritillaria cirrhosa* control, to determine whether the sample that is being tested contains the expected product.

In another study, Chen et al. [32] isolated the genomic DNA of *Coptis chinensis* (*Ranunculus, Coptis Salisb.*) as a template, and then analyzed the many thermal cycle parameters in the ISSR (inter-simple sequence repeat) reaction system using two factors and single factors, as well as the effect and influence of some major components of amplification to identify the optimal conditions. Finally, the reaction system and amplification procedure that are suitable for the ISSR analysis of *Coptis chinensis* were established under optimum conditions. The establishment of this optimization system provides a standardized procedure for the identification and genetic diversity analysis of *Coptis chinensis* by the ISSR marker technique in the future.

The greatest advantage of a gene chip is high throughput. The process of marking and hybridizing probes for many genes can be completed in a single experiment; the degree of automation is high, and the data are objective and reliable. However, its greatest shortcoming is that the gene chip cannot be used to find new things; it can only be used to find things that have been found before (and have been printed on the chip).

2.2. Protein Chip

Protein chip technology is mainly used for protein analysis and exploration. The methodology involves the use of advanced microelectronics technology to analyze the surface of some carriers in order to establish a system that is suitable for microbiological research [33–35]. First, some known proteins are fixed to the carrier in order; then, the molecules can specifically interact with known molecules based on molecular properties, after which the molecules are ready for purification and subsequent treatment. Finally, the protein components can be more quickly and accurately screened [36,37].

Li et al. [38] used an NP10 chip and a protein chip combined with SELDI-TOF MS (surface-enhanced laser desorption/inionation-time of flight-mass spectra) to analyze the peptide composition and protein of tortoise shell glue. The peptide component/protein quality fingerprint of tortoise shell glue was obtained. This technology can be used for the digital analysis of tortoise shell glue. Wang et al. [39] obtained the protein/peptide from the dried and processed *Pheretima asiatica*, and analyzed the protein/peptide information with the surface-enhanced laser desorption/ionization time-of-flight mass spectrometry technique. Twenty-nine peaks of protein/peptide molecular weight were obtained. Among them, crude and dried *Pheretima asiatica* had 17 peaks of meaningful molecular weight, and processed *Pheretima asiatica* had 12 peaks. There was an amino acid residue among multiple adjacent protein peaks, and multi-groups of molecular weight of crude and processed *Pheretima asiatica* were extremely similar, indicating that these peptides may represent the same peptides. Molecular weight fingerprinting is obtained from the crude and processed *Pheretima asiatica* protein/peptide by laser desorption/ionization time-of-flight mass spectrometry; it can serve as a digitalized quality and control standard of *Pheretima asiatica*, and provide references for the further isolation, purification, and verification of the proteins/peptides associated with *Pheretima asiatica* function.

Protein chip technology has the following advantages: it is low-cost because it requires only a few samples and reagents in the process, it is more accurate and sensitive than conventional ELISA (enzyme linked immune sorbent assay), it is easy to operate due to automation, it is highly accurate, a large number of proteins can be quickly analyzed, and it has few sample requirements. The sample to be tested can be detected and analyzed by a protein chip only by simple processing. However, this technology has some shortcomings compared to gene chip technology: the purification of protein is more difficult, there is a lack of mature protein technology, the technology is inseparable from the function and re-modification of other proteins, and proteins are more variable among themselves. Therefore, protein chips are used less frequently than gene chips.

3. DNA Barcoding

Ever since Canadian scientist Paul Herbert [40] proposed DNA barcoding technology back in 2003, this diagnostic technology has been widely used in the field of species identification. DNA barcoding uses a short, unique DNA sequence to identify a species. Since traditional Chinese medicines are derived from a variety of animals, plants, and minerals, this technology can be used for the accurate identification and evaluation of medicinal ingredients. This technique is widely used in the field of biological and medicinal material identification because it is not influenced by the morphology of the sample or the surrounding environment [41]. The application of this technology has broadened widely, and has reached multiple fields and industries, such as ecology, development, and evolution, Chinese herbal medicine identification, and genetic identification [42]. A non-trivial proportion of the public fungal DNA sequences are compromised in terms of quality and reliability, contributing noise and bias to sequence-borne inferences. R. Henrik Nilsson et al. [43] discussed various aspects and pitfalls of sequence quality assessment. Based on their observations, they provided a set of guidelines to assist in the manual quality management of newly generated, near-full-length (Sanger-derived) fungal internal transcribed spacer (ITS) sequences, and to some extent also sequences of shorter read lengths, other genes or markers, and groups of organisms. A flow chart describing the process of the DNA barcoding molecular identification of Chinese herbal medicines is shown in Figure 2.

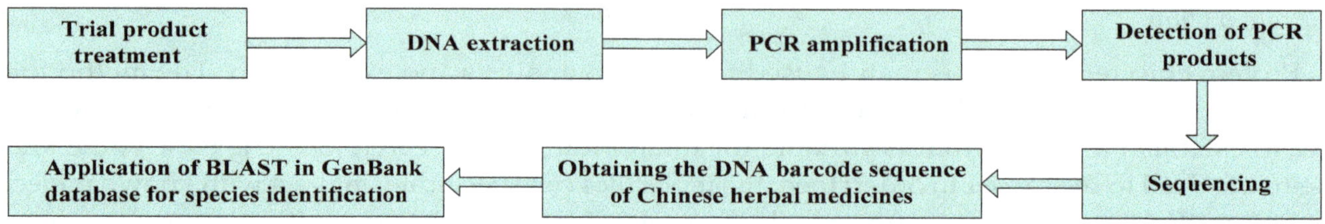

Figure 2. Molecular identification process of DNA barcoding in Chinese medicinal materials.

According to the selection of different DNA sequences, DNA barcoding can be roughly divided into four categories: (1) mitochondrial DNA barcode, (2) ribosomal DNA barcode, (3) chloroplast DNA barcode, and (4) DNA barcode combination identification. A suitable DNA barcode should meet the following conditions: high throughput so that it can be routinely sequenced in plant species, suitable for producing a high-sequence mass coverage of bidirectional sequences with minimal unsequenced bases, high resolution so that most species can be distinguished, and DNA fragments should be short enough so that degraded DNA can be amplified.

3.1. Mitochondrial DNA Barcode

Mitochondrial DNA barcode technology was the first technique to be used in the classification and identification of animal mitochondria. Mitochondrial COI (cytochrome oxidase) gene barcode technology has been an important method in identifying animal Chinese herbal medicine [44]. Hebert [40] eventually selected COI sequences, because COI sequences ensured sufficient variation and were easy to be amplified by universal primers, and there were few deletions and insertions in its own DNA sequence, so it was suitable for the analysis of closely related taxa.

Shi Linchun et al. [45], in order to distinguish *periostracum serpentis*(the skin that the snake shed, PS) from its adulterants, PCR amplified and sequenced COI sequences of 68 samples from 13 species. Furthermore, the DNA barcoding gap and phylogenetic cluster analysis were carried out. The results showed that three specimens of *periostracum serpentis Elaphe taeniura* (Cope), *E. carinata* (Guenther), and *Zaocys humnades* (Cantor)—had DNA barcode gaps, and they were separated into independent branches on the neighbor-joining (NJ) system clustering tree. As a DNA barcode, COI can not only identify three basic genera and species of Chinese herbal snakeskin, it can also distinguish between snakeskin and their easily confused products. This shows that DNA barcoding can be used for the identification of the snake shedding that is found in Chinese medicinal products. Zhang Hongyin et al. used the same method to identify the pseudo products of *centipede* [46], *deer medicine* [47], and *Gekko gecko* [48]. The results showed that the COI gene can be used as an effective marker for identifying at least metazoan Chinese medicinal ingredients.

The COI gene sequence exists in the vast majority of animal cells. There are only one set of genomic chromosomes in a cell, whereas there are hundreds of mitochondria per cell; that is why mitochondrial DNA is more easily recovered. The mitochondrial DNA mutation rate is 10 times greater than that of genomic DNA. According to this characteristic, it is easier for us to accurately differentiate species. However, one disadvantage of this technology is that it requires a lot of tedious work in the identification of species, and its identification results are not produced directly or quickly.

3.2. Ribosomal DNA Barcode

The nuclear ribosomal ITS (internal transcribed spacer) region contains the ITS1 intergenic region, the ITS2 intergenic region, and the 5.8S gene (ITS1-5.8S-ITS2) ranging in size from 400 bp to over 1000 bp. ITS has a high capacity for species identification and technical scalability. Ribosomal DNA is a polygene family, and ITS exists in highly repetitive ribosomal DNA [49,50]. Due to its fast evolution speed and short length, this genetic marker is widely used in the field of angiosperm branching analysis in plant systems and fungal metabarcoding [51].

Li et al. [52] used ITS primers to amplify the ITS sequences of *Hedyotis diffusa Willd* (*Rubiaceae, Cerastium*) and *Corymbose Hedyotis Herb* (*Oldenlandia corymbosa* L.), and found that there are obvious differences in the ITS sequences between the two plants. They then used the established phylogenetic tree to analyze the common *Herba Hedyotis* that is available on the market, and found that only the medicinal herbs purchased from Guangzhou were genuine; those purchased from Boston and Hong Kong were actually *Corymbose Hedyotis Herb*. Moreover, there was confusion regarding the samples that were used in previous studies of medicinal materials, indicating that it is difficult to distinguish between the two types of medicinal materials based on their shape and characteristics. Thus, the identification of Chinese medicinal materials assisted by bioinformatics is more accurate than the traditional identification method. Based on ITS2 barcode technology, Yu Junlin et al. [53] explored 17 samples of *Bupleurum longiradiatum Turcz* and 31 samples of *Bupleurum* species. They examined the intraspecific variation of two species by analyzing and studying two categories, and identified an accurate distinction between two major species. In order to ensure the accuracy of the study, the samples used in the study have been validated using BLAST (Basic Local Alignment Search Tool) following the DNA barcoding identification system for Chinese medicinal materials (http://www.tcmbarcode.cn). The ITS2 barcode sequence can accurately identify *Bupleurum* and *B. longiradiatum Turcz*. The minimum kimura 2-parameter(K2P) distance between *Bupleurum* and *B. longiradiatum Turcz* is far greater than the maximum K2P distance within the species of *Bupleurum*, and the NJ tree shows that the *B. longiradiatum Turcz* constitutes a single branch, which can be distinguished from *Bupleurum*. Therefore, *Bupleurum* and *B. longiradiatum Turcz* can be distinguished consistently and accurately using the ITS2 barcode. Shi Yuhu et al. [54] used the ITS2 sequence as a barcode to identify the herbal tea ingredient *Plumeriarubra* and its adulterants. Genomic DNAs from 48 samples were extracted; the ITS2 sequences were amplified and sequenced bidirectionally; and then they were assembled and obtained using CodonCode Aligner (https://www.codoncode.com/). The sequences were aligned using ClustalW, the genetic distances were computed by the K2P model, and the NJ phylogenetic tree was constructed using MEGA5.0. The results showed that the length of the ITS2 sequence of *P. rubra* were 244 bp. The intraspecific genetic distances (0–0.0166) were much smaller than interspecific ones between *P. rubra* and its adulterants (0.3208–0.6504). The NJ tree indicated that *P. rubra* and its adulterants could be distinguished clearly. Therefore, using the ITS2 barcode can accurately and effectively distinguish the herbal tea ingredient *P. rubra* from its adulterants, which provides a new molecular method to identify *P. rubra* and ensure its safety in use.

ITS sequences are usually used to distinguish some species that are closely related to each other. It may be difficult to identify the different geographical distribution and host types of some fungi. Not all 18 S and 28 rRNA databases have been established, and many of the data are still lacking, which results in the inability to rapidly identify the required ITS fragments, thus inevitably affecting the application of ITS methods.

3.3. Chloroplast DNA Barcode

psbA-trnH intervals in common flowering plants are between 340–660 bp. This sequence interval was compared with nine other genetic markers (matK, rbcL, and ITS are included) with a discrimination efficiency of 83% and an amplification efficiency of 100% [55–58]. Zhang Yaqin et al. [59] used psbA-trnH sequence technology to study and explore *Pyrrosia* in order to distinguish the appearance and form of counterfeit herbs that are similar to those of *Pyrrosia*. They collected partial sequences of psbA-trnH in the chloroplast genes of *Pyrrosias* and some counterfeit products. Next, the group used two conventional methods based on the partial sequence of the collected psbA-trnH: the minimum distance method and the similarity search method. It was found that the former method could not distinguish between two *Pyrrosia* species; however, the use of the psbA-trnH sequence clearly distinguishes some other counterfeit products mixed with other kinds of *Pyrrosias*. This study shows that the psbA-trnH sequence can effectively distinguish and identify genuine and counterfeit products of pteridophytes.

As one of the fastest evolving regions in plant chloroplasts, the matK gene is about 1500 bp [60]. Hilu and Lahaye [61] also pointed out that it can be implemented in a single fragment when choosing a barcode of a plant. The fragment is then added according to the complexity of the group. Genievskaya Y et al. [62] identified species defined by morphological traits using sequences of the nuclear ribosomal DNA ITS1-5.8S-ITS2 region and matK.

The polymorphic sequence positions in Kazakh populations and GenBank (Benson et al. 2017) references were acquired by comparison with GenBank sequences, which identified a difference between local populations of sand rice (*Agriophyllum squarrosum*). ITS and matK sequence analysis revealed a segregation of *Agriophyllum squarrosum* (L.) *Moq* from *A. minus* into separate branches in maximum-likelihood dendrograms. ITS analysis can be used to characterize the populations of *A. squarrosum* growing far away from each other. The data obtained in this study laid the foundation for the further study of *A. squarrosum* populations, and summed up the advantages and disadvantages of this technology. Wang Xiaoming et al. [63] used the matK sequence method in their analysis and exploration of *Herba Abri* (*Leguminous*, *Abrus* L.), and summarized the advantages and disadvantages of this technology in the applications of the DNA barcode in this plant. The modified CTAB (Cetyltriethyl Ammnonium Bromide) method was used to extract the total DNA of nine kinds of *Herba Abri* from different regions, and the matK sequence was amplified using the universal primers of leguminous plants. After that, both K2P genetic distance calculation and the creation of the NJ tree showed that the ITS2 sequences can be used as the DNA barcode sequence of the *Fabaceae* plants. The results showed that the total length of the matK sequence was 889–895 bp; the genetic distance between different plants was far greater than the genetic distance within the populations of the same plant, and even the smallest interspecific genetic distance still exceeded the maximum intraspecific genetic distance. Therefore, matK sequence technology can serve as a DNA barcode for leguminous plants.

rbcL is a fragment of the coding region of a chloroplast gene. The rbcL fragment has a low species resolution, but it is of relatively high species resolution in angiosperms [64–66]. Chen Jianxiong et al. [67] collected nine samples of Chinese lobelia of different origins and seven samples of *Mazus japonicus*, and extracted the total DNA from all of the samples collected. The rbcL fragment in the chloroplast DNA of the sample was sequenced, and Clustal X 2.1 software (University College Dublin, Dublin, Ireland) was used for multiple sequence alignments. The NJ clustering feature of MEGA 5.0 software (Center of Evolutionary Functional Genomics Biodesign Institute Arizona State University, Phoenix, AZ, USA) was then used for cluster analysis. They designed specific primers identified by SNP micropoints for two groups of samples, established a specific PCR identification method, and used SYBR Green I (Molecular Probes) dye to establish a rapid detection method for two kinds of Chinese herbs. They successfully identified the Chinese lobelia and *Mazus japonicus*. The results showed that rbcL has strong molecular identification ability and rbcL is easy to amplify and compare. So, rbcL was selected as the DNA barcode for species identification. Huang Qionglin et al. [68] identified *Nervilia fordii (Hance) Schltr.* and its adulterants byrbcLsequencing. They used a commercial kit to extract genomic DNA from fresh leaves; PCR amplification and sequencing were conducted with a pair of universal primers. DNAMAN (https://www.lynnon.com/); Clustal X (University College Dublin, Dublin, Ireland) and MEGA 4.0 software (Center of Evolutionary Functional Genomics Biodesign Institute Arizona State University, Phoenix, AZ, USA) were used for sequence alignment, genetic distance analysis, and clustering analysis. They acquired 502 bp sequences of the rbcL gene from *N. fordii* and its adulterants. Three types of *N. fordii* showed completely consistent sequence data, and differences in five sites were shown between *N. fordii* and *N. plicata*. The interspecific variations were larger than the intraspecific ones. In the cluster dendrogram, all of the species were monophyletic and distinguished from the others. The results showed that the rbcL gene can be used as a DNA barcode to identify *N. fordii* and its counterfeit.

3.4. DNA Barcode Combination Identification

Studies have found that it is difficult to identify some species accurately if only one genetic marker is used in the identification of Chinese herbal medicines. Therefore, it is not possible to rely on a single genetic marker to identify all of the species, especially for those species with complex genetic backgrounds [69–72]. Therefore, a combination of DNA barcodes can be used to accurately identify species.

The CBOL Plant Working Group [73] found that rbcL and matK were suitable barcodes for plants, because these two barcodes were used to successfully identify 550 species from 907 samples, with a success rate of roughly 72%. Moreover, the barcode can become a basic criterion for the molecular identification of species, and it can also provide references for discovering hidden species. Yong et al. [74] tested the universality of rbcL, matK, trnH-psbA, and ITS as DNA barcodes for tree species, and then examined the accuracy of the phylogenetic inference and species identification in three tropical cloud forests (Table 1). Their results suggested that rbcL and trnH-psbA should be adopted as the standard DNA barcode for tree species in tropical cloud forests. The success rates of identifying four fragments were all higher than 41.00%, demonstrating that these fragments are candidates for use in species identification. They used random fragment combinations of rbcL, matK, and trnH-psbA to infer phylogenetic relationships, and established the optimal evolutionary tree with high supporting values in tropical cloud forests.

Table 1. Results of PCR amplification success rate and DNA sequencing rate for rbcL, matK, trnH-psbA, and internal transcribed spacer (ITS), respectively.

	RbcL	MatK	TrnH-psbA	ITS
Success rates of PCR amplification	75.26% ± 3.65%	57.24% ± 4.42%	79.28% ± 7.08%	50.31% ± 6.64%
Rates of DNA sequencing	63.84% ± 4.32%	50.82% ± 4.36%	72.87% ± 11.37%	45.15% ± 8.91%

Priyanka et al. [75] studied the steno-endemic species of the genus *Decalepis* (*Decalepisar ayalpathra Venter*, which is locally known as Amirthapala, is a steno-endemic species in the eastern and western ghats of peninsular India), and found the corresponding DNA barcode, which can be used to monitor and stop the illegal trade of these endangered species. They used rbcL, matK, psbA-trnH, ITS, and ITS2 as DNA barcode candidates. The average intraspecific variation was 0–0.27%, which was less than the distance to the nearest neighbor (0.4–11.67%) with ITS and matK. Finally, they combined rbcL, matK, andITS to produce 100% species resolution using the PAUP (http://paup.phylosolutions.com/, Phylogenetic Analysis Using PAUP) and BOLD (http://v3.boldsystems.org/, Bold systems) methods with the least number of marker combinations to support a character-based approach. They found that the most advantageous barcode datasets were achieved by combining rbcL, matK, ITS, mat, ITS, and ITS2, with a consistency index (CI) of 85% and 90%, respectively. They included 2106 characters in the former dataset for parsimony analysis, among which 103 were parsimony informative, and 18 variable characters were parsimony-uninformative. The 1836 total characters of the latter dataset contributed 146 informative characters. Therefore, the rbcL, matK, and ITS combination is considered to be the best choice for species resolution in the genus *Decalepis*. DNA barcoding has greatly improved species identification and resolution.

Compared with use of single DNA barcodes for identifying species, combining DNA barcodes can greatly improve species resolution. However, researchers need to find the right combination of DNA barcodes. The use of DNA barcode combinations for species identification is more complicated and costlier than use of a single barcode. Therefore, different identification methods should be selected for different species, so as to achieve quick and convenient identification results.

3.5. *The Limitations of DNA Barcoding*

Although DNA barcoding has many advantages in the identification of Chinese herbal medicine, it still has some limitations. In fungi, the ITS region has been sequenced for something similar to 1% of the estimated number of extant species. This makes it tricky to compare newly generated sequences to the entries in GenBank. First, traditional Chinese herbal medicines are usually made of dead animals and plants. Due to collection, processing, storage time, and storage conditions, the DNA macromolecules in the test sample may have been destroyed and degraded prior to analysis. Second, it is difficult to obtain the template that is needed to identify DNA molecules when only small fragments are retained in the sample. Third, although DNA barcoding technology can effectively identify and evaluate plant and animal material in medicinal ingredients, the technology cannot identify the mineral components. Finally, DNA barcoding technology requires technical expertise and is costly.

In order to solve these problems, we should promote basic research, enrich the genome sequencing data, and find small fragments of molecular markers. When dealing with medicinal materials, we should strictly follow standard methods and conditions to avoid damaging the DNA. Results can be validated by using multiple fragments or molecular markers to obtain more accurate DNA fragments, and we should simply test the plant or animal tissues or extracts upfront, before they are used to make medicines. Previous researchers used ClustalW to multiple sequence alignment, but this method is a tool that is long-since obsolete. The user should go for a recent (and readily updated) one, rather than relying on old programs. We recommend any of MAFFT (Katoh et al. 2010), Muscle (Edgar et al. 2004), and PRANK (Löytynoja et al. 2005) for large or otherwise non-trivial sequence datasets [76]. For phylogenetic analysis, NJ is a long-since obsolete tool; Bayesian inference in MrBayes is a whole lot more powerful. In future research, researchers can use these better tools to improve identification.

4. Future Perspectives

Chinese medicine reveals a rich history of our people's long-term struggle against diseases, and has made great contributions to the prosperity and rebirth of the Chinese nation. Starting from Shennong's Herbal Classic (https://en.wikipedia.org/wiki/Shennong_Ben_Cao_Jing), Chinese medicine plays a very important role in our historical society. However, after a period of instability, the development of traditional Chinese medicine is regrettably at a standstill. With improvements in people's standards of living and access to technology, medical practitioners and researchers in China and abroad have taken a new interest in Chinese herbal medicine. However, this increased interest in traditional Chinese medicine has led to the emergence of counterfeit and low-quality Chinese herbal medicine in the market, which has greatly reduced the efficacy of Chinese herbal medicine. Therefore, it is essential that we develop a fast and accurate method to identify counterfeit and low-quality Chinese herbal medicines.

Chinese herbal medicines can be identified based on three aspects: appearance, chemical composition, and molecular characteristics. Many medicinal herbs are similar in appearance and are not easy to differentiate, which increases the risk of false identification. Since many Chinese herbal medicines are made up of many kinds of materials that are subsequently ground into a powder, it is very difficult to identify Chinese medicinal ingredients according to chemical constituents. Bioinformatics has brought new opportunities for the identification of traditional Chinese herbal medicine ingredients using high throughput and big data. At present, there are mainly two methods of bioinformatics identification of Chinese herbal medicine: biochip technology and DNA barcoding. The use of DNA barcoding is more extensive, because this method has different identification methods for different kinds of Chinese herbal medicines, making it a more targeted approach. Moreover, the combination identification method of DNA barcoding allows for more species to be identified accurately. The first step toward the DNA barcoding of Chinese herbal medicine is to extract different DNA fragments from different species and select the most suitable DNA fragment, followed by PCR amplification to detect PCR products. Finally, the target bands should be bidirectionally sequenced by DNA sequencing. This method is not without its limitations, and cannot be used to identify all of the species. Therefore, future studies should combine traditional Chinese medicine identification methods, biochip technology, DNA barcoding technology, and high-throughput sequencing-driven metabarcoding to identify Chinese medicinal ingredients and obtain more accurate and rapid identification results. We believe that through the continuous efforts of researchers, the bioinformatics-assisted identification of Chinese herbal medicines will make substantial advancements. It will one day be possible to quickly identify counterfeit and inferior medicinal ingredients from a large number of mixed Chinese medicinal materials, thus standardizing the market of Chinese herbal medicines.

References

1. Zhao, Y.; Hellum, B.H.; Liang, A.H.; Nilsen, O.G. The in vitro inhibition of human cyp1a2, cyp2d6 and cyp3a4 by tetrahydropalmatine, neferine and berberine. *Phytother. Res.* **2012**, *26*, 277–283. [CrossRef] [PubMed]
2. Chen, R.; Dong, J.; Cui, X.; Wang, W.; Yasmeen, A.; Deng, Y.; Zeng, X.M.; Tang, Z. DNA based identification of medicinal materials in chinese patent medicines. *Sci. Rep.* **2012**, *2*, 958. [CrossRef] [PubMed]
3. Han, J.; Song, J.; Yao, H.; Chen, X.; Chen, S. Comparison of DNA barcoders in identifying medicinal materials. *China J. Chin. Mater. Med.* **2012**, *37*, 1056–1061.
4. Qian, Z.; Lin, C.R.; Xiao, P.G. Identification and standard of traditional Chinese Medicine. *China J. Chin. Mater. Med.* **2014**, *39*, 2153–2154.
5. Read, T.D.; Salzberg, S.L.; Pop, M.; Shumway, M.; Umayam, L.; Jiang, L.X.; Holtzapple, E.; Busch, J.D.; Smith, K.L.; Schupp, J.M.; et al. Comparative genome sequencing for discovery of novel polymorphisms in bacillus anthracis. *Science* **2002**, *296*, 2028–2033. [CrossRef] [PubMed]
6. Giacomelli, L.; Covani, U. Bioinformatics and data mining studies in oral genomics and proteomics: New trends and challenges. *Open Dent. J.* **2010**, *4*, 67–71. [CrossRef] [PubMed]
7. Wang, J.X.; Li, M.; Yu, Z.G. Nonlinear science and network methods for prediction problems in bioinformatics and systems biology. *Curr. Bioinform.* **2016**, *11*, 154–155. [CrossRef]
8. Prejzendanc, T.; Wasik, S.; Blazewicz, J. Computer representations of bioinformatics models. *Curr. Bioinform.* **2016**, *11*, 551–560. [CrossRef]
9. Gao, J.Z.; Cui, W.; Sheng, Y.J.; Ruan, J.S.; Kurgan, L. Psionplus: Accurate sequence-based predictor of ion channels and their types. *PLoS ONE* **2016**, *11*, e0152964. [CrossRef] [PubMed]
10. Hou, D.; Song, J.; Shi, L.; Ma, X.; Xin, T.; Han, J.; Xiao, W.; Sun, Z.; Cheng, R.; Yao, H. Stability and accuracy assessment of identification of traditional chinese materia medica using DNA barcoding: A case study on flos lonicerae japonicae. *BioMed. Res. Int.* **2013**, *2013*, 549037. [CrossRef] [PubMed]
11. Sun, Z.Y.; Chen, S.L. Identification of cortex herbs using the DNA barcode nrits2. *J. Nat. Med.* **2013**, *67*, 296–302. [CrossRef] [PubMed]
12. Cao, S.P.; Guo, L.N.; Luo, H.M.; Yuan, H.; Chen, S.Y.; Zheng, J.; Lin, R.C. Application of coi barcode sequence for the identification of snake medicine (zaocys). *Mitochondrial DNA Part A* **2016**, *27*, 483–489. [CrossRef] [PubMed]
13. Ma, J.W.; Gu, H. A novel method for predicting protein subcellular localization based on pseudo amino acid composition. *BMB Rep.* **2010**, *43*, 670–676. [CrossRef] [PubMed]
14. Avin, F.A.; Bhassu, S.; Tan, Y.S.; Shahbazi, P.; Vikineswary, S. Molecular divergence and species delimitation of the cultivated oyster mushrooms: Integration of igs1 and its. *Sci. World J.* **2014**, *10*, 793414. [CrossRef] [PubMed]
15. Liu, Y.; Zeng, X.; He, Z.; Zou, Q. Inferring microrna-disease associations by random walk on a heterogeneous network with multiple data sources. *Trans. Comput. Biol. Bioinform.* **2017**, *14*, 905–915. [CrossRef] [PubMed]
16. Wang, S.; Huang, G.; Hu, Q.; Zou, Q. A network-based method for the identification of putative genes related to infertility. *Biochim. Biophys. Acta Gen. Subj.* **2016**, *1860*, 2716–2724. [CrossRef] [PubMed]
17. Zou, Q.; Li, J.; Hong, Q.; Lin, Z.; Wu, Y.; Shi, H.; Ju, Y. Prediction of microrna-disease associations based on social network analysis methods. *BioMed Res. Int.* **2015**, *2015*, 810514. [CrossRef] [PubMed]

18. Zou, Q.; Li, J.; Wang, C.; Zeng, X. Approaches for recognizing disease genes based on network. *BioMed. Res. Int.* **2014**, *2014*, 416323. [CrossRef] [PubMed]
19. Chen, G.X.; Zhao, J.C.; Zhao, X.; Zhao, P.S.; Duan, R.J.; Nevo, E.; Ma, X.F. A psammophyte *Agriophyllum squarrosum* (L.) moq.: A potential food crop. *Genet. Resour. Crop Evol.* **2014**, *61*, 669–676. [CrossRef]
20. Tamura, K.; Stecher, G.; Peterson, D.; Filipski, A.; Kumar, S. Mega6: Molecular evolutionary genetics analysis version 6.0. *Mol. Biol. Evol.* **2013**, *30*, 2725–2729. [CrossRef] [PubMed]
21. Smith, D.R. Mutation rates in plastid genomes: They are lower than you might think. *Genome Biol. Evol.* **2015**, *7*, 1227–1234. [CrossRef] [PubMed]
22. Darriba, D.; Taboada, G.L.; Doallo, R.; Posada, D. Jmodeltest 2: More models, new heuristics and parallel computing. *Nat. Methods* **2012**, *9*, 772. [CrossRef] [PubMed]
23. Bucher, C.; Torhorst, J.; Bubendorf, L.; Schraml, P.; Kononen, J.; Moch, H.; Mihatsch, M.; Kallioniemi, O.; Sauter, G. Tissue microarrays (tissue chips) for high-throughput cancer genetics: Linking molecular changes to clinical endpoints. *Am. J. Hum. Genet.* **1999**, *65*, A10.
24. Li, P.; Guo, M.; Wang, C.; Liu, X.; Zou, Q. An overview of snp interactions in genome-wide association studies. *Brief. Funct. Genom.* **2015**, *14*, 143–155. [CrossRef] [PubMed]
25. Yao, H.; Song, J.Y.; Liu, C.; Luo, K.; Han, J.P.; Li, Y.; Pang, X.H.; Xu, H.X.; Zhu, Y.J.; Xiao, P.G.; et al. Use of its2 region as the universal DNA barcode for plants and animals. *PLoS ONE* **2010**, *5*, e13102. [CrossRef] [PubMed]
26. Mei, Q.L.; Zhang, H.X.; Liang, C. A discriminative feature extraction approach for tumor classification using gene expression data. *Curr. Bioinform.* **2016**, *11*, 561–570. [CrossRef]
27. Hossen, M.B.; Siraj-Ud-Doulah, M. Identification of robust clustering methods in gene expression data analysis. *Curr. Bioinform.* **2017**, *12*, 558–562. [CrossRef]
28. Zhang, Y.B.; Wang, J.; Wang, Z.T.; But, P.P.H.; Shaw, P.C. DNA microarray for identification of the herb of dendrobium species from chinese medicinal formulations. *Planta Med.* **2003**, *69*, 1172–1174. [PubMed]
29. Liu, Z.Q.; Hao, M.G. Application of the AFLP combining with SCAR to identify the geoherbalism of traditional Chinese medicines. *Chin. J. Mod. Chin. Med.* **2005**, *1*, 1.
30. Li, H.; Li, Z.; Song, H. Study on quality standard of biyanling tablet. *China Pharm.* **2012**, *21*, 31–32.
31. Gong, B.Q.; Luo, G.H.; Wang, W. Research Progress on the Identification Methods of Fritillaria. *J. Anhui Agric. Sci.* **2009**, *37*, 1603–1604.
32. Chen, D.; Li, L.; Lu, C.; Zhong, G.; Qu, X.; Peng, R. Study on optimization of issr reaction conditions for coptis chinensis franch. *Bull. Bot. Res.* **2007**, *27*, 77–81.
33. Xin, T.Y.; Li, X.J.; Yao, H.; Lin, Y.L.; Ma, X.C.; Cheng, R.Y.; Song, J.Y.; Ni, L.H.; Fan, C.Z.; Chen, S.L. Survey of commercial rhodiola products revealed species diversity and potential safety issues. *Sci. Rep.* **2015**, *5*, 8337. [CrossRef] [PubMed]
34. Miao, C.P.; Li, X.H.; Jiang, D.M. Spatial variability of agriophyllum squarrosum across scales and along the slope on an active sand dune in semi-arid china. *Arid Land Res. Manag.* **2013**, *27*, 231–244. [CrossRef]
35. Zhao, P.S.; Zhang, J.W.; Qian, C.J.; Zhou, Q.; Zhao, X.; Chen, G.X.; Ma, X.F. Snp discovery and genetic variation of candidate genes relevant to heat tolerance and agronomic traits in natural populations of sand rice (agriophyllum squarrosum). *Front. Plant Sci.* **2017**, *8*, 536. [CrossRef] [PubMed]
36. Wang, J.; Xia, N.H. Quantitative Analysis of Morphological Characters of Ardisia crenata Complex (Primulaceae). *J. Trop. Subtrop. Bot.* **2013**, *21*, 543–548.
37. Mishra, P.; Kumar, A.; Nagireddy, A.; Mani, D.N.; Shukla, A.K.; Tiwari, R.; Sundaresan, V. DNA barcoding: An efficient tool to overcome authentication challenges in the herbal market. *Plant Biotechnol. J.* **2016**, *14*, 8–21. [CrossRef] [PubMed]
38. Li, D.Z.; Gao, L.M.; Li, H.T.; Wang, H.; Ge, X.J.; Liu, J.Q.; Chen, Z.D.; Zhou, S.L.; Chen, S.L.; Yang, J.B.; et al. Comparative analysis of a large dataset indicates that internal transcribed spacer (its) should be incorporated into the core barcode for seed plants. *Proc. Natl. Acad. Sci. USA* **2011**, *108*, 19641–19646. [PubMed]
39. Wang, R.G.; Li, C.M.; Wang, L.Y.; Liu, X.L.; Liu, H.P.; You, Z.L. Biological characteristics of dried and processed pheretima asiatica by the interaction of protein/peptide in traditional chinese medicine with protein chips as carriers. *J. Clin. Rehabil. Tissue Eng. Res.* **2008**, *12*, 2489–2492.
40. Hebert, P.D.N.; Cywinska, A.; Ball, S.L.; DeWaard, J.R. Biological identifications through DNA barcodes. *Proc. R. Soc. B Biol. Sci.* **2003**, *270*, 313–321. [CrossRef] [PubMed]

41. Miao, M.; Warren, A.; Song, W.B.; Wang, S.; Shang, H.M.; Chen, Z.G. Analysis of the internal transcribed spacer 2 (its2) region of scuticociliates and related taxa (ciliophora, oligohymenophorea) to infer their evolution and phylogeny. *Protist* **2008**, *159*, 519–533. [CrossRef] [PubMed]
42. Coleman, A.W. Pan-eukaryote its2 homologies revealed by rna secondary structure. *Nucleic Acids Res.* **2007**, *35*, 3322–3329. [CrossRef] [PubMed]
43. Nilsson, R.H.; Tedersoo, L.; Abarenkov, K.; Ryberg, M.; Kristiansson, E.; Hartmann, M.; Schoch, C.L.; Nylander, J.A.A.; Bergsten, J.; Porter, T.M. Five simple guidelines for establishing basic authenticity and reliability of newly generated fungal its sequences. *Mycokeys* **2012**, *4*, 37–63. [CrossRef]
44. Trias-Blasi, A.; Vorontsova, M. Plant identification is key to conservation. *Nature* **2015**, *521*, 161. [CrossRef] [PubMed]
45. Shi, L.C.; Chen, J.; Liu, D.; Zhang, Y.H.; Jia, J.; Zhang, H.; Yao, H. Molecular Identification of Serpentis Periostracum and Its Adulterants Based on COI Sequence. *World Sci. Technol. Modemization Tradit. Chin. Med.* **2014**, *2*, 284–287.
46. Zhang, H.; Chen, J.; Jia, J.; Liu, D.; Shi, L.; Zhang, H.; Song, J.; Yao, H. Identification of scolopendra subspinipes mutilans and its adulterants using DNA barcode. *China J. Chin. Mater. Med.* **2014**, *39*, 2208–2211.
47. Liu, D.; Qian, Q.; Zhang, H.; Zeng, D.; Jia, J.; Zhang, H. Molecular Identification of the Traditional Chinese Medicine of the Deers Using COI Barcode Sequence. *World Sci. Technol. Modemization Tradit. Chin. Med.* **2014**, *2*, 274–278.
48. Zhang, H.; Shi, L.; Liu, D.; Jia, J.; Chen, J.; Zhang, H. Identification of Gekko geeko Linnaeus and Adulterants Using the COI Barcode. *World Sci. Technol.-Mod. Tradit. Chin. Med.* **2014**, *2*, 269–273.
49. Chen, S.L.; Yao, H.; Han, J.P.; Liu, C.; Song, J.Y.; Shi, L.C.; Zhu, Y.J.; Ma, X.Y.; Gao, T.; Pang, X.H.; et al. Validation of the its2 region as a novel DNA barcode for identifying medicinal plant species. *PLoS ONE* **2010**, *5*, e8613. [CrossRef] [PubMed]
50. Holtken, A.M.; Schroder, H.; Wischnewski, N.; Degen, B.; Magel, E.; Fladung, M. Development of DNA-based methods to identify cites-protected timber species: A case study in the meliaceae family. *Holzforschung* **2012**, *66*, 97–104.
51. Erickson, D.L.; Jones, F.A.; Swenson, N.G.; Pei, N.; Bourg, N.A.; Chen, W.; Davies, S.J.; Ge, X.J.; Hao, Z.; Howe, R.W.; et al. Comparative evolutionary diversity and phylogenetic structure across multiple forest dynamics plots: A mega-phylogeny approach. *Front. Genet.* **2014**, *5*, 358. [CrossRef] [PubMed]
52. Li, M.; Jiang, R.W.; Hon, P.M.; Cheng, L.; Li, L.L.; Zhou, J.R.; Shaw, P.C.; But, P.P.H. Authentication of the anti-tumor herb baihuasheshecao with bioactive marker compounds and molecular sequences. *Food Chem.* **2010**, *119*, 1239–1245. [CrossRef]
53. Yu, J.; Zhao, S.; Ren, M.; Qian, Q.; Pang, X. Identification of bupleurum chinense and b. Longiradiatum based on its2 barcode. *China J. Chin. Mater. Med.* **2014**, *39*, 2160–2163.
54. Shi, Y.H.; Sun, W.; Fang, G.H.; Deng, R.B.; Xu, W.L.; Huang, X.D.; Weng, S.Q.; Li, C.Y.; Chen, S.L. Identification of herbal tea ingredient Plumeria rubra and its adulterants using DNA barcoding. *China J. Chin. Mater. Med.* **2014**, *39*, 2199–2203.
55. Liu, J.; Yan, H.F.; Newmaster, S.G.; Pei, N.C.; Ragupathy, S.; Ge, X.J. The use of DNA barcoding as a tool for the conservation biogeography of subtropical forests in china. *Divers. Distrib.* **2015**, *21*, 188–199. [CrossRef]
56. Pei, N.C.; Erickson, D.L.; Chen, B.F.; Ge, X.J.; Mi, X.C.; Swenson, N.G.; Zhang, J.L.; Jones, F.A.; Huang, C.L.; Ye, W.H.; et al. Closely-related taxa influence woody species discrimination via DNA barcoding: Evidence from global forest dynamics plots. *Sci. Rep.* **2015**, *5*, 15127. [CrossRef] [PubMed]
57. Wang, X.; Long, W.; Yang, X.; Xiong, M.; Kang, Y.; Huang, J.; Wang, X.; Hong, X.; Zhou, Z.; Lu, Y.; et al. Patterns of plant diversity within and among three tropical cloud forest communities in hainan island. *Chin. J. Plant Ecol.* **2016**, *40*, 469–479.
58. Burgess, K.S.; Fazekas, A.J.; Kesanakurti, P.R.; Graham, S.W.; Husband, B.C.; Newmaster, S.G.; Percy, D.M.; Hajibabaei, M.; Barrett, S.C.H. Discriminating plant species in a local temperate flora using the rbcl plus matk DNA barcode. *Methods Ecol. Evol.* **2011**, *2*, 333–340. [CrossRef]
59. Zhang, Y.; Shi, Y.; Song, M.; Lin, Y.; Ma, X.; Sun, W.; Xiang, L.; Liu, X. Identification of pyrrosiae folium and its adulterants based on psba-trnh sequence. *China J. Chin. Mater. Med.* **2014**, *39*, 2222–2226.
60. Lu, M.M.; Ci, X.Q.; Yang, G.P.; Li, J. DNA barcoding of subtropical forest trees—A study from ailao mountains nature reserve, Yunnan, China. *Plant Divers. Resour.* **2013**, *35*, 733–741.

61. Lahaye, R.; Van der Bank, M.; Bogarin, D.; Warner, J.; Pupulin, F.; Gigot, G.; Maurin, O.; Duthoit, S.; Barraclough, T.G.; Savolainen, V. DNA barcoding the floras of biodiversity hotspots. *Proc. Natl. Acad. Sci. USA* **2008**, *105*, 2923–2928. [CrossRef] [PubMed]
62. Genievskaya, Y.; Abugalieva, S.; Zhubanysheva, A.; Turuspekov, Y. Morphological description and DNA barcoding study of sand rice (agriophyllum squarrosum, chenopodiaceae) collected in kazakhstan. *BMC Plant Biol.* **2017**, *17*, 177. [CrossRef] [PubMed]
63. Wang, X.; Niu, X.; Wei, N.; Ji, K. Its sequence used as DNA barcode applied in chinese herb abrus cantoniensis. *Genom. Appl. Biol.* **2012**, *31*, 603–608.
64. Huang, X.C.; Ci, X.Q.; Conran, J.G.; Li, J. Application of DNA barcodes in asian tropical trees—A case study from xishuangbanna nature reserve, Southwest China. *PLoS ONE* **2015**, *10*, e0129295. [CrossRef] [PubMed]
65. Aldrich, J.; Cherney, B.W.; Merlin, E.; Christopherson, L. The role of insertions/deletions in the evolution of the intergenic region between psba and trnh in the chloroplast genome. *Curr. Genet.* **1988**, *14*, 137–146. [CrossRef] [PubMed]
66. Fazekas, A.J.; Burgess, K.S.; Kesanakurti, P.R.; Graham, S.W.; Newmaster, S.G.; Husband, B.C.; Percy, D.M.; Hajibabaei, M.; Barrett, S.C.H. Multiple multilocus DNA barcodes from the plastid genome discriminate plant species equally well. *PLoS ONE* **2008**, *3*, e2802. [CrossRef] [PubMed]
67. Chen, J.X.; Wei, Y.C.; Huang, Z.H.; Xu, H.L.; Lu, W.; Liang, Y.C. Rapid PCR Differentiation between Lobelia chinensis and Mzaus pumilus. *Fujian J. Agric. Sci.* **2017**, *32*, 730–733.
68. Huang, Q.; Liang, L.; Rui, H.E.; Zhan, R.; Chen, W. Molecular identification of nervilia fordii (hance) schltr. And its adulterants by rbcl gene. *Chin. J. Trop. Crop.* **2012**, *33*, 1630–1634.
69. Devey, D.S.; Chase, M.W.; Clarkson, J.J. A stuttering start to plant DNA barcoding: Microsatellites present a previously overlooked problem in non-coding plastid regions. *Taxon* **2009**, *58*, 7–15.
70. Simeone, M.C.; Piredda, R.; Papini, A.; Vessella, F.; Schirone, B. Application of plastid and nuclear markers to DNA barcoding of euro-mediterranean oaks (quercus, fagaceae): Problems, prospects and phylogenetic implications. *Bot. J. Linn. Soc.* **2013**, *172*, 478–499. [CrossRef]
71. Saarela, J.M.; Sokoloff, P.C.; Gillespie, L.J.; Consaul, L.L.; Bull, R.D. DNA barcoding the canadian arctic flora: Core plastid barcodes (rbcl plus matk) for 490 vascular plant species. *PLoS ONE* **2013**, *8*, e77982. [CrossRef] [PubMed]
72. Letcher, S.G. Phylogenetic structure of angiosperm communities during tropical forest succession. *Proc. R. Soc. B Biol. Sci.* **2010**, *277*, 97–104. [CrossRef] [PubMed]
73. Hollingsworth, P.M.; Forrest, L.L.; Spouge, J.L.; Hajibabaei, M.; Ratnasingham, S.; van der Bank, M.; Chase, M.W.; Cowan, R.S.; Erickson, D.L.; Fazekas, A.J.; et al. A DNA barcode for land plants. *Proc. Natl. Acad. Sci. USA* **2009**, *106*, 12794–12797.
74. Kang, Y.; Deng, Z.Y.; Zang, R.G.; Long, W.X. DNA barcoding analysis and phylogenetic relationships of tree species in tropical cloud forests. *Sci. Rep.* **2017**, *7*, 12564. [CrossRef] [PubMed]
75. Mishra, P.; Kumar, A.; Sivaraman, G.; Shukla, A.K.; Kaliamoorthy, R.; Slater, A.; Velusamy, S. Character-based DNA barcoding for authentication and conservation of iucn red listed threatened species of genus decalepis (apocynaceae). *Sci. Rep.* **2017**, *7*, 14910. [CrossRef] [PubMed]
76. Hyde, K.D.; Udayanga, D.; Manamgoda, D.S.; Tedersoo, L.; Larsson, E.; Abarenkov, K.; Bertrand, Y.J.K.; Oxelman, B.; Hartmann, M.; Kauserud, H.; et al. Incorporating molecular data in fungal systematics: A guide for aspiring researchers. *Quant. Biol.* **2013**, *3*, 1–32. [CrossRef]

9

Theoretical Prediction of the Complex P-Glycoprotein Substrate Efflux based on the Novel Hierarchical Support Vector Regression Scheme

Chun Chen [1], Ming-Han Lee [1], Ching-Feng Weng [2] and Max K. Leong [1,2,*]

[1] Department of Chemistry, National Dong Hwa University, Shoufeng, Hualien 97401, Taiwan; 610212011@ems.ndhu.edu.tw (C.C.); 610512018@gms.ndhu.edu.tw (M.-H.L.)

[2] Department of Life Science and Institute of Biotechnology, National Dong Hwa University, Shoufeng, Hualien 97401, Taiwan; cfweng@gms.ndhu.edu.tw

[*] Correspondence: leong@gms.ndhu.edu.tw.

Academic Editors: Xiangxiang Zeng, Alfonso Rodríguez-Patón and Quan Zou

Abstract: P-glycoprotein (P-gp), a membrane-bound transporter, can eliminate xenobiotics by transporting them out of the cells or blood–brain barrier (BBB) at the expense of ATP hydrolysis. Thus, P-gp mediated efflux plays a pivotal role in altering the absorption and disposition of a wide range of substrates. Nevertheless, the mechanism of P-gp substrate efflux is rather complex since it can take place through active transport and passive permeability in addition to multiple P-gp substrate binding sites. A nonlinear quantitative structure–activity relationship (QSAR) model was developed in this study using the novel machine learning-based hierarchical support vector regression (HSVR) scheme to explore the perplexing relationships between descriptors and efflux ratio. The predictions by HSVR were found to be in good agreement with the observed values for the molecules in the training set (n = 50, r^2 = 0.96, q^2_{CV} = 0.94, RMSE = 0.10, s = 0.10) and test set (n = 13, q^2 = 0.80–0.87, RMSE = 0.21, s = 0.22). When subjected to a variety of statistical validations, the developed HSVR model consistently met the most stringent criteria. A mock test also asserted the predictivity of HSVR. Consequently, this HSVR model can be adopted to facilitate drug discovery and development.

Keywords: P-glycoprotein; efflux ratio; in silico; machine learning; hierarchical support vector regression; absorption; distribution; metabolism; excretion; toxicity

---

1. Introduction

Permeability glycoprotein also known as P-glycoprotein (P-gp), which belongs to the ATP-binding cassette (ABC) superfamily of transporters, can actively transport a wide range of structurally and mechanistically diverse endogenous and xenobiotic chemical agents across the cell membrane at the energy expense of ATP hydrolysis [1]. P-gp, a 170-kDa plasma membrane protein encoded by the multidrug resistance gene (*MDR1/ABCB1*), is expressed at high levels in various tissues such as blood–brain-barriers (BBB), gastrointestinal tract (GIT), liver, kidney, and placenta [2–6]. In addition, P-gp plays significant roles in cell and tissue detoxification and elimination of harmful substances per se [1]. For example, the accumulation of neurotoxic amyloid-β (Aβ) peptides in the brain represents a pathogenic hallmark of Alzheimer's disease (AD), which is the most common form of dementia in aging populations [7]. It has been found that the decreased clearance rather than production of Aβ is the primary formation of the deleterious Aβ plaques in the brain [8]. The decreased elimination of Aβ from the brain into the blood can be partially attributed to the dysfunction of P-gp function, leading to the progression of AD [9–11]. Furthermore, it has been shown that Aβ can downregulate the P-gp expression along with other transporters and consequently lead to further accelerated neurodegeneration [12]. Hence, it has been suggested to increase Aβ clearance from the brain by restoring P-gp function of BBB to reduce Aβ brain accumulation as a new strategy in the medical treatment of the early stages of AD [13,14].

Additionally, P-gp efflux can profoundly implicate the role of drug absorption, distribution, metabolism, excretion, and toxicity (ADME/Tox) [15] that can clinically alter the administrated drug efficacy or even lead to various adverse side-effects due to drug–drug interaction (DDI) in the case of polypharmacy [16]. For instance, rifampin can interact with the P-gp substrate digoxin, leading to a lower accumulation of digoxin, as demonstrated by a clinical study [17]. Moreover, it is of particular interest to observe the subtle role played by P-gp in the central nervous system (CNS) since P-gp can affect the BBB penetration and pharmacological activities of administrated drugs [18]. The CNS-related side-effects of non-CNS drugs can be eliminated by P-gp because of their limited BBB penetration [19,20]. For instance, the P-gp substrate loperamide, which is a long-acting anti-diarrheal agent by agonizing the μ-opioid receptor, does not cause any CNS side-effects when administrated alone due to the blockage of the BBB penetration by P-gp [21]. When co-administrated with the P-gp inhibitor quinidine, loperamide produces adverse respiratory depression without significant alteration of the plasma accumulation due to its central opioid effect [22]. Conversely, P-gp can restrict or even eliminate the entry of CNS-targeted drugs into the brain, resulting in the reduction of the clinical efficacy [23].

In addition to normal tissues and organs, various types of tumor can over-express P-gp, producing multidrug resistance (MDR) [24], in which a single drug causes a non-drug resistant cell or cell line to become cross-resistant to other pharmacologically unrelated drugs due to the increase of administrated drug efflux and the decrease of intracellular drug accumulation [25]. As a result, P-gp efflux remains a major obstacle in the success of various kinds of cancer treatment [26] as well as infectious diseases [3,27]. For instance, brain tumor is one of the leading forms of malignancy and one of highest causes of cancer-related mortality among young adults aged less than 40 years and children [28] and glioma is the most common type of primary brain cancer with limited survival time and rate [29]. The CNS penetration of cediranib, which is a tyrosine kinase inhibitor for the treatment of glioma, is severely limited by the P-gp active efflux [30]. Co-administration of P-gp inhibitors is conceptually plausible and yet infeasible to circumvent MDR because of ineffective P-gp inhibitors in practical clinical applications [31,32]. Alternatively, P-gp can be considered as an anti-target in pharmaceutical research [33] especially in the field of CNS-targeted therapeutics [34,35]. Nevertheless, not all of marketed drugs have to be P-gp non-substrates provided that their therapeutic index is large with respect to the P-gp efflux ratio (ER) [36,37]. For instance, risperidone and 9-hydroxyl risperidone are clinically approved therapeutic agents for the treatment of schizophrenia even though they are P-gp substrates [38]. Accordingly, it is conceivable to expect that quantitative measure, viz. P-gp substrate efflux ratio, is more clinically relevant than qualitative classification, viz. substrate/non-substrate classification.

Of various in vitro assays to measure the efflux ratio [39–42], the monolayer efflux assay is the most relevant to drug distribution and the most commonly used in practice [20], in which the polarized epithelial cells, such as Madin–Darby canine kidney (MDCK) cells, are transfected with the *MDR1* gene, followed by measuring the ratios between basolateral-to-apical (B→A) apparent permeability (P_{app}) and apical-to-basolateral (A→B) P_{app} [43].

$$\mathrm{ER} = \frac{P_{app}(\mathrm{B} \rightarrow \mathrm{A})}{P_{app}(\mathrm{A} \rightarrow \mathrm{B})}, \quad (1)$$

$$P_{app} = \frac{1}{AC_0} \cdot \frac{dQ}{dt}, \quad (2)$$

where P_{app} is evaluated using the membrane surface area (A), initial dosing concentration of the test molecule (C_0) in the donor compartment, and the amount of molecule transported per time (dQ/dt) in the receiver compartment [44]. Normally, molecules with ER > 2 are classified as P-gp substrates [39].

In contrast to in vitro and in vivo assays, in silico approaches are usually swift, inexpensive, less labor intensive, and less time-consuming for drug discovery and ADME/Tox profiling [45,46]. In fact, numerous P-gp classification structure–activity relationship (CSAR) models have been published elsewhere [47–68], whereas in silico quantitative studies of efflux ratio are scant [69–71]. Nevertheless, it is highly challenging to accurately model P-gp–substrate interactions [72] since P-gp is highly promiscuous per se as the result of the fact that P-gp can undergo substantial conformational changes upon binding with various ligands as illustrated by Figure 1 of Leong et al. [73]. In addition, P-gp has multiple substrate binding sites, as reported [72,74–77]. The mechanism of P-gp substrate efflux is far more complicated than P-gp–substrate interactions since P-gp substrate efflux can take place through various routes in that substrates can be actively transported by P-gp from the cytoplasm into the extracellular environment in an energy-dependent manner or through a protein channel positioned between the inner and outer leaflets of the lipid membrane, as illustrated by Figure 2 of Edwards [78]. In addition to active transport, P-gp substrates can also passively diffuse from the cytoplasm into the extracellular environment through transcellular diffusion and/or paracellular route, as illustrated by Figure 1 of Balimane et al. [79]. Notably, the P-gp substrate vinblastine, for instance, can be both passively diffused and actively transported [80]. As such, those modeling schemes employed by previously published investigations can only render the direct protein-ligand interactions and they are not suitable to model the efflux ratio. Conversely, quantitative structure–activity relationship (QSAR) schemes, which are a mathematic means to establish the relationship between biological activity and chemical characteristics, provide better approaches to model the efflux ratio since they can take into account any mechanisms that can occur through complex routes [81].

The complexity of P-gp mediated efflux can be problematic once the delicate roles played by those associated chemical features, viz. descriptors in QSAR models, are considered. For instance, inhibitors, modulators, and substrates can interact with P-gp using the hydrophobicity, hydrogen-bond acceptor (HBA), and hydrogen-bond donor (HBD) features [47,73,82]. Accordingly, hydrophobicity, HBA, and HBD can simultaneously enhance and reduce the P-gp efflux, and it is plausible to expect extremely nonlinear relationships between those chemical features and efflux ratio, suggesting that those linear models can yield significant prediction errors once applied to the test samples that are very different from their training patterns.

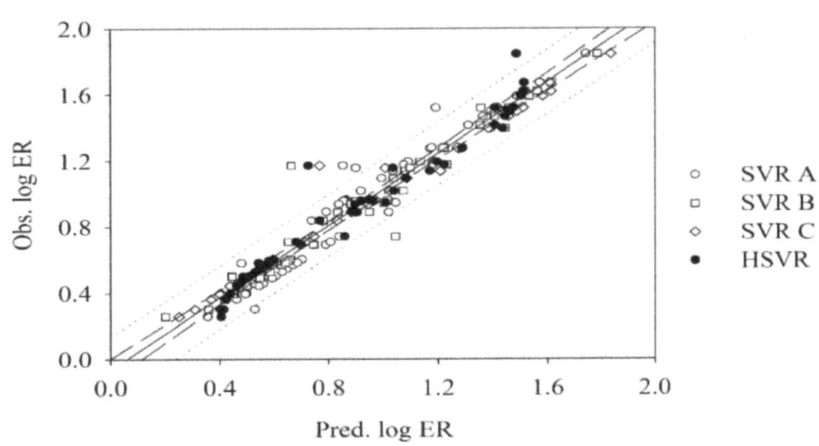

Figure 1. Observed log ER vs. the log ER predicted by SVR A (open circle), SVR B (open square), SVR C (open diamond), and HSVR (solid circle) for the molecules in the training set. The solid line, dashed line, and dotted lines correspond to the HSVR regression of the data, 95% confidence.

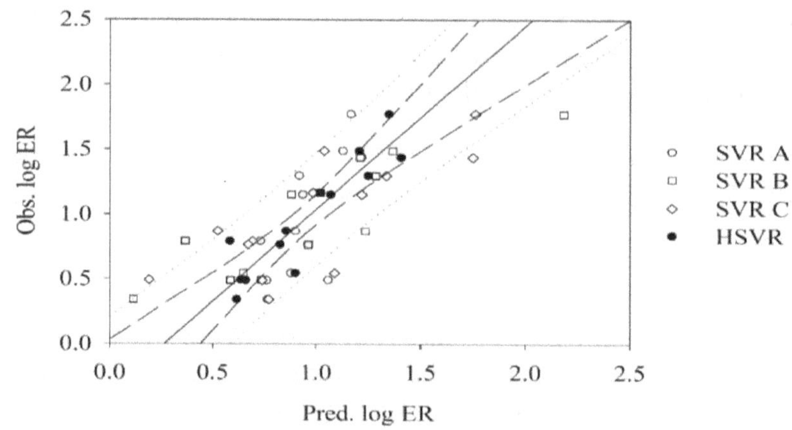

Figure 2. Observed log ER vs. the log ER predicted by SVR A (open circle), SVR B (open square), SVR C (open diamond), and HSVR (solid circle) for the molecules in the test set. The solid line, dashed line, and dotted lines correspond to the HSVR regression of the data, 95% confidence interval for the HSVR regression, and 95% confidence interval for the prediction, respectively.

Thus, it seems extremely difficult, if not completely impossible, to develop a sound in silico model to predict the P-gp substrate efflux ratio to compressively take into account those critical factors mentioned above. A solution to such challenge, however, can be obtained by the novel hierarchical support vector regression (HSVR) scheme proposed by Leong et al. [83] because HSVR can render the complex and varied dependencies of descriptors. As such, HSVR can simultaneously possess the advantageous characteristics of a local model and a global model, viz. broader coverage of applicability domain and higher level of predictivity, respectively. Furthermore, HSVR is designated to circumvent the "mesa effect" [84] in that the performance of a developed model deteriorates dramatically when applied to extrapolated predictions as demonstrated elsewhere [85,86]. In other words, HSVR is insensitive to outliers as compared with the other predictive models that is of critical importance to a predictive model [87]. Herein, the objective of this investigation was to develop an accurate, fast, and predictive in silico model based on the HSVR scheme to predict the P-gp substrate efflux ratio to facilitate drug discovery to design molecules with a preferable ADME/Tox profile.

2. Results

2.1. Data Compilation

More than 550 compounds were collected after comprehensive literature search. data curation was carefully carried out by eliminating those compounds: (i) with only qualitative array results (i.e., substrate or non-substrate); (ii) without specific ER values; or (iii) chemical structures. In addition, cells used to express P-gp protein also play a significant role in determining ER values. For instance, the measured ER values of astemizole were 2.16 and 0.6 assayed in MDCK and human colon adenocarcinoma (Caco-2) cells, respectively [51,88]. Of various assayed cells, 63 molecules tested in MDCK cells were selected from various sources [23,39,88–101] since it constituted the largest amount of data. The data size is seemingly small since several CSAR models have been derived based on rather large amounts of data. For instance, Li et al. [66] built various predictive models based on 423 P-gp substrates and 399 non-substrates compiled from numerous sources. Nevertheless, their data were generated from different assay conditions (e.g., different cell lines), leading to high levels of data heterogeneity. QSAR models, conversely, are vulnerable to data inhomogeneity [102]. Additionally, some molecules such as selenium-containing ones [103] were excluded because their topological descriptors, for instance, cannot be enumerated. Those ER values were discarded when they were not consistent with their measured P_{app} (B→A) and P_{app} (A→B) values [104]. Recently, the efflux ratios of more than 4000 Amgen in-house compounds were measured [105]. It is plausible to expect that the great sample amount and data consistency can furnish a good ER pool. Unfortunately, those chemical structures are proprietary, leading to the fact that there are only limited quantitative data with chemical structures available in the public domain to date. Those factors partially contribute to the fact that there is no genuine QSAR model has been published.

As such, only very limited data samples with available chemical structures and consistent assay conditions were recruited in this study to maximize the structural diversity and to maintain data homogeneity after purging inappropriate data based on above-mentioned criteria. Table S1 lists the SMILES strings, CAS registry numbers, efflux ratio values, and literature references of all molecules collected in this study.

2.2. Data Partition

Of all molecules adopted in this study, 50 and 13 molecules were randomly assigned to the training set and test set, respectively, with a ca. 4:1 ratio as suggested [106]. Figure S1 displays the projection of all molecules enrolled in this investigation in chemical space, spanned by the first three principal components (PCs), explaining 94.6% of the variance in the original data. As illustrated, both datasets exhibited high levels of similarity in the chemical space. Furthermore, the high levels of biological and chemical similarity between both datasets can also be validated by Figure S2, which shows the histograms of log ER, molecular weight (MW), polar surface area (PSA), number of HBA, and number of HBD in density form for all molecules in the training set and test set. Thus, it can be asserted that there was no substantial bias in datasets.

2.3. SVRE

Of all generated SVR models using various combinations of descriptors and runtime parameters, three SVR models, denoted by SVR A, SVR B, and SVR C, were assembled to construct the SVR ensemble, which was further subjected to regression by another SVR to generate the HSVR model. Table S2 summarizes the optimal runtime parameters of SVR A, SVR B, and SVR C. These three SVR models, which adopted 4, 6, and 3 descriptors (Table 1), respectively, were selected based on their individual performances on all molecules and statistical analyses in the training set and test set. Table S1 lists the predicted log ER values. Tables 2 and 3 summarize the associated statistical analyses of these three SVR models in the training set and test set, respectively. Figures 1 and 2 display the scatter plots of observed versus the predicted log ER values by SVR A, SVR B, and SVR C for the molecules in the training set and test set, respectively.

Table 1. Descriptor selected as the input of SVR models in the ensemble and their description.

Descriptor	SVR A	SVR B	SVR C	Description
SA			x[†]	Total surface area
n_{N+O}	x	x		Number of nitrogen and oxygen atoms
V_m	x	x	x	Molecule volume
PSA	x	x		Polar surface area
HBD	x	x		Number of hydrogen bond donating groups
n_{Rot}		x	x	Number of rotatable bonds
n_{Ar}		x		Number of aromatic rings

[†] Selected.

Figure 1 shows that the predictions by SVR A, SVR B, and SVR C are in good agreement with the observed values for most of the molecules in the training set as further manifested by their small RMSDs, average deviations, standard deviations (s), and larger r^2 parameters (Table 2). Of 50 training samples, SVR A, SVR B, and SVR C gave rise to 28, 3, and 2 predictions, which deviated from the experimental values by more than 0.10, respectively. It can be further observed in Figure 1 that most of the points predicted by SVR C generally lie on or are closer to the regression line when compared with

SVR A and SVR B. As a result, SVR C produced the lowest MAE (0.02), s (0.06), and RMSE (0.06) and the highest r^2 parameter (0.98), suggesting that SVR C performed better than SVR A and SVR B for the molecules in the training set. Nevertheless, the predictions of quinidine (**48**) by SVR A, SVR B and SVR C unanimously yielded the maximum residuals of 0.32, 0.51 and 0.40, respectively, denoting that SVR A executed better than SVR B and SVR C.

Table 2. Statistic evaluations, namely correlation coefficient (r^2), maximum residual (Δ_{Max}), mean absolute error (MAE), standard deviation (s), RMSE, and 10-fold cross-validation correlation coefficient (q^2_{CV}) evaluated by SVR A, SVR B, SVR C, and HSVR in the training set.

	SVR A	SVR B	SVR C	HSVR
r^2	0.95	0.95	0.98	0.96
Δ_{Max}	0.32	0.51	0.40	0.45
MAE	0.11	0.07	0.02	0.06
s	0.12	0.10	0.06	0.10
RMSE	0.12	0.10	0.06	0.10
q^2_{CV}	0.01	0.01	0.07	0.94

The predictions by SVR A, SVR B, and SVR C in the test set are also in good agreement with the experimental values (Figure 2). Nevertheless, most of the residuals obtained by the three SVR models in the test set are more than 0.15 (11, 11, and 8, respectively). It can be further observed in Table 3 that the mean absolute errors computed by SVR A, SVR B, and SVR C unequivocally increase from 0.11, 0.07, and 0.02 in the training set to 0.29, 0.22, and 0.24 in test set, respectively. The other statistical parameters also suggest that the performances of these three models in the SVRE slightly decline from the training set to the test set (Tables 2 and 3). The maximum residual computed by SVR C in the test set was yielded from the prediction of cimetidine (**13**) with an absolute residual of 0.55, which were only 0.34 and 0.10 by SVR A and SVR B, respectively. Similarly, vinblastine (**58**) was best predicted by SVR C with an absolute residual of 0.01, whereas SVR A and SVR B gave rise to absolute errors of 0.60 and 0.41, respectively.

Table 3. Statistic evaluations, correlation coefficients q^2, q^2_{F1}, q^2_{F2}, and q^2_{F3}, concordance correlation coefficient (CCC), maximal absolute residual (Δ_{Max}), mean absolute error (MAE), standard deviation (s), and RMSE evaluated by SVR A, SVR B, SVR C, and HSVR in the test set.

	SVR A	SVR B	SVR C	HSVR
q^2	0.54	0.75	0.60	0.83
q^2_{F1}	0.39	0.67	0.55	0.80
q^2_{F2}	0.39	0.67	0.54	0.80
q^2_{F3}	0.38	0.66	0.54	0.80
CCC	0.45	0.86	0.78	0.87
Δ_{Max}	0.60	0.42	0.55	0.42
MAE	0.29	0.22	0.24	0.17
s	0.35	0.26	0.30	0.22
RMSE	0.34	0.25	0.29	0.21

Furthermore, SVR A, SVR B, and SVR C yielded the q^2 values of 0.54, 0.75, and 0.60 in the test and the cross-validation correlation coefficients q^2_{CV} of 0.01, 0.01, and 0.07 in the training set, respectively (Tables 2 and 3). When subjected to the Y-scrambling test, SVR A, SVR B, and SVR C gave rise to the $\langle r_s^2 \rangle$ values of 0.02, 0.03, and 0.03, respectively (Table 1). The almost zero values of $\langle r_s^2 \rangle$ as well as substantial differences between corresponding r^2 and $\langle r_s^2 \rangle$ signify that those three SVR models in the ensemble are not the result of chance correlation [107]. Conversely, the substantial differences between r^2 and q^2 and between r^2 and q^2_{CV} imply the over-fitting characteristics of these three models that actually can be further manifested by their small q^2_{F1}, q^2_{F2}, q^2_{F3}, and CCC values (Table 3). As a result, it is plausible to expect that these models are local models per se, which have limited coverage of applicability domain (vide infra) [108].

2.4. HSVR

The HSVR model was produced by the regression of the SVR ensemble based on the predictions of all molecules and statistical evaluations in the training set (Table S1 and Table 2). Table S2 lists the optimal runtime conditions for the final SVR model. It can be observed in Figure 1 that the HSVR model showed better prediction accuracy than SVR A, SVR B, and SVR C for the molecules in the training set because the distances between the predictions by HSVR and regression line are generally between the largest ones and smallest ones produced by its SVR counterparts in the ensemble. However, HSVR executed better than any of SVR models in the ensemble in some cases. The predictions of desloratadine (**19**) by SVR A, SVR B, SVR C, and HSVR, for instance, yielded absolute residuals of 0.10, 0.06, 0.01, and 0.00, respectively. Statistically, HSVR performed better than SVR A and SVR B, whereas SVR C, in turn, functioned negligibly better than HSVR, as manifested by those parameters listed in Table 2. For example, SVR A, SVR B, SVR C, and HSVR yielded the r^2 values of 0.95, 0.95, 0.98, and 0.96, respectively.

When applied to the test samples, HSVR only showed insignificant performance decreases from the training set to the test set. For instance, RMSE increased from 0.10 in the training set to 0.21 in the test set (Tables 2 and 3). However, the maximum residual declined from 0.45 in the training set to 0.42 in the test set. Figure 2 displays that HSVR showed better performance than SVR A, SVR B, and SVR C in the test set. The performance predominance of HSVR can be further manifested by those statistical parameters listed in Table 3. For instance, SVR A, SVR B, SVR C, and HSVR gave rise to MAE values of 0.29, 0.22, 0.24, and 0.17, respectively. Similar observation that HSVR generated smaller absolute residuals than its counterparts in the ensemble can also be found in the test set. The absolute prediction error of paliperidone (**41**), for instance, was 0.14 given rise by HSVR, whereas SVR A, SVR B, and SVR C produced residuals of 0.57, 0.25, and 0.30, respectively. When compared with its counterparts in the ensemble, HSVR generally produced consist and small errors in both training set and test set as manifested by those parameters associated with error listed in Tables 2 and 3, suggesting that HSVR has broader coverage of applicability domain. Additionally, HSVR yielded the smallest differences between r^2 and q^2_{CV}(0.02) and between r^2 and q^2 (0.13), indicating that HSVR was a well-trained model or no over-fitting effect was observed because it will otherwise produce at least one significant difference among those parameters. Similarly, the possibility of chance correlation of HSVR can be eliminated by Y-scrambling since it also produced an almost zero $\langle r_s^2 \rangle$ (0.03) and marked difference between r^2 and $\langle r_s^2 \rangle$ (Table 2) [107].

2.5. Predictive Evaluations

Figure 3 displays the scatter plots of the residual vs. the log ER values predicted by HSVR for the molecules in the training set and test set. It can be observed that the residuals are approximately evenly distributed on both sides of x-axis along the range of predicted values in both datasets, suggesting that there is no systematic error associated with the HSVR model [102]. The unbiased predictions can be further exhibited by its almost negligible average residuals that were −0.02 and −0.02 in the training set and test set, respectively (Table S1).

The predictivity of generated HSVR model was further evaluated by the validation requirements proposed by Golbraikh et al. [109], Ojha et al. [110], Roy et al. [111], and Chirico and Gramatica [112] (Equations (18)–(21)) in the training set and test set. Table 4 summarizes the results, from which it can be observed that HSVR maintained similar high levels of performance in the training set and test set. Additionally, HSVR fulfilled all validation requirements, indicating that this predictive model is highly accurate and predictive.

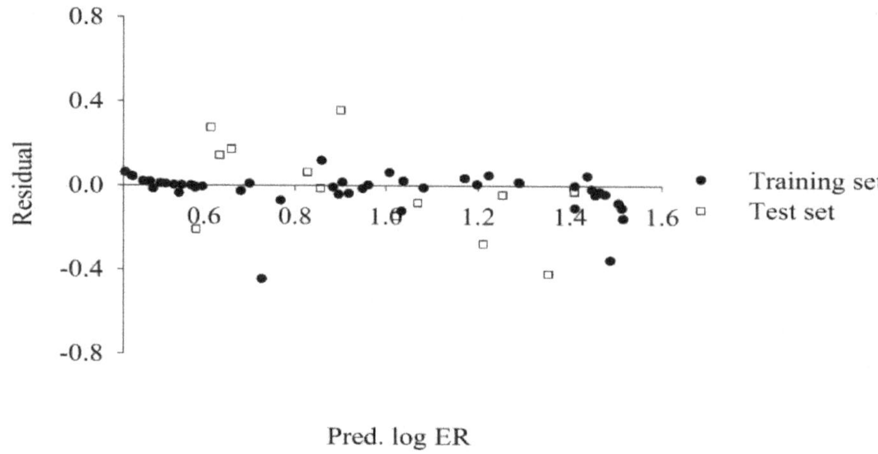

Figure 3. Residual vs. the log ER predicted by HSVR in the training set (solid circle) and test set (open square).

Table 4. Validation verification of HSVR based on prediction performance of those molecules in the training set and test set.

	Training Set	Test Set
n	50	13
r_o^2	0.95	0.77
k	1.03	1.05
r'^2_o	0.94	0.52
r_m^2	0.90	0.72
r'^2_m	0.85	0.60
$\langle r_m^2 \rangle$	0.88	0.66
Δr_m^2	0.05	0.12
$r^2, q^2_{\mathrm{CV}}, q^2, q^2_{\mathrm{Fn}} \geq 0.70$	x	x
$\lvert r^2 - q^2_{\mathrm{CV}} \rvert < 0.10$	x	N/A
$(r^2 - r_o^2)/r^2 < 0.10$ and $0.85 \leq k \leq 1.15$	x	x
$\lvert r_o^2 - r'^2_o \rvert < 0.30$	x	x
$r_m^2 \geq 0.65$	x	x
$\langle r_m^2 \rangle \geq 0.65$ and $\Delta r_m^2 < 0.20$	x	x
CCC ≥ 0.85	N/A †	x

† Not applicable.

2.6. Mock Test

To mimic real world challenges, the developed HSVR model was further tested on the P-gp substrates assayed by Crivori et al. [51]. Of all marketed drugs measured by Crivori et al., 12 were also enrolled in this study, yielding a good way to calibrate the testing system. However, these molecules were measured in Caco-2 cells, whereas all of the molecules adopted in this study were tested in MDCK cells, suggesting that those compounds assayed by Crivori et al. are not qualified as the second external or test set since those validation criteria (vide supra) are not applicable to these compounds. To eliminate the discrepancy between both assay systems, the linear correlation between both assay systems for those common molecules was first inspected and the obtained scatter plot is illustrated in Figure 4. It can be observed that the experimental values in both systems were modestly correlated with each other well with an r value of 0.78. Thus, it is plausible to examine the HSVR model with those novel P-gp substrates assayed in Caco-2 cells.

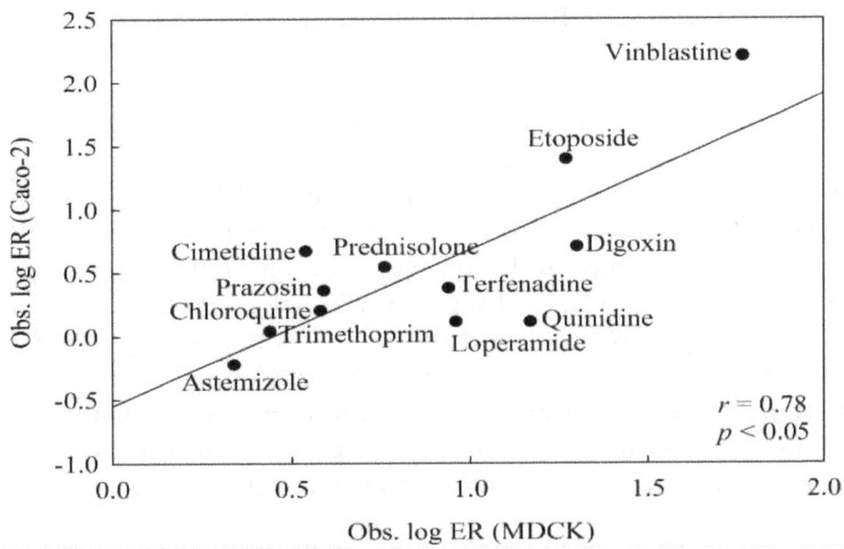

Figure 4. The observed log ER values (Caco-2) vs. the observed log ER values (MDCK).

Figure 5 displays the tested results of the nine novel drugs. It can be observed that the r value between experimental log ER obtained in the Caco-2 cells and predicted log ER in the MDCK cells was 0.77. The negligible difference between both numbers (0.78 vs. 0.77) suggests that the predictions by the HSVR model can almost reproduce the experimental observations and this mock test unequivocally assured the predictive capability of HSVR.

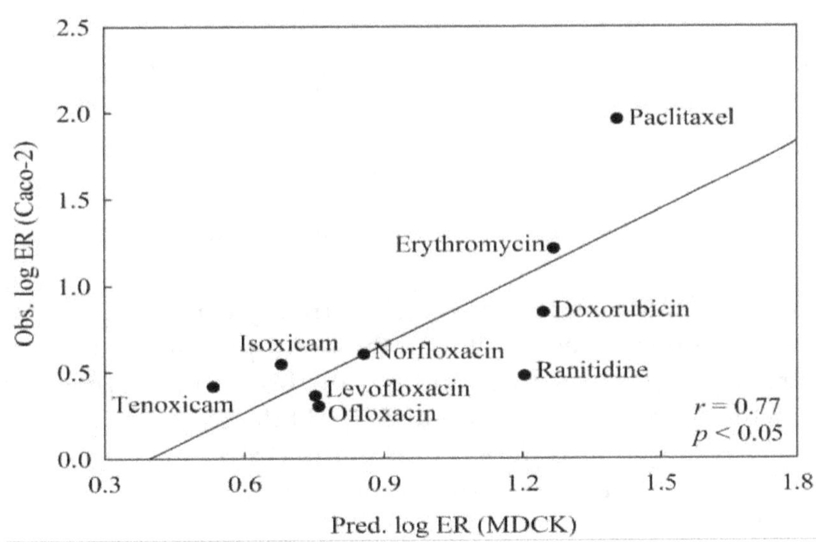

Figure 5. The observed log ER values (Caco-2) vs. the predicted log ER values (MDCK).

3. Discussion

Collectively, seven descriptors were adopted in this study. Intrinsically, the sample-to-descriptor ratio was ca. 7:1, which is significantly larger than 5, viz. the minimal requirement to lessen the probability of chance correlations in a predictive model [113]. However, the process of P-gp substrate efflux is complex since it can take place thought various routes (vide supra). As such, different descriptors were adopted by different classification models. Of various descriptors selected by qualitative predictive models, hydrophobic, HBA, and HBD are the most frequently selected chemical features, as illustrated by the model proposed by Penzotti et al. [47]. However, the analysis of Amgen in-house compounds can reveal that HBD and topological PSA (tPSA) are the predominant factors associated with ER [105].

Figure 6 displays the average log ER for each histogram bin of HBD for all molecules selected in this study. It can be observed that the average log ER value initially increased with HBD when HBD was no more than 6 and then subsequently decreased when HBD was more than 6. Such positive dependence of log ER on HBD is, in fact, consistent with the analysis made by Hitchcock et al. [105]. However, those Amgen in-house compounds had HBD of no more than 5, leading to an only positive relationship between log ER and HBD. Such discrepancy in both systems can be conceivably attributed to the fact that the initial P-gp substrate binding can be enhanced by HBD as illustrated by the pharmacophore models of Penzotti et al. [47], whereas the consequent transport of the substrates into the extracellular environment can be hampered by too many HBDs, plausibly because of the increase in water desolvation energy [114] and the decrease in membrane fluidity [115]. As such, a nonlinear relationship between HBD and log ER was yielded consequently.

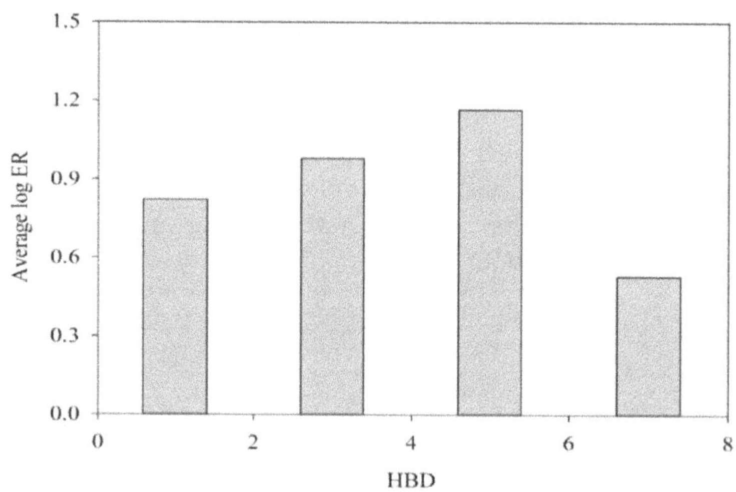

Figure 6. Average log ER vs. the distribution of HBD.

It has been observed that hydrophobicity, which normally can be represented by log P, plays an important role in P-gp–substrate interaction due to the hydrophobic nature of the substrate binding pocket, resulting in stronger P-gp substrate binding for those more hydrophobic substrates [116]. Nevertheless, the interaction between substrates and lipid bilayer as well as the release of substrates into the extracellular environment also depend on the hydrophobicity of substrates (vide supra), leading to a nonlinear relationship between log P and log ER. Figure 7 displays the average log ER for histogram bin of log P for all molecules enlisted in this study. It can be observed that the average log ER initially increased with log P when log P was smaller and decreased with log P when log P became higher. Such observation is qualitatively similar to the trend of P_{app} (A$\rightarrow$B) found by Hitchcock et al. [105].

Nevertheless, it is unusual to observe that log P was not included in this study, whereas the number of aromatic rings (n_{Ar}) was enlisted in this study. Such inconsistency can be realized by the fact that the average log P values increased with n_{Ar} for all of molecules included in this study, as illustrated in Figure 8, which displays the average log P versus the distribution of n_{Ar}. As such, it is plausible to replace log P by n_{Ar}. Furthermore, it has been found that aromatic ring moieties are important in substrate recognition and efflux modulation [117,118]. More importantly, the empirical observation has indicated that models with the selection of n_{Ar} unanimously showed better performance than those with the selection of log P (data not shown).

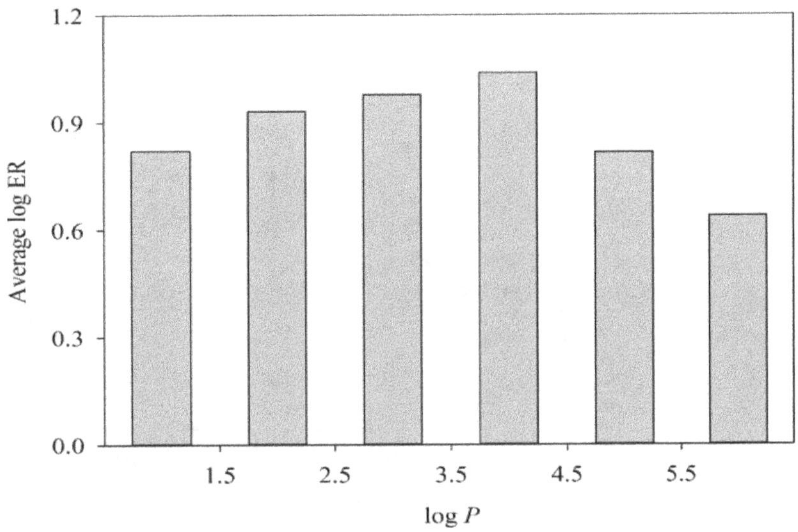

Figure 7. Average log ER vs. the distribution of log P.

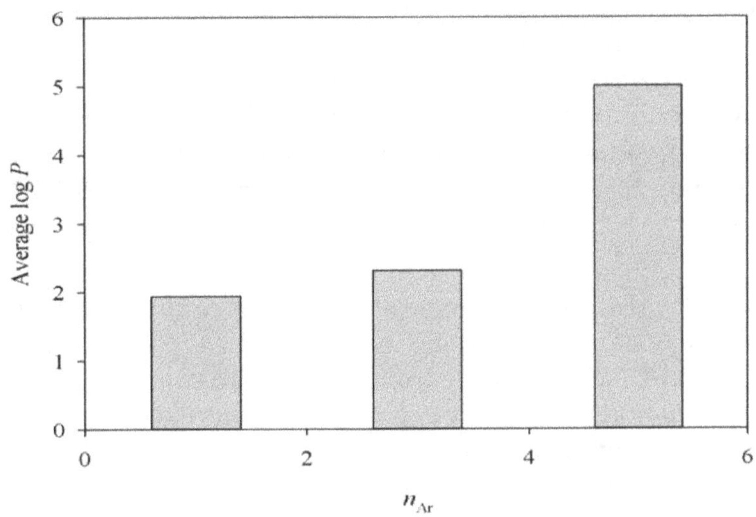

Figure 8. Average log P vs. the distribution of number of aromatic ring (n_{Ar}).

The significant role of HBA in the P-gp–substrate interaction has been manifested by molecular docking simulations [71] as well as numerous qualitative models. Additionally, it has been suggested that HBA can enhance P-gp-mediated efflux [56]. Nevertheless, it is unusual to observe that none of SVR models in the ensemble has adopted HBA, plausibly because the descriptor number of nitrogen and oxygen atoms ($n_{\mathrm{N+O}}$) correlated well with HBA as demonstrated by Figure 9, which displays $n_{\mathrm{N+O}}$ versus HBA. In fact, Desai et al. [56] adopted $n_{\mathrm{N+O}}$ instead of HBA as the substrate classification criterion. Furthermore, empirical model development has shown that models with the selection of $n_{\mathrm{N+O}}$ executed better than those with the selection of HBA (data not shown). As a result, the descriptor $n_{\mathrm{N+O}}$ was selected in lieu of HBA.

The descriptor tPSA is a modified version to swiftly calculate the polar surface area only based on the additive polar surface areas [119]. The recursive partitioning (RP) model of Joung et al. [68] indicated the significant role of PSA in classifying molecules as P-gp substrates/non-substrates. Moreover, Hitchcock et al. also found the profound contribution of tPSA to P-gp mediated efflux (vide supra). Accordingly, the more sophisticated version of PSA was adopted in this study since it can function as polarity and hydrogen-bonding features [66].

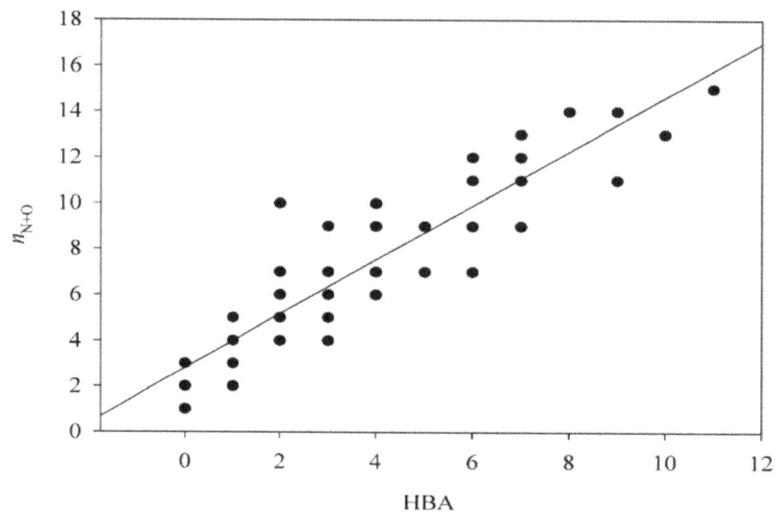

Figure 9. The number of nitrogen and oxygen (n_{N+O}) vs. HBA.

It has been observed that the substrate size, which can be characterized by molecular weight (MW), molecular volume (V_m), and total surface area (SA), can have a large impact on P-gp–substrate interaction as well as passive permeability [120]. Nevertheless, it has been suggested that both V_m and SA can be better metrics to estimate the actual molecular size [121], and MW, conversely, was closely associated with V_m with an r^2 values of 0.98 for the molecules enlisted in this study. In fact, it has been postulated that V_m rather than MW is a better metric to associate with ER [122]. Accordingly, V_m and SA were adopted to render the size effects, whereas MW was discarded to reduce the probability of spurious correlations.

It has been found that P-gp substrates generally have more rotatable bonds than non-substrates since more flexible molecules can be more easily to adopt favorable orientation to interact with P-gp [49,66,123]. In fact, non-CNS drugs are more flexible than their CNS counterparts [23] since molecules with more conformational flexibility can favor the internal H-bond formation, which, in turn, can enhance the passive membrane permeability [124]. As such, substrate conformational flexibility, which can be characterized by the number of rotatable bond (n_{Rot}), can facilitate not only the active transport but also passive permeability of P-gp substrates, and n_{Rot} was adopted in this investigation.

Gunaydin et al. [69] only took into account the contribution of the differences between free energy in water (G_{H2O}) and that in chloroform (G_{CHCl3}), viz. $\Delta G_{H2O-CHCl3}$, since it was hypothesized that P-gp undergoes a conformation change from the intercellular-facing state to extracellular-facing state upon binding with substrates. As such, the transported substrates experience from a lipophilic environment into a hydrophilic one. In addition to $\Delta G_{H2O-CHCl3}$, the contribution of $\Delta G_{DMSO-CHCl3}$ was also computed in this study to mimic the assay conditions. Nevertheless, neither of the solvation free energy differences was selected in this study due to their insignificant contribution to ERs (data not shown), plausibly because the P-gp conformation change can only account for a small part of the whole complicated efflux process and, additionally, passive permeability is not resulted from the P-gp conformation change. The predictive model of Gunaydin et al. [102], nevertheless, was derived only based on 12 marketed drugs that cannot comprehensively render the complex efflux. As such, more descriptors will be required in case of more diverse samples.

Didziapetris et al. [63] proposed the "rule-of-fours," which states that molecules with: (i) $n_{N+O} \geq 8$; (ii) MW > 400; and (iii) acid pK_a > 4 are likely to be P-gp substrates. Of all molecule with ER > 2 selected in this study, viz. substrates, approximately 32%, 52%, and 100% can meet the criteria $n_{N+O} \geq 8$, MW > 400, and acid pK_a > 4, respectively, and only 29% can completely fulfill those three criteria. Actually, Li et al. [66] also found that only ca. 34% of samples can simultaneously meet those three criteria. Furthermore, it is not unusual to observe that different rules have been proposed to classify molecules into P-gp substrates/non-substrates. Desai et al. [56], for instance, have proposed the molecules with TPSA >100 Å^2 and most basic pK_a > 8 have higher probability to be substrates. The inconsistency in various proposed rules can be plausibly explained as those rules were derived only based on linear analyses of those P-gp substrates/non-substrates. However, such bisection is not always true, as manifested by the naïve Bayesian classifiers built by Li et al. [66]. In addition, the size and hydrophobicity of substrates can affect the substrate-membrane interactions nonlinearly [125]. Further complexity can be raised once the P-gp substrate efflux is considered instead of P-gp substrate/non-substrate classification since the P-gp substrate efflux can take place through various routes (vide supra), leading to nonlinear relationships between some descriptors and log ER, such as HBD and log P (Figures 6 and 7). Numerous attempts have been made in this study to develop various partial least square (PLS) models to accommodate the novel 2-QSAR scheme [86] and no satisfactory models were produced (data not shown). Conversely, the accurate and predictive HSVR can comprehensively describe such nonlinear dependence of log ER on descriptors.

Moreover, it has been observed that P-gp and other ABC members, namely breast cancer resistant protein (BCRP/ABCG2) and multidrug resistance-associated protein 4 (ABCC4/MRP4), play a critical role in BBB permeability [126], which can take place via various routes [127] in addition to the already complicated P-gp mediated efflux. As such, it is plausible to expect that it is extremely difficult to develop a sound in silico model to predict BBB permeability if not entirely impossible [128]. The development of an accurate in silico model in this study to predict the P-gp substrate efflux can pave the way to establish a sound theoretical model to predict the BBB permeability in the future. Most molecules adopted in this study are marketed drugs for treating various illnesses, such as HIV infection, allergy symptoms, rheumatoid arthritis, hypertension, diarrhea, and different types of cancer in addition to assorted CNS-related disorders (Supplementary Materials Table S1). The broad spectrum of therapeutic agents unequivocally indicates that the data samples are structurally diverse, which can be further manifested by the fact that the average minimum distance between two molecules, viz. the distance between two nearest neighbors, in the chemical space was 2.06 with an standard deviation of 1.39 and the maximum distance between two collected samples was 29.57 (Supplementary Materials Figure S1), giving rise to an ratio of ca. 1:14. As such, it is plausible to expect that developed HSVR should have a larger coverage of applicability domain accordingly, which is an important characteristic for a predictive model in practical application. More importantly, the derived HSVR model and published P-gp substrate/non-substrate classification models can work in a synergistic fashion, in which the latter can be used to identify those P-gp substrates and the former can be deployed to predict their efflux ratios.

4. Materials and Methods

4.1. Data Compilation

A sound predictive model can only be built based on good quality of sample data [102]. To compile quality data for this study, a comprehensive literature search was conducted to retrieve efflux ratio values from various sources to maximize the structural diversity. If there were two or more available efflux ratio data for a given compound and in close range, the average values were then taken to warrant better consistency. Further data curation was carried out by cautiously inspecting molecular structures to remove those molecules without definite stereochemistry.

4.2. Molecular Descriptors

All of the molecules enlisted in this study were subjected to full geometry optimization using the density functional theory (DFT) B3LYP method with the basis set 6-31G(d,p) by the *Gaussian 09* package (Gaussian, Wallingford, CT) in the dimethyl sulfoxide (DMSO) solvent system using the polarizable continuum model (PCM) [129,130] to mimic the experimental conditions. These geometries were confirmed to be real minima on the potential energy surface by force calculations when no imaginary frequency was obtained. Additionally, atomic charges were also calculated by the molecular electrostatic potential-based method of Merz and Kollman [131] and the highest occupied molecular orbital energy (E_{HOMO}), lowest unoccupied molecular orbital energy (E_{LUMO}), free energy (ΔG), and dipole (μ) were also retrieved from the optimization calculations since those quantum mechanics descriptors have been adopted previously. As such, it is of necessity to employ a more sophisticated quantum mechanics method to optimize those selected molecules and to calculate their associated descriptors.

The *Discovery Studio* package (BIOVIA, San Diego, CA) and *E-Dragon* (available at the web site http://www.vcclab.org/lab/edragon/) were also utilized to calculate more than 200 one-, two-, and three-dimensional molecular descriptors of those optimized molecules. These descriptors can be classified as electronic descriptors, spatial descriptors, structural descriptors, thermodynamic descriptors, topological descriptors, and E-state indices.

Data filtering was initially performed by removing those descriptors missing for at least one sample or showing little or no discrimination against all samples. Furthermore, only one descriptor should be kept among those descriptors with intercorrelation values of $r^2 > 0.8$ to reduce the probability of spurious correlations as postulated by Topliss and Edwards [113]. It is not uncommon to observe that certain descriptors with broader ranges outweigh those with narrower ranges because of substantial variations in magnitudes. Nevertheless, such problem can be resolved when the non-descriptive descriptors, viz. real variable descriptors, are normalized with the following equation [132]

$$\chi_{ij} = (x_{ij} - \langle x_j \rangle) / \left[\sum_{i=1}^{n} (x_{ij} - \langle x_j \rangle)^2 / (n-1)\right]^{1/2} \quad (3)$$

where x_{ij} and χ_{ij} represent the original and normalized jth descriptors of the ith compound, respectively; $\langle x_j \rangle$ stands for the mean value of the original jth descriptor; and n is the number of samples.

Descriptor selection plays a pivotal role in determining the performance of predictive models [133]. More descriptors will be needed once there are more training samples with more diverse structures [102]. Conversely, it is highly possible to yield an over-trained model when there are too many selected descriptors [134]. The descriptor selection was initially executed by genetic function approximation (GFA) using the QSAR module of *Discovery Studio* due to its effectiveness and efficiency [135]. Further descriptor selection was carried out by the recursive feature elimination (RFE) method, in which the predictive model was repeatedly generated by all but one of descriptors. The descriptors were then ranked according to their contributions to the predictive performance; and the descriptor with least contribution was discarded [136].

4.3. Data Partition

The collected molecules were divided into two datasets, namely the training set and test set, to develop and to verify the predictive models using the Kennard–Stone (KS) algorithm [137] implemented in *MATLAB* (The Mathworks, Natick, MA, USA) with an approximate 4:1 ratio as suggested [106]. It has been suggested that a sound model can be derived only based on chemically and biologically similar training samples and test samples [138]. As such, the data distribution was carefully examined to ensure the high levels of biological and chemical similarity in both datasets.

4.4. Hierarchical Support Vector Regression

Support vector machine (SVM) proposed by Vapnik et al. [139] was initially designated for use in classification and consequently modified for regression problems by nonlinearly mapping the input data into a higher-dimension space, in which a linear regression is performed [140]. SVM regression takes into account both the training error and the model complexity as compared with the traditional regression algorithms, which develop predictive models by minimizing the training error. As such, SVM performs better than traditional regression methods because of its advantageous characteristics, namely dimensional independence, limited number of freedom, excellent generalization capability, global optimum, and easy to implement [141].

Similar to other linear or machine learning (ML)-based QSAR techniques, SVM has to tradeoff between the characteristics of a global model, viz. broader coverage of applicability domain (AD), and a local model, viz. higher level of predictivity [108]. This seeming dilemma, nevertheless, can be plausibly resolved using the hierarchical support vector regression (HSVR) scheme, which was initially proposed by Leong et al. and was derived from SVM [83], because HSVR can simultaneously take into consideration both seemingly mutually exclusive characteristics. Practically speaking, it has been demonstrated that HSVR outperformed a number of ML-based models, namely artificial neural network (ANN), genetic algorithm (GA), and SVM [85].

The detail of HSVR has been mentioned elsewhere [83]. Briefly, a panel of SVR models was built by the *LIBSVM* package (software available at http://www.csie.ntu.edu.tw/~cjlin/libsvm) based on various descriptor combinations, and each SVR model represented a local model. The model generation and verification were executed using the modules *svm-train* and *svm-predict*, respectively, implemented in the *LIBSVM* package. The regression modes, namely, ε-SVR and γ-SVR, were adopted, and radial basis function (RBF) was employed as the kernel due to its simplicity and better performance when compared with the others [142]. The runtime parameters, namely regression modes ε-SVR and ν-SVR, the associated ε and ν, cost C, and the kernel width γ, were scanned by the systemic grid search algorithm using an in-house Perl script [143], in which all parameters were changed independently in a parallel fashion.

Two SVR models were initially adopted to develop an SVR ensemble (SVRE), which, in turn, was further subjected to regression by another SVR to yield the final HSVR model. The two-member SVREs were continuously assembled until the HSVR model performed well. Otherwise, the three- or even four-member ensembles were built by adding one or more SVR models, respectively, if all two-member ensembles failed to perform well. The descriptor selection and ensemble assembly were predominantly governed by the principle of Occam's razor [144] by adopting the fewest descriptors and SVR models.

4.5. Predictive Evaluation

The predictivity of a generated model was evaluated by several statistic metrics. The coefficients r^2 and q^2 in the training set and external set, respectively, for the linear least square regression were computed by the following equation

$$r^2,\, q^2 = 1 - \sum_{i=1}^{n}(\hat{y}_i - y_i)^2 / \sum_{i=1}^{n}(y_i - \langle\hat{y}\rangle)^2 \tag{4}$$

where $\hat{y}_i$ and y_i are the predicted and observed values, respectively; and $\langle\hat{y}\rangle$ and n stand for the average predicted value and the number of samples in the dataset, respectively.

Furthermore, the residual Δ_i, which is the difference between y_i and $\hat{y}_i$, was calculated

$$\Delta_i = y_i - \hat{y}_i \tag{5}$$

The root mean square error (RMSE) and the mean absolute error (MAE) for n samples in the dataset were computed

$$\mathrm{RMSE} = \left[\sum_{i=1}^{n} \Delta_i^2 / n\right]^{1/2}, \tag{6}$$

$$\mathrm{MAE} = \frac{1}{n}\sum_{i=1}^{n}|\Delta_i|, \tag{7}$$

The produced model was further subjected to 10-fold cross-validation instead of the widely used leave-one-out due to its better performance [145], giving rise to the correlation coefficient of 10-fold cross validation q^2_{CV}. In addition to cross-validation, the developed models were also internally validated by the *Y*-scrambling test [102], which was carried out by randomly permuting the log ER values, viz. *Y* values, to refit the previously developed models while the descriptors were remained unaltered, giving rise to the correlation coefficient r_s^2. The observed log ER values were scrambled 25 times as suggested [107] to produce the average correlation coefficient $\langle r_s^2 \rangle$. Furthermore, various modified versions of r^2 proposed by Ojha et al. [110] were also computed

$$r_m^2 = r^2\left(1 - \sqrt{|r^2 - r_o^2|}\right), \tag{8}$$

$$r'^2_m = r^2\left(1 - \sqrt{\left|r^2 - r'^2_o\right|}\right), \tag{9}$$

$$\left\langle r_m^2 \right\rangle = \left(r_m^2 + r'^2_m\right)/2, \tag{10}$$

$$\Delta r_m^2 = \left|r_m^2 - r'^2_m\right|, \tag{11}$$

where the correlation coefficient r_o^2 and the slope of the regression line k were calculated from the regression line (predicted vs. observed values) through the origin, whereas r'^2_o was calculated from the regression line (observed vs. predicted values) through the origin.

Moreover, the correlation coefficients q_{F1}^2, q_{F2}^2, and q_{F3}^2 and concordance correlation coefficient (CCC) proposed by Shi et al. [146], Schüürmann et al. [147], Consonni et al. [148], and Chirico and Gramatica [149] were also computed by *QSARINS* [150,151] to evaluate the model performance in the external dataset

$$q_{\mathrm{F1}}^2 = 1 - \sum_{i=1}^{n_{\mathrm{EXT}}} (y_i - \hat{y}_i)^2 / \sum_{i=1}^{n_{\mathrm{EXT}}} (y_i - \langle y_{\mathrm{TR}} \rangle)^2, \tag{12}$$

$$q_{\mathrm{F2}}^2 = 1 - \sum_{i=1}^{n_{\mathrm{EXT}}} (y_i - \hat{y}_i)^2 / \sum_{i=1}^{n_{\mathrm{EXT}}} (y_i - \langle y_{\mathrm{EXT}} \rangle)^2, \tag{13}$$

$$q_{\mathrm{F3}}^2 = 1 - \left[\sum_{i=1}^{n_{\mathrm{EXT}}} (y_i - \hat{y}_i)^2 / n_{\mathrm{EXT}}\right] / \left[\sum_{i=1}^{n_{\mathrm{TR}}} (y_i - \langle y_{\mathrm{TR}} \rangle)^2 / n_{\mathrm{TR}}\right], \tag{14}$$

$$\mathrm{CCC} = \frac{2\sum_{i=1}^{n_{\mathrm{EXT}}} (y_i - \langle y_{\mathrm{EXT}} \rangle)(\hat{y}_i - \langle \hat{y}_{\mathrm{EXT}} \rangle)}{\sum_{i=1}^{n_{\mathrm{EXT}}} (y_i - \langle y_{\mathrm{EXT}} \rangle)^2 + (\hat{y}_i - \langle \hat{y}_{\mathrm{EXT}} \rangle)^2 + n_{\mathrm{EXT}}(\langle y_{\mathrm{EXT}} \rangle - \langle \hat{y}_{\mathrm{EXT}} \rangle)^2} \tag{15}$$

where n_{TR} and n_{EXT} are the numbers of samples in the training set and external set, respectively; $\langle \hat{y}_{\mathrm{TR}} \rangle$ is the average predicted value in the training set; and $\langle y_{\mathrm{EXT}} \rangle$ and $\langle \hat{y}_{\mathrm{EXT}} \rangle$ are the average observed and predicted values in the external set, respectively.

Various criteria for those statistical parameters have been proposed to gauge the model predictivity [152]. For instance, Chirico and Gramatica considered that both q_{F3}^2 and *CCC* are the best validation parameters to measure the predictivity [149], whereas Roy et al. suggested that $\langle r_m^2 \rangle$ and Δr_m^2 are the most stringent metrics [111]. Recently, Todeschini et al. demonstrated that q_{F3}^2 is the most reliable metric [112]. The parameter q_{F2}^2 has been adopted by Organization for Economic Co-operation and Development (OECD) to assess the performance of QSAR models [147].

More importantly, a model can be considered as predictive if it can meet the most stringent criteria collectively proposed by Golbraikh et al. [109], Ojha et al. [110], Roy et al. [111], and Chirico and Gramatica [112].

$$r^2, q_{\mathrm{CV}}^2, q^2, q_{\mathrm{Fn}}^2 \geq 0.70, \tag{16}$$

$$\left|r^2 - q_{\mathrm{CV}}^2\right| < 0.10, \tag{17}$$

$$\left(r^2 - r_o^2\right)/r^2 < 0.10 \text{ and } 0.85 \leq k \leq 1.15, \tag{18}$$

$$\left| r_o^2 - r'^2_o \right| < 0.30, \tag{19}$$

$$r_m^2 \geq 0.65, \tag{20}$$

$$\left\langle r_m^2 \right\rangle \geq 0.65 \, \text{and} \, \Delta r_m^2 < 0.20, \tag{21}$$

$$CCC \geq 0.85, \tag{22}$$

where r in Equations (18)–(21) represents the parameters r and q in the training set and external set, respectively; and q_{Fn} stands for q_{F1}, q_{F2}, and q_{F3}.

5. Conclusions

P-gp substrate efflux can be a major obstacle in the success of CNS-targeted therapeutic delivery as well as a critical pharmacokinetic factor for causing DDIs. On the other hand, the CNS-related side-effects of non-CNS drugs can be reduced by P-gp mediated efflux. As such, P-gp substrate efflux is of critical importance to drug discovery and development regardless of CNS drugs or non-CNS drugs. An in silico model to predict the P-gp substrate efflux can be valuable to drug discovery and development. Nevertheless, P-gp substrate efflux is a complex process that can take place through various routes, namely active transport and passive permeability, leading to different descriptor combinations as well as different relationships to render these variations in different mechanisms. In this study, a QSAR predictive model derived from the novel hierarchical support vector regression (HSVR) scheme, which can simultaneously possess the advantageous characteristics of a local model and a global model, viz. broader coverage of applicability domain and higher level of predictivity, respectively, was developed to envisage the P-gp substrate efflux ratio. The developed HSVR showed great prediction accuracy for the 50 and 13 molecules in the training set and test set, respectively, with excellent predictivity and statistical significance. When mock tested by a group of molecules to mimic real challenges, the derived HSVR model also executed accordantly well. Furthermore, the HSVR model can elucidate the discrepancies among all published P-gp substrate classifiers, indicating its superiority. Hence, it can be affirmed that this HSVR model can be adopted as an accurate and reliable predictive tool, even in the high throughput fashion, to facilitate drug discovery and development by designing drug candidates with a more desirable pharmacokinetic profile.

Supplementary Materials: Table S1. Selected compounds for this study, their names, SMILES strings, CAS numbers, observed log ER values and predicted values by SVR A, SVR B, and HSVR, data partitions, and references; Table S2. Optimal runtime parameters for the SVR models; Figure S1. Molecular distribution for the samples in the training set (solid circle) and test set (open square) in the chemical space spanned by three principal components; Figure S2. Histograms of: (A) observed log ER; (B) molecular weight (MW); (C) polar surface area (PSA); (D) number of hydrogen bond acceptor (HBA); and (E) number of hydrogen bond donor (HBD) in density form for all molecules in the training set and test set.

Author Contributions: C.C., C.F.W., and M.K.L. conceived and designed the study; C.C., M.H.L., and M.K.L. performed the experiments and analyzed the data; and C.C., C.F.W., and M.K.L. wrote the paper.

Acknowledgments: This work was financially supported by the Ministry of Science and Technology, Taiwan. Some calculations were performed at the National Center for High-Performance Computing, Taiwan. The authors are grateful to Prof. Paola Gramatica for providing free license of *QSARINS* and Yi-Lung Ding for helping data analysis.

References

1. Schinkel, A.H.; Jonker, J.W. Mammalian drug efflux transporters of the atp binding cassette (abc) family: An overview. *Adv. Drug Deliv. Rev.* **2003**, *55*, 3–29. [CrossRef]
2. Thiebaut, F.; Tsuruo, T.; Hamada, H.; Gottesman, M.M.; Pastan, I.; Willingham, M.C. Cellular localization of the multidrug-resistance gene product p-glycoprotein in normal human tissues. *Proc. Natl. Acad. Sci. USA* **1987**, *84*, 7735–7738. [CrossRef] [PubMed]

3. Kim, R.B.; Fromm, M.F.; Wandel, C.; Leake, B.; Wood, A.J.; Roden, D.M.; Wilkinson, G.R. The drug transporter p-glycoprotein limits oral absorption and brain entry of hiv-1 protease inhibitors. *J. Clin. Investig.* **1998**, *101*, 289–294. [CrossRef] [PubMed]
4. Cordon-Cardo, C.; O'Brien, J.P.; Casals, D.; Rittman-Grauer, L.; Biedler, J.L.; Melamed, M.R.; Bertino, J.R. Multidrug-resistance gene (p-glycoprotein) is expressed by endothelial cells at blood-brain barrier sites. *Proc. Natl. Acad. Sci. USA* **1989**, *86*, 695–698. [CrossRef] [PubMed]
5. Schinkel, A.H. P-glycoprotein, a gatekeeper in the blood-brain barrier. *Adv. Drug Deliv. Rev.* **1999**, *36*, 179–194. [CrossRef]
6. Vähäkangas, K.; Myllynen, P. Drug transporters in the human blood-placental barrier. *Br. J. Pharmacol.* **2009**, *158*, 665–678. [CrossRef] [PubMed]
7. Gosselet, F.; Saint-Pol, J.; Candela, P.; Fenart, L. Amyloid-β peptides, alzheimer's disease and the blood-brain barrier. *Curr. Alzheimer Res.* **2013**, *10*, 1015–1033. [CrossRef] [PubMed]
8. Mawuenyega, K.G.; Sigurdson, W.; Ovod, V.; Munsell, L.; Kasten, T.; Morris, J.C.; Yarasheski, K.E.; Bateman, R.J. Decreased clearance of cns β-amyloid in alzheimer's disease. *Science* **2010**, *330*, 1774. [CrossRef] [PubMed]
9. van Assema, D.M.E.; Lubberink, M.; Bauer, M.; van der Flier, W.M.; Schuit, R.C.; Windhorst, A.D.; Comans, E.F.I.; Hoetjes, N.J.; Tolboom, N.; Langer, O.; et al. Blood–brain barrier p-glycoprotein function in alzheimer's disease. *Brain* **2012**, *135*, 181–189. [CrossRef] [PubMed]
10. Jedlitschky, G.; Vogelgesang, S.; Kroemer, H.K. Mdr1-p-glycoprotein (abcb1)-mediated disposition of amyloid-β peptides: Implications for the pathogenesis and therapy of alzheimer's disease. *Clin. Pharmacol. Ther.* **2010**, *88*, 441–443. [CrossRef] [PubMed]
11. Cascorbi, I.; Flüh, C.; Remmler, C.; Haenisch, S.; Faltraco, F.; Grumbt, M.; Peters, M.; Brenn, A.; Thal, D.R.; Warzok, R.W.; et al. Association of atp-binding cassette transporter variants with the risk of alzheimer's disease. *Pharmacogenomics* **2013**, *14*, 485–494. [CrossRef] [PubMed]
12. Brenn, A.; Grube, M.; Peters, M.; Fischer, A.; Jedlitschky, G.; Kroemer, H.K.; Warzok, R.W.; Vogelgesang, S. Beta-amyloid downregulates mdr1-p-glycoprotein (abcb1) expression at the blood-brain barrier in mice. *Int. J. Alzheimers Dis.* **2011**, *2011*. [CrossRef] [PubMed]
13. Neuwelt, E.A.; Bauer, B.; Fahlke, C.; Fricker, G.; Iadecola, C.; Janigro, D.; Leybaert, L.; Molnár, Z.; O'Donnell, M.E.; Povlishock, J.T.; et al. Engaging neuroscience to advance translational research in brain barrier biology. *Nat. Rev. Neurosci.* **2011**, *12*, 169–182. [CrossRef] [PubMed]
14. Wolf, A.; Bauer, B.; Hartz, A. Abc transporters and the alzheimer's disease enigma. *Front. Psychiatry* **2012**, *3*. [CrossRef] [PubMed]
15. Selick, H.E.; Beresford, A.P.; Tarbit, M.H. The emerging importance of predictive adme simulation in drug discovery. *Drug Discov. Today* **2002**, *7*, 109–116. [CrossRef]
16. Montanari, F.; Ecker, G.F. Prediction of drug–abc-transporter interaction—recent advances and future challenges. *Adv. Drug Deliv. Rev.* **2015**, *86*, 17–26. [CrossRef] [PubMed]
17. Greiner, B.; Eichelbaum, M.; Fritz, P.; Kreichgauer, H.P.; von Richter, O.; Zundler, J.; Kroemer, H.K. The role of intestinal p-glycoprotein in the interaction of digoxin and rifampin. *J. Clin. Investig.* **1999**, *104*, 147–153. [CrossRef] [PubMed]
18. Padowski, J.M.; Pollack, G.M. Influence of time to achieve substrate distribution equilibrium between brain tissue and blood on quantitation of the blood–brain barrier p-glycoprotein effect. *Brain Res.* **2011**, *1426*, 1–17. [CrossRef] [PubMed]
19. Bagal, S.; Bungay, P. Restricting cns penetration of drugs to minimise adverse events: Role of drug transporters. *Drug Discov. Today Technol.* **2014**, *12*, e79–e85. [CrossRef] [PubMed]
20. Hochman, J.H.; Ha, S.N.; Sheridan, R.P. Establishment of p-glycoprotein structure–transport relationships to optimize cns exposure in drug discovery. In *Blood-Brain Barrier in Drug Discovery: Optimizing Brain Exposure of Cns Drugs and Minimizing Brain Side Effects for Peripheral Drugs*; Di, L., Kerns, E.H., Eds.; John Wiley & Sons, Inc.: Hoboken, NJ, USA, 2015; pp. 113–124.
21. Schinkel, A.H.; Wagenaar, E.; Mol, C.A.; van Deemter, L. P-glycoprotein in the blood-brain barrier of mice influences the brain penetration and pharmacological activity of many drugs. *J. Clin. Investig.* **1996**, *97*, 2517–2524. [CrossRef] [PubMed]
22. Aszalos, A. Drug–drug interactions affected by the transporter protein, p-glycoprotein (abcb1, mdr1): II. Clinical aspects. *Drug Discov. Today* **2007**, *12*, 838–843. [CrossRef] [PubMed]

23. Doan, K.M.M.; Humphreys, J.E.; Webster, L.O.; Wring, S.A.; Shampine, L.J.; Serabjit-Singh, C.J.; Adkison, K.K.; Polli, J.W. Passive permeability and p-glycoprotein-mediated efflux differentiate central nervous system (cns) and non-cns marketed drugs. *J. Pharmacol. Exp. Ther.* **2002**, *303*, 1029–1037. [CrossRef] [PubMed]
24. Hennessy, M.; Spiers, J.P. A primer on the mechanics of p-glycoprotein the multidrug transporter. *Pharmacol. Res.* **2007**, *55*, 1–15. [CrossRef] [PubMed]
25. Gottesman, M.M.; Pastan, I. Biochemistry of multidrug resistance mediated by the multidrug transporter. *Ann. Rev. Biochem.* **2003**, *62*, 385–427. [CrossRef] [PubMed]
26. Breier, A.; Gibalova, L.; Seres, M.; Barancik, M.; Sulova, Z. New insight into p-glycoprotein as a drug target. *Anticancer Agents Med. Chem.* **2013**, *13*, 159–170. [CrossRef] [PubMed]
27. Ambudkar, S.V.; Dey, S.; Hrycyna, C.A.; Ramachandra, M.; Pastan, I.; Gottesman, M.M. Biochemical, cellular, and pharmacological aspects of the multidrug transporter. *Annu. Rev. Pharmacol. Toxicol.* **1999**, *39*, 361–398. [CrossRef] [PubMed]
28. Siegel, R.L.; Miller, K.D.; Jemal, A. Cancer statistics, 2017. *CA Cancer J. Clin.* **2017**, *67*, 7–30. [CrossRef] [PubMed]
29. Clarke, J.; Penas, C.; Pastori, C.; Komotar, R.J.; Bregy, A.; Shah, A.H.; Wahlestedt, C.; Ayad, N.G. Epigenetic pathways and glioblastoma treatment. *Epigenetics* **2013**, *8*, 785–795. [CrossRef] [PubMed]
30. Wang, T.; Agarwal, S.; Elmquist, W.F. Brain distribution of cediranib is limited by active efflux at the blood-brain barrier. *J. Pharmacol. Exp. Ther.* **2012**, *341*, 386–395. [CrossRef] [PubMed]
31. Palmeira, A.; Sousa, E.H.; Vasconcelos, M.M.; Pinto, M. Three decades of p-gp inhibitors: Skimming through several generations and scaffolds. *Curr. Med. Chem.* **2012**, *19*, 1946–2025. [CrossRef] [PubMed]
32. van Hoppe, S.; Schinkel, A.H. What next? Preferably development of drugs that are no longer transported by the abcb1 and abcg2 efflux transporters. *Pharmacol. Res.* **2017**, 122–144. [CrossRef] [PubMed]
33. Crivori, P. Computational models for p-glycoprotein substrates and inhibitors. In *Antitargets: Prediction and Prevention of Drug Side Effects*; Vaz, R.J., Klabunde, T., Eds.; Wiley-VCH Verlag GmbH & Co. KGaA: Weinheim, Germany, 2008; Volume 38, pp. 367–397.
34. Terasaki, T.; Hosoya, K.I. The blood-brain barrier efflux transporters as a detoxifying system for the brain. *Adv. Drug Deliv. Rev.* **1999**, *36*, 195–209. [CrossRef]
35. Garg, P.; Verma, J. In silico prediction of blood brain barrier permeability: An artificial neural network model. *J. Chem. Inf. Model.* **2006**, *46*, 289–297. [CrossRef] [PubMed]
36. Kalvass, J.C.; Maurer, T.S.; Pollack, G.M. Use of plasma and brain unbound fractions to assess the extent of brain distribution of 34 drugs: Comparison of unbound concentration ratios to in vivo p-glycoprotein efflux ratios. *Drug Metab. Dispos.* **2007**, *35*, 660–666. [CrossRef] [PubMed]
37. Di, L.; Rong, H.; Feng, B. Demystifying brain penetration in central nervous system drug discovery. *J. Med. Chem.* **2013**, *56*, 2–12. [CrossRef] [PubMed]
38. Inoue, T.; Osada, K.; Tagawa, M.; Ogawa, Y.; Haga, T.; Sogame, Y.; Hashizume, T.; Watanabe, T.; Taguchi, A.; Katsumata, T.; et al. Blonanserin, a novel atypical antipsychotic agent not actively transported as substrate by p-glycoprotein. Prog. Neuropsychopharmacol. *Biol. Psychiatry* **2012**, *39*, 156–162.
39. Polli, J.W.; Wring, S.A.; Humphreys, J.E.; Huang, L.; Morgan, J.B.; Webster, L.O.; Serabjit-Singh, C.S. Rational use of in vitro p-glycoprotein assays in drug discovery. *J. Pharmacol. Exp. Ther.* **2001**, *299*, 620–628. [PubMed]
40. Hochman, J.H.; Yamazaki, M.; Ohe, T.; Lin, J.H. Evaluation of drug interactions with p-glycoprotein in drug discovery: In vitro assessment of the potential for drug-drug interactions with p-glycoprotein. *Curr. Drug MeTable* **2002**, *3*, 257–273. [CrossRef]
41. Schwab, D.; Fischer, H.; Tabatabaei, A.; Poli, S.; Huwyler, J. Comparison of in vitro p-glycoprotein screening assays: Recommendations for their use in drug discovery. *J. Med. Chem.* **2003**, *46*, 1716–1725. [CrossRef] [PubMed]
42. Zhang, Y.; Bachmeier, C.; Miller, D.W. In vitro and in vivo models for assessing drug efflux transporter activity. Adv. *Drug Deliv. Rev.* **2003**, *55*, 31–51. [CrossRef]
43. Sugano, K.; Shirasaka, Y.; Yamashita, S. Estimation of michaelis–menten constant of efflux transporter considering asymmetric permeability. *Int. J. Pharm.* **2011**, *418*, 161–167. [CrossRef] [PubMed]
44. Storch, C.H.; Nikendei, C.; Schild, S.; Haefeli, W.E.; Weiss, J.; Herzog, W. Expression and activity of p-glycoprotein (mdr1/abcb1) in peripheral blood mononuclear cells from patients with anorexia nervosa compared with healthy controls. *Int. J. Eating Disord.* **2008**, *41*, 432–438. [CrossRef] [PubMed]

45. Egan, W.J. Computational models for adme. In *Annual Reports in Medicinal Chemistry*; John, E.M., Ed.; Academic Press: San Diego, CA, USA, 2007; Volume 42, pp. 449–467.
46. Sliwoski, G.; Kothiwale, S.; Meiler, J.; Lowe, E.W. Computational methods in drug discovery. *Pharmacol. Rev.* **2014**, *66*, 334–395. [CrossRef] [PubMed]
47. Penzotti, J.E.; Lamb, M.L.; Evensen, E.; Grootenhuis, P.D.J. A computational ensemble pharmacophore model for identifying substrates of p-glycoprotein. *J. Med. Chem.* **2002**, *45*, 1737–1740. [CrossRef] [PubMed]
48. Gombar, V.K.; Polli, J.W.; Humphreys, J.E.; Wring, S.A.; Serabjit-Singh, C.S. Predicting p-glycoprotein substrates by a quantitative structure-activity relationship model. *J. Pharm. Sci.* **2004**, *93*, 957–968. [CrossRef] [PubMed]
49. Xue, Y.; Yap, C.W.; Sun, L.Z.; Cao, Z.W.; Wang, J.F.; Chen, Y.Z. Prediction of p-glycoprotein substrates by a support vector machine approach. *J. Chem. Inf. Comput. Sci.* **2004**, *44*, 1497–1505. [CrossRef] [PubMed]
50. Wang, Y.H.; Li, Y.; Yang, S.L.; Yang, L. Classification of substrates and inhibitors of p-glycoprotein using unsupervised machine learning approach. *J. Chem. Inf. Model.* **2005**, *45*, 750–757. [CrossRef] [PubMed]
51. Crivori, P.; Reinach, B.; Pezzetta, D.; Poggesi, I. Computational models for identifying potential p-glycoprotein substrates and inhibitors. *Mol. Pharma.* **2006**, *3*, 33–44. [CrossRef]
52. de Cerqueira Lima, P.; Golbraikh, A.; Oloff, S.; Xiao, Y.; Tropsha, A. Combinatorial qsar modeling of p-glycoprotein substrates. *J. Chem. Inf. Model.* **2006**, *46*, 1245–1254. [CrossRef] [PubMed]
53. Huang, J.; Ma, G.; Muhammad, I.; Cheng, Y. Identifying p-glycoprotein substrates using a support vector machine optimized by a particle swarm. *J. Chem. Inf. Model.* **2007**, *47*, 1638–1647. [CrossRef] [PubMed]
54. Li, W.-X.; Li, L.; Eksterowicz, J.; Ling, X.B.; Cardozo, M. Significance analysis and multiple pharmacophore models for differentiating p-glycoprotein substrates. *J. Chem Inf. Model.* **2007**, *47*, 2429–2438. [CrossRef] [PubMed]
55. Wang, Z.; Chen, Y.; Liang, H.; Bender, A.; Glen, R.C.; Yan, A. P-glycoprotein substrate models using support vector machines based on a comprehensive data set. *J. Chem. Inf. Model.* **2011**, *51*, 1447–1456. [CrossRef] [PubMed]
56. Desai, P.V.; Sawada, G.A.; Watson, I.A.; Raub, T.J. Integration of in silico and in vitro tools for scaffold optimization during drug discovery: Predicting p-glycoprotein efflux. Mol. *Pharmaceutics* **2013**, *10*, 1249–1261. [CrossRef] [PubMed]
57. Ecker, G.F.; Stockner, T.; Chiba, P. Computational models for prediction of interactions with abc-transporters. *Drug Discov. Today* **2008**, *13*, 311–317. [CrossRef] [PubMed]
58. Adenot, M. A practical approach to computational models of the blood–brain barrier. In *Handbook of Neurochemistry and Molecular Neurobiology: Neural Membranes and Transport*; Lajtha, A., Reith, M.E.A., Eds.; Springer: New York, NY, USA, 2007; pp. 109–150.
59. Ivanciuc, O. Artificial immune systems in drug design: Recognition of p-glycoprotein substrates with airs (artificial immune recognition system). Internet Electron. *J. Mol. Des.* **2006**, *5*, 542–554.
60. Bikadi, Z.; Hazai, I.; Malik, D.; Jemnitz, K.; Veres, Z.; Hari, P.; Ni, Z.; Loo, T.W.; Clarke, D.M.; Hazai, E.; et al. Predicting p-glycoprotein-mediated drug transport based on support vector machine and three-dimensional crystal structure of p-glycoprotein. *PLoS ONE* **2011**, *6*, e25815. [CrossRef] [PubMed]
61. Erić, S.; Kalinić, M.; Ilić, K.; Zloh, M. Computational classification models for predicting the interaction of drugs with p-glycoprotein and breast cancer resistance protein. *SAR QSAR Environ. Res.* **2014**, *25*, 939–966. [CrossRef] [PubMed]
62. Pan, X.; Mei, H.; Qu, S.; Huang, S.; Sun, J.; Yang, L.; Chen, H. Prediction and characterization of p-glycoprotein substrates potentially bound to different sites by emerging chemical pattern and hierarchical cluster analysis. *Int. J. Pharm.* **2016**, *502*, 61–69. [CrossRef] [PubMed]
63. Didziapetris, R.; Japertas, P.; Avdeef, A.; Petrauskas, A. Classification analysis of p-glycoprotein substrate specificity. *J. Drug Target.* **2003**, *11*, 391–406. [CrossRef] [PubMed]
64. Broccatelli, F. Qsar models for p-glycoprotein transport based on a highly consistent data set. *J. Chem. Inf. Model.* **2012**, *2*, 2462–2470. [CrossRef] [PubMed]
65. Poongavanam, V.; Haider, N.; Ecker, G.F. Fingerprint-based in silico models for the prediction of p-glycoprotein substrates and inhibitors. *Bioorg. Med. Chem.* **2012**, *20*, 5388–5395. [CrossRef] [PubMed]
66. Li, D.; Chen, L.; Li, Y.; Tian, S.; Sun, H.; Hou, T. Admet evaluation in drug discovery. 13. Development of in silico prediction models for p-glycoprotein substrates. *Mol. Pharm.* **2014**, *11*, 716–726. [PubMed]

67. Estrada, E.; Molina, E.; Nodarse, D.; Uriarte, E. Structural contributions of substrates to their binding to p-glycoprotein. A topsmode approach. *Curr. Pharm. Des.* **2010**, *16*, 2676–2709. [CrossRef] [PubMed]
68. Joung, J.Y.; Kim, H.; Kim, H.M.; Ahn, S.K.; Nam, K.-Y.; No, K.T. Prediction models of p-glycoprotein substrates using simple 2d and 3d descriptors by a recursive partitioning approach. *Bull. Korean Chem. Soc.* **2012**, *33*, 1123–1127. [CrossRef]
69. Gunaydin, H.; Weiss, M.M.; Sun, Y. De novo prediction of p-glycoprotein-mediated efflux liability for druglike compounds. *ACS Med. Chem. Lett.* **2013**, *4*, 108–112. [CrossRef] [PubMed]
70. Dolghih, E.; Jacobson, M.P. Predicting efflux ratios and blood-brain barrier penetration from chemical structure: Combining passive permeability with active efflux by p-glycoprotein. *ACS Chem. Neurosci.* **2012**, *4*, 361–367. [CrossRef] [PubMed]
71. Dolghih, E.; Bryant, C.; Renslo, A.R.; Jacobson, M.P. Predicting binding to p-glycoprotein by flexible receptor docking. *PLoS Comput. Biol.* **2011**, *7*, e1002083. [CrossRef] [PubMed]
72. Subramanian, N.; Condic-Jurkic, K.; O'Mara, M.L. Structural and dynamic perspectives on the promiscuous transport activity of p-glycoprotein. *Neurochem. Int.* **2016**, *98*, 146–152. [CrossRef] [PubMed]
73. Leong, M.K.; Chen, H.B.; Shih, Y.H. Prediction of promiscuous p-glycoprotein inhibition using a novel machine learning scheme. *PLoS ONE* **2012**, *7*, e33829. [CrossRef] [PubMed]
74. Garrigues, A.; Loiseau, N.; Delaforge, M.; Ferté, J.; Garrigos, M.; André, F.; Orlowski, S. Characterization of two pharmacophores on the multidrug transporter p-glycoprotein. *Mol. Pharmacol.* **2002**, *62*, 1288–1298. [CrossRef] [PubMed]
75. Chufan, E.E.; Sim, H.M.; Ambudkar, S.V. Molecular basis of the polyspecificity of p-glycoprotein (abcb1): Recent biochemical and structural studies. In *Advances in Cancer Research: Abc Transporters and Cancer*; John, D.S., Toshihisa, I., Eds.; Academic Press: San Diego, CA, USA, 2015; Volume 125, pp. 71–96.
76. Ferreira, R.J.; Ferreira, M.J.U.; dos Santos, D.J.V.A. Molecular docking characterizes substrate-binding sites and efflux modulation mechanisms within p-glycoprotein. *J. Chem. Inf. Model.* **2013**, *53*, 1747–1760. [CrossRef] [PubMed]
77. Aller, S.G.; Yu, J.; Ward, A.; Weng, Y.; Chittaboina, S.; Zhuo, R.; Harrell, P.M.; Trinh, Y.T.; Zhang, Q.; Urbatsch, I.L.; et al. Structure of p-glycoprotein reveals a molecular basis for poly-specific drug binding. *Science* **2009**, *323*, 1718–1722. [CrossRef] [PubMed]
78. Edwards, G. Ivermectin: Does p-glycoprotein play a role in neurotoxicity? *Filaria J.* **2003**, *2* (Suppl. 1), S8. [CrossRef] [PubMed]
79. Balimane, P.V.; Han, Y.H.; Chong, S. Current industrial practices of assessing permeability and p-glycoprotein interaction. *AAPS J.* **2006**, *8*, E1–E13. [CrossRef] [PubMed]
80. Roger, P.; Sahla, M.E.; Mäkelä, S.; Gustafsson, J.Å.; Baldet, P.; Rochefort, H. Decreased expression of estrogen receptor β protein in proliferative preinvasive mammary tumors. *Cancer Res.* **2001**, *61*, 2537–2541. [PubMed]
81. Speck-Planche, A.; Cordeiro, M.N.D.S. Multi-target qsar approaches for modeling protein inhibitors. Simultaneous prediction of activities against biomacromolecules present in gram-negative bacteria. *Curr. Top. Med. Chem.* **2015**, *15*, 1801–1813. [CrossRef] [PubMed]
82. Ferreira, R.J.; dos Santos, D.J.V.A.; Ferreira, M.J.U.; Guedes, R.C. Toward a better pharmacophore description of p-glycoprotein modulators, based on macrocyclic diterpenes from euphorbia species. *J. Chem. Inf. Model.* **2011**, *51*, 1315–1324. [CrossRef] [PubMed]
83. Leong, M.K.; Chen, Y.M.; Chen, T.H. Prediction of human cytochrome p450 2b6-substrate interactions using hierarchical support vector regression approach. *J. Comput. Chem.* **2009**, *30*, 1899–1909. [CrossRef] [PubMed]
84. Caudill, M. Using neural networks: Hybrid expert networks. *AI Expert* **1990**, *5*, 49–54.
85. Leong, M.K.; Lin, S.W.; Chen, H.B.; Tsai, F.Y. Predicting mutagenicity of aromatic amines by various machine learning approaches. *Toxicol. Sci.* **2010**, *116*, 498–513. [CrossRef] [PubMed]
86. Ding, Y.L.; Lyu, Y.C.; Leong, M.K. In silico prediction of the mutagenicity of nitroaromatic compounds using a novel two-qsar approach. *Toxicol. In Vitro* **2017**, *40*, 102–114. [CrossRef] [PubMed]
87. Gnanadesikan, R.; Kettenring, J.R. Robust estimates, residuals, and outlier detection with multiresponse data. *Biometrics* **1972**, *28*, 81–124. [CrossRef]
88. Carrara, S.; Reali, V.; Misiano, P.; Dondio, G.; Bigogno, C. Evaluation of in vitro brain penetration: Optimized pampa and mdckii-mdr1 assay comparison. *Int. J. Pharm.* **2007**, *345*, 125–133. [CrossRef] [PubMed]
89. Chen, C.; Hanson, E.; Watson, J.W.; Lee, J.S. P-glycoprotein limits the brain penetration of nonsedating but not sedating h1-antagonists. *Drug Metab. Dispos.* **2003**, *31*, 312–318. [CrossRef] [PubMed]

90. Eriksson, U.G.; Dorani, H.; Karlsson, J.; Fritsch, H.; Hoffmann, K.-J.; Olsson, L.; Sarich, T.C.; Wall, U.; Schützer, K.-M. Influence of erythromycin on the pharmacokinetics of ximelagatran may involve inhibition of p-glycoprotein-mediated excretion. *Drug Metab. Dispos.* **2006**, *34*, 775–782. [CrossRef] [PubMed]
91. Feng, B.; Mills, J.B.; Davidson, R.E.; Mireles, R.J.; Janiszewski, J.S.; Troutman, M.D.; de Morais, S.M. In vitro p-glycoprotein assays to predict the in vivo interactions of p-glycoprotein with drugs in the central nervous system. *Drug Metab. Dispos.* **2008**, *36*, 268–275. [CrossRef] [PubMed]
92. Gertz, M.; Harrison, A.; Houston, J.B.; Galetin, A. Prediction of human intestinal first-pass metabolism of 25 cyp3a substrates from in vitro clearance and permeability data. *Drug Metab. Dispos.* **2010**, *38*, 1147–1158. [CrossRef] [PubMed]
93. Huang, L.; Wang, Y.; Grimm, S. Atp-dependent transport of rosuvastatin in membrane vesicles expressing breast cancer resistance protein. *Drug Metab. Dispos.* **2006**, *34*, 738–742. [CrossRef] [PubMed]
94. Luo, S.; Pal, D.; Shah, S.J.; Kwatra, D.; Paturi, K.D.; Mitra, A.K. Effect of hepes buffer on the uptake and transport of p-glycoprotein substrates and large neutral amino acids. *Mol. Pharm.* **2010**, *7*, 412–420. [CrossRef] [PubMed]
95. Taub, M.E.; Podila, L.; Ely, D.; Almeida, I. Functional assessment of multiple p-glycoprotein (p-gp) probe substrates: Influence of cell line and modulator concentration on p-gp activity. *Drug Metab. Dispos.* **2005**, *33*, 1679–1687. [CrossRef] [PubMed]
96. Troutman, M.D.; Thakker, D.R. Novel experimental parameters to quantify the modulation of absorptive and secretory transport of compounds by p-glycoprotein in cell culture models of intestinal epithelium. *Pharm. Res.* **2003**, *20*, 1210–1224. [CrossRef] [PubMed]
97. Wager, T.T.; Chandrasekaran, R.Y.; Hou, X.; Troutman, M.D.; Verhoest, P.R.; Villalobos, A.; Will, Y. Defining desirable central nervous system drug space through the alignment of molecular properties, in vitro adme, and safety attributes. *ACS Chem. Neurosci.* **2010**, *1*, 420–434. [CrossRef] [PubMed]
98. Callegari, E.; Malhotra, B.; Bungay, P.J.; Webster, R.; Fenner, K.S.; Kempshall, S.; LaPerle, J.L.; Michel, M.C.; Kay, G.G. A comprehensive non-clinical evaluation of the cns penetration potential of antimuscarinic agents for the treatment of overactive bladder. *Br. J. Clin. Pharmacol.* **2011**, *72*, 235–246. [CrossRef] [PubMed]
99. Obradovic, T.; Dobson, G.; Shingaki, T.; Kungu, T.; Hidalgo, I. Assessment of the first and second generation antihistamines brain penetration and role of p-glycoprotein. *Pharm. Res.* **2007**, *24*, 318–327. [CrossRef] [PubMed]
100. Liu, Q.; Wang, C.; Meng, Q.; Huo, X.; Sun, H.; Peng, J.; Ma, X.; Sun, P.; Liu, K. Mdr1 and oat1/oat3 mediate the drug-drug interaction between puerarin and methotrexate. *Pharm. Res.* **2014**, *31*, 1120–1132. [CrossRef] [PubMed]
101. Kim, W.Y.; Benet, L.Z. P-glycoprotein (p-gp/mdr1)-mediated efflux of sex-steroid hormones and modulation of p-gp expression in vitro. *Pharm. Res.* **2004**, *21*, 1284–1293. [CrossRef] [PubMed]
102. Cherkasov, A.; Muratov, E.N.; Fourches, D.; Varnek, A.; Baskin, I.I.; Cronin, M.; Dearden, J.; Gramatica, P.; Martin, Y.C.; Todeschini, R.; et al. Qsar modeling: Where have you been? Where are you going to? *J. Med. Chem.* **2014**, *57*, 4977–5010. [CrossRef] [PubMed]
103. McIver, Z.A.; Kryman, M.W.; Choi, Y.; Coe, B.N.; Schamerhorn, G.A.; Linder, M.K.; Davies, K.S.; Hill, J.E.; Sawada, G.A.; Grayson, J.M.; et al. Selective photodepletion of malignant t cells in extracorporeal photopheresis with selenorhodamine photosensitizers. *Bioorg. Med. Chem.* **2016**, *24*, 3918–3931. [CrossRef] [PubMed]
104. Lee, W.; Crawford, J.J.; Aliagas, I.; Murray, L.J.; Tay, S.; Wang, W.; Heise, C.E.; Hoeflich, K.P.; La, H.; Mathieu, S.; et al. Synthesis and evaluation of a series of 4-azaindole-containing p21-activated kinase-1 inhibitors. Bioorg. *Med. Chem. Lett.* **2016**, *26*, 3518–3524. [CrossRef] [PubMed]
105. Hitchcock, S.A. Structural modifications that alter the p-glycoprotein efflux properties of compounds. *J. Med. Chem.* **2012**, *55*, 4877–4895. [CrossRef] [PubMed]
106. Tropsha, A.; Gramatica, P.; Gombar, V.K. The importance of being earnest: Validation is the absolute essential for successful application and interpretation of qspr models. *QSAR Comb. Sci.* **2003**, *22*, 69–77. [CrossRef]
107. Rücker, C.; Rücker, G.; Meringer, M. Y-randomization and its variants in qspr/qsar. *J. Chem. Inf. Model.* **2007**, *47*, 2345–2357. [CrossRef] [PubMed]

108. Netzeva, T.I.; Worth, A.; Aldenberg, T.; Benigni, R.; Cronin, M.T.D.; Gramatica, P.; Jaworska, J.S.; Kahn, S.; Klopman, G.; Marchant, C.A.; et al. Current status of methods for defining the applicability domain of (quantitative) structure-activity relationships : The report and recommendations of ecvam workshop 52. *Altern. Lab. Anim.* **2005**, *33*, 1–19.
109. Golbraikh, A.; Shen, M.; Xiao, Z.Y.; Xiao, Y.D.; Lee, K.H.; Tropsha, A. Rational selection of training and test sets for the development of validated qsar models. *J. Comput.-Aided Mol. Des.* **2003**, *17*, 241–253. [CrossRef] [PubMed]
110. Ojha, P.K.; Mitra, I.; Das, R.N.; Roy, K. Further exploring rm2 metrics for validation of qspr models. *Chemometr. Intell. Lab. Syst.* **2011**, *107*, 194–205. [CrossRef]
111. Roy, K.; Mitra, I.; Kar, S.; Ojha, P.K.; Das, R.N.; Kabir, H. Comparative studies on some metrics for external validation of qspr models. *J. Chem. Inf. Model.* **2012**, *52*, 396–408. [CrossRef] [PubMed]
112. Chirico, N.; Gramatica, P. Real external predictivity of qsar models. Part 2. New intercomparable thresholds for different validation criteria and the need for scatter plot inspection. *J. Chem. Inf. Model.* **2012**, *52*, 2044–2058. [CrossRef] [PubMed]
113. Topliss, J.G.; Edwards, R.P. Chance factors in studies of quantitative structure-activity relationships. *J. Med. Chem.* **1979**, *22*, 1238–1244. [CrossRef] [PubMed]
114. Desai, P.V.; Raub, T.J.; Blanco, M.J. How hydrogen bonds impact p-glycoprotein transport and permeability. Bioorg. *Med. Chem. Lett.* **2012**, *22*, 6540–6548. [CrossRef] [PubMed]
115. Teixeira, V.H.; Vila-Viçosa, D.; Baptista, A.M.; Machuqueiro, M. Protonation of dmpc in a bilayer environment using a linear response approximation. *J. Chem. Theory Comput.* **2014**, *10*, 2176–2184. [CrossRef] [PubMed]
116. Clay, A.T.; Sharom, F.J. Lipid bilayer properties control membrane partitioning, binding, and transport of p-glycoprotein substrates. *Biochemistry* **2013**, *52*, 343–354. [CrossRef] [PubMed]
117. Raub, T.J. P-glycoprotein recognition of substrates and circumvention through rational drug design. *Mol. Pharmaceutics* **2006**, *3*, 3–25. [CrossRef]
118. Suzuki, T.; Fukazawa, N.; San-nohe, K.; Sato, W.; Yano, O.; Tsuruo, T. Structure-activity relationship of newly synthesized quinoline derivatives for reversal of multidrug resistance in cancer. *J. Med. Chem.* **1997**, *40*, 2047–2052. [CrossRef] [PubMed]
119. Prasanna, S.; Doerksen, R.J. Topological polar surface area: A useful descriptor in 2d-qsar. *Curr. Med. Chem.* **2009**, *16*, 21–41. [CrossRef] [PubMed]
120. Ferté, J. Analysis of the tangled relationships between p-glycoprotein-mediated multidrug resistance and the lipid phase of the cell membrane. *Eur. J. Biochem.* **2000**, *267*, 277–294. [CrossRef] [PubMed]
121. Johnson, T.W.; Dress, K.R.; Edwards, M. Using the golden triangle to optimize clearance and oral absorption. *Bioorg. Med. Chem. Lett.* **2009**, *19*, 5560–5564. [CrossRef] [PubMed]
122. Pettersson, M.; Hou, X.; Kuhn, M.; Wager, T.T.; Kauffman, G.W.; Verhoest, P.R. Quantitative assessment of the impact of fluorine substitution on p-glycoprotein (p-gp) mediated efflux, permeability, lipophilicity, and metabolic stability. *J. Med. Chem.* **2016**, *59*, 5284–5296. [CrossRef] [PubMed]
123. Jabeen, I.; Wetwitayaklung, P.; Klepsch, F.; Parveen, Z.; Chiba, P.; Ecker, G.F. Probing the stereoselectivity of p-glycoprotein-synthesis, biological activity and ligand docking studies of a set of enantiopure benzopyrano[3,4-b][1,4]oxazines. *Chem. Commun.* **2011**, *47*, 2586–2588. [CrossRef] [PubMed]
124. Rezai, T.; Bock, J.E.; Zhou, M.V.; Kalyanaraman, C.; Lokey, R.S.; Jacobson, M.P. Conformational flexibility, internal hydrogen bonding, and passive membrane permeability: Successful in silico prediction of the relative permeabilities of cyclic peptides. *J. Am. Chem. Soc.* **2006**, *128*, 14073–14080. [CrossRef] [PubMed]
125. Rauch, C.; Paine, S.W.; Littlewood, P. Can long range mechanical interaction between drugs and membrane proteins define the notion of molecular promiscuity? Application to p-glycoprotein-mediated multidrug resistance (mdr). Biochim. Biophys. *Acta-Gen. Subj.* **2013**, *1830*, 5112–5118. [CrossRef] [PubMed]
126. Declèves, X.; Jacob, A.; Yousif, S.; Shawahna, R.; Potin, S.; Scherrmann, J.-M. Interplay of drug metabolizing cyp450 enzymes and abc transporters in the blood-brain barrier. *Curr. Drug MeTable* **2011**, *12*, 732–741. [CrossRef]
127. Passeleu-Le Bourdonnec, C.; Carrupt, P.-A.; Scherrmann, J.; Martel, S. Methodologies to assess drug permeation through the blood–brain barrier for pharmaceutical research. *Pharm. Res.* **2013**, *30*, 2729–2756. [CrossRef] [PubMed]
128. Leong, M.K. In silico prediction of the blood-brain barrier permeation: Are we there yet? *Med. Chem.* **2015**, *5*, 130. [CrossRef]

129. Cammi, R.; Tomasi, J. Remarks on the use of the apparent surface charges (asc) methods in solvation problems: Iterative versus matrix-inversion procedures and the renormalization of the apparent charges. *J. Comput. Chem.* **1995**, *16*, 1449–1458. [CrossRef]
130. Miertuš, S.; Scrocco, E.; Tomasi, J. Electrostatic interaction of a solute with a continuum. A direct utilizaion of ab initio molecular potentials for the prevision of solvent effects. *Chem. Phys.* **1981**, *55*, 117–129. [CrossRef]
131. Besler, B.H.; Merz, K.M.J.; Kollman, P.A. Atomic charges derived from semiempirical methods. *J. Comput. Chem.* **1990**, *11*, 431–439. [CrossRef]
132. Kettaneh, N.; Berglund, A.; Wold, S. Pca and pls with very large data sets. *Comput. Stat. Data Anal.* **2005**, *48*, 69–85. [CrossRef]
133. Tseng, Y.J.; Hopfinger, A.J.; Esposito, E.X. The great descriptor melting pot: Mixing descriptors for the common good of qsar models. *J. Comput. Aided Mol. Des.* **2012**, *26*, 39–43. [CrossRef] [PubMed]
134. Burden, F.R.; Ford, M.G.; Whitley, D.C.; Winkler, D.A. Use of automatic relevance determination in qsar studies using bayesian neural networks. *J. Chem. Inf. Comput. Sci.* **2000**, *40*, 1423–1430. [CrossRef] [PubMed]
135. Rogers, D.; Hopfinger, A.J. Application of genetic function approximation to quantitative structure-activity relationships and quantitative structure-property relationships. *J. Chem. Inf. Comput. Sci.* **1994**, *34*, 854–866. [CrossRef]
136. Guyon, I.; Weston, J.; Barnhill, S.; Vapnik, V. Gene selection for cancer classification using support vector machines. *Mach. Learn.* **2002**, *46*, 389–422. [CrossRef]
137. Kennard, R.W.; Stone, L.A. Computer aided design of experiments. *Technometrics* **1969**, *11*, 137–148. [CrossRef]
138. Tropsha, A. Recent trends in statistical qsar modeling of environmental chemical toxicity. In *Molecular, Clinical and Environmental Toxicology. Volume 3: Environmental Toxicology*; Luch, A., Ed.; Springer Basel: New York, NY, USA, 2012; Volume 101, pp. 381–411.
139. Cortes, C.; Vapnik, V. Support-vector networks. *Mach. Learn.* **1995**, *20*, 273–297. [CrossRef]
140. Vapnik, V.; Golowich, S.; Smola, A. *Support Vector Method for Function Approximation, Regression Estimation, and Signal Processing, Advances in Neural Information Processing Systems 9*; Mozer, M., Jordan, M.I., Petsche, T., Eds.; MIT Press: Cambridge, MA, USA, 1997; pp. 281–287.
141. Schölkopf, B.; Smola, A. *Learning with Kernels: Support Vector Machines, Regularization, Optimization, and Beyond*, 1st ed.; MIT Press: Cambridge, MA, USA, 2002.
142. Kecman, V. *Learning and Soft Computing :Support Vector Machines, Neural Networks, and Fuzzy Logic Models*; MIT Press: Cambridge, MA, USA, 2001; p. 576.
143. Leong, M.K.; Syu, R.G.; Ding, Y.L.; Weng, C.F. Prediction of n-methyl-d-aspartate receptor glun1-ligand binding affinity by a novel svm-pose/svm-score combinatorial ensemble docking scheme. *Sci. Rep.* **2017**, *7*, 40053. [CrossRef] [PubMed]
144. Dearden, J.C.; Cronin, M.T.D.; Kaiser, K.L.E. How not to develop a quantitative structure–activity or structure–property relationship (qsar/qspr). *SAR QSAR Environ. Res.* **2009**, *20*, 241–266. [CrossRef] [PubMed]
145. Breiman, L.; Spector, P. Submodel selection and evaluation in regression. The x-random case. *Int. Stat. Rev.* **1992**, *60*, 291–319. [CrossRef]
146. Shi, L.M.; Fang, H.; Tong, W.; Wu, J.; Perkins, R.; Blair, R.M.; Branham, W.S.; Dial, S.L.; Moland, C.L.; Sheehan, D.M. Qsar models using a large diverse set of estrogens. *J. Chem. Inf. Comput. Sci.* **2001**, *41*, 186–195. [CrossRef] [PubMed]
147. Schüürmann, G.; Ebert, R.U.; Chen, J.; Wang, B.; Kühne, R. External validation and prediction employing the predictive squared correlation coefficient-test set activity mean vs training set activity mean. *J. Chem. Inf. Model.* **2008**, *48*, 2140–2145. [CrossRef] [PubMed]
148. Consonni, V.; Ballabio, D.; Todeschini, R. Comments on the definition of the q2 parameter for qsar validation. *J. Chem. Inf. Model.* **2009**, *49*, 1669–1678. [CrossRef] [PubMed]
149. Chirico, N.; Gramatica, P. Real external predictivity of qsar models: How to evaluate it? Comparison of different validation criteria and proposal of using the concordance correlation coefficient. *J. Chem. Inf. Model.* **2011**, *51*, 2320–2335. [CrossRef] [PubMed]
150. Gramatica, P.; Chirico, N.; Papa, E.; Cassani, S.; Kovarich, S. Qsarins: A new software for the development, analysis, and validation of qsar mlr models. *J. Comput. Chem.* **2013**, *34*, 2121–2132. [CrossRef]

151. Gramatica, P.; Cassani, S.; Chirico, N. Qsarins-chem: Insubria datasets and new qsar/qspr models for environmental pollutants in qsarins. *J. Comput. Chem.* **2014**, *35*, 1036–1044. [CrossRef] [PubMed]

152. Gramatica, P.; Sangion, A. A historical excursus on the statistical validation parameters for qsar models: A clarification concerning metrics and terminology. *J. Chem. Inf. Model.* **2016**, *56*, 1127–1131. [CrossRef] [PubMed]

10

ParaBTM: A Parallel Processing Framework for Biomedical Text Mining on Supercomputers

Yuting Xing [1], Chengkun Wu [1,*], Xi Yang [1], Wei Wang [1], En Zhu [1] and Jianping Yin [2]

[1] School of Computer Science, National University of Defense Technology, Changsha, Hunan 410073, China; xingyuting16@nudt.edu.cn (Y.X.); yangxi1016@nudt.edu.cn (X.Y.); g.webywang@gmail.com (W.W.); enzhu@nudt.edu.cn (E.Z.)

[2] School of Computer Science and Network Security, Dongguan University of Technology, Dongguan, Guangdong 523808, China; jpyin@dgut.edu.cn

* Correspondence: chengkun_wu@nudt.edu.cn.

Abstract: A prevailing way of extracting valuable information from biomedical literature is to apply text mining methods on unstructured texts. However, the massive amount of literature that needs to be analyzed poses a big data challenge to the processing efficiency of text mining. In this paper, we address this challenge by introducing parallel processing on a supercomputer. We developed paraBTM, a runnable framework that enables parallel text mining on the Tianhe-2 supercomputer. It employs a low-cost yet effective load balancing strategy to maximize the efficiency of parallel processing. We evaluated the performance of paraBTM on several datasets, utilizing three types of named entity recognition tasks as demonstration. Results show that, in most cases, the processing efficiency can be greatly improved with parallel processing, and the proposed load balancing strategy is simple and effective. In addition, our framework can be readily applied to other tasks of biomedical text mining besides NER.

Keywords: biomedical text mining; big data; Tianhe-2; parallel computing; load balancing

---

1. Introduction

With the rapid development of biotechnology, the amount of biomedical literature is growing exponentially. For instance, PubMed (https://www.ncbi.nlm.nih.gov/pubmed/), the most recognized biomedical literature database, indexes over 28 million entries for biomedical literature. Most of that information is presented in the form of unstructured texts. It is almost impossible for any domain expert to digest such a massive amount of information within a short period of time. Therefore, automated tools are essential for a systematic understanding of literature. To deal with the literature big data challenge, text mining methods are commonly applied to extract relevant knowledge from vast amounts of literature, and this has become a prominent trend in recent years [1].

Typical tasks of biomedical text mining include named entity recognition and relation extraction. One of the most fundamental tasks of biomedical text mining is named entity recognition (NER). Its task is to recognize target entities that represents key concepts from unstructured biomedical texts, such as proteins, genes, mutations, diseases, etc. There are some existing start-of-art biomedical tools that use text mining methods to identify some specific types of entities, such as mutations [2,3], genes [4,5], and diseases [6,7]. Most of these tools can achieve satisfactory recognition performance (F score over 80%) on standard datasets.

Relation extraction (RE) is a process that typically follows NER and aims to discover semantic connections between entities. Nowadays, there are a number of RE tools using different methods to identify biomedical entity interactions [8,9], such as drug–gene relationships [10–12], gene–disease relationships [9,13,14] and protein–protein interaction [15]. Some of them can achieve high F scores (over 80%) on several annotated datasets.

NER and RE are the preliminary steps in mining information from literature. With the uncovered facts, it is possible to construct a complex knowledge graph, which can assist new knowledge discovery and hypotheses generation. In order to achieve this goal, it is necessary to process as many articles as possible. However, text mining procedures are time consuming. BioContext, for instance, an integrated text mining system for large-scale extraction and contextualization of biomolecular events, took nearly 3 months to complete a full run of the system, which analyzed 20 million MEDLINE abstracts and several hundred thousand PMC open access full-texts using 100 concurrent processes [16]. In addition, some text mining tools, like GNormPlus [17] require a substantial amount of memory ($\geq$5 GB), due to the necessity of loading a large gene dictionary and complementary data structures. Consequently, commodity servers cannot fulfil the computation and storage demands of large-scale text mining. Cloud-based solutions in Map-Reduce mode can partially fulfil computational resource demands. However, practically speaking, many text mining components were written in different languages, and they are dependent on a complex collection of third-party libraries, which prevents them from being readily transplanted into a high-level framework, like Hadoop and Spark. In addition, we dived into the details of load balancing, which cannot be readily supported by Map-Reduce.

An alternative solution to address this computational challenge is to harness the power of high performance computers. High performance computers (HPC) like Tianhe-2 [18] represent high-end computing infrastructures that have traditionally been used to solve compute-intensive scientific and engineering problems. The system configuration of Tianhe-2 is listed in Table 1.

Table 1. The system configuration of Tianhe-2.

Items	Content
Manufacturer	NUDT
Cores	3,120,000
Memory	1,024,000 GB
CPU	Intel Xeon E5-2692v2 12 C 2.2 GHz
Interconnect	TH Express-2
Linpack Performance(Rmax)	33,862.7 TFlop/s
Theoretical Peak(Rpeak)	54,902.4 TFlop/s
HPCG [TFlop/s]	580.109
Operating System	Kylin Linux
MPI	MPICH2 with a customized GLEX channel

Although the software stack on Tianhe-2 is designed and optimized for compute-intensive tasks, its high-performance architecture does provide the capability and capacity of big data processing. Nonetheless, to employ Tianhe-2 for big data processing is not a trivial task, which requires expert knowledge of the system architecture and parallel programming. The programming model is MPI-based (message passing interface) [18], which adds an extra dimension of complexity to normal programming languages like C/C++, Python, Java, etc. Most existing text mining tools are implemented without parallel processing. Therefore, it is necessary to develop an enabling framework that can support parallel text mining without the need to rewrite the original code. In this paper, we develop a parallel processing framework for text mining on the Tianhe-2 supercomputer. The framework integrates text mining tools as plugins. It unifies the input–output stream, implements the parallel processing across multiple compute nodes using the MPI model, and it applies a carefully crafted load balancing strategy to improve the parallelization efficiency. Without a loss of generality, we demonstrate the effectiveness of our framework using multiple NER tools as the demonstration plugins, which can recognize genes, mutations and diseases appearing in biomedical literature. More sophisticated tools of biomedical text mining can be readily integrated into the framework. In the remaining of this paper, we will introduce how paraBTM works and evaluate its performance on Tianhe-2.

2. Results and Discussion

To verify the effectiveness of the parallel framework, we constructed a corpus named 60K, which consists of 60,000 randomly selected articles from PubMed. For NER plugins, we chose three state-of-the-art tools (GNormPlus [17], tmVar [2], DNorm [7]), developed by NCBI (National Centre for Biotechnology Information). We measured the performance in terms of the total processing time and the average processing time (across all processes), and the total time includes the time of initialization and the actual processing time of different plugins.

The 60 K corpus is presented in the NXML format, which is a standard format provided by NCBI. Titles, abstracts, and full-texts from NXML files are extracted and re-written in the PubTator format. All input and output files processed by paraBTM should follow the PubTator format and the PubTator format starts with:

<PMID> | t | <Title of the paper>
<PMID> | a | <The rest of the paper>

The output file will be appended with annotated information like named entities followed in a tab-separated way.

A basic fact is that the time overhead of text mining is not proportional to the number of input articles. We verified this via a single process run over several groups of randomly selected articles. The result is depicted in Figure 1. Here, different colors represent different processing plugins. Related numbers are also listed in Table 2.

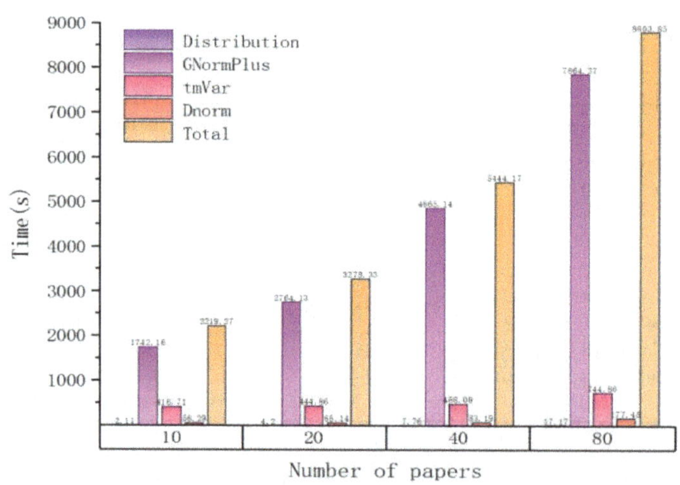

Figure 1. The time cost of processing different numbers of input articles in serial.

Table 2. Time distribution of processing different numbers of input articles in serial.

Number of Papers	Time(s)				
	Distribution	GNormPlus	tmVar	Dnorm	Total
10	2.11	1742.16	416.71	58.29	2219.27
20	4.2	2764.13	444.86	65.14	3278.33
40	7.76	4865.14	488.08	83.19	5444.17
80	17.17	7864.37	744.86	177.45	8803.85

As the number of input files increases, the time cost also increases but not linearly. For example, when the number of input articles is equal to 10, it takes about 36 min for tagging entities, and when the number of articles increases to 100, the spent time is about 3 h (180 min). This can be attributed to another important observation, that is, the total processing time is approximately proportional to the total input size (sum of file lengths as measured by number of characters), which is illustrated in

Figure 2 (size unit is MB, mega-bytes) and Table 3. The workload of each plugin can be better estimated by the total length of input files, which is the basis for our load balancing strategy in the following part.

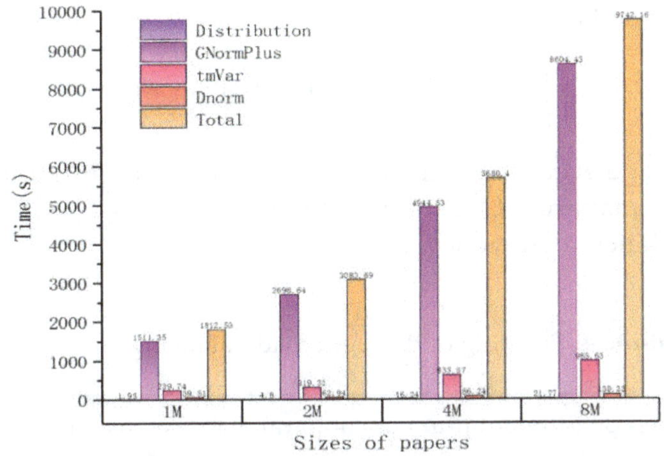

Figure 2. The time cost of processing different input sizes in serial.

Table 3. Time cost distribution of processing different input sizes in serial.

Sizes of Papers (M)	Time (s)				
	Distribution	GNormPlus	tmVar	Dnorm	Total
1	1.93	1511.35	239.74	59.51	1812.53
2	4.80	2696.64	319.31	62.94	3083.69
4	16.24	4944.35	633.57	86.24	5680.40
8	21.77	8604.43	985.63	130.33	9742.16

Figure 3 shows the time spent on paraBTM processing with different numbers of parallel processes on an input dataset of 16 MBs (including 175 articles) which is composed of articles randomly selected from the 60 K corpus. Parallel processing greatly reduces the processing time and different load-balancing strategies do affect the parallel efficiency. paraBTM costs about 500 s (under the Short-Board balancing strategy) when 64 processes are employed, which is around 1/16 the processing time of 2 processes. To note, each process needs to carry out initialization for every plugin, which means you cannot reduce the total processing time any further if the initialization time cost becomes the majority part.

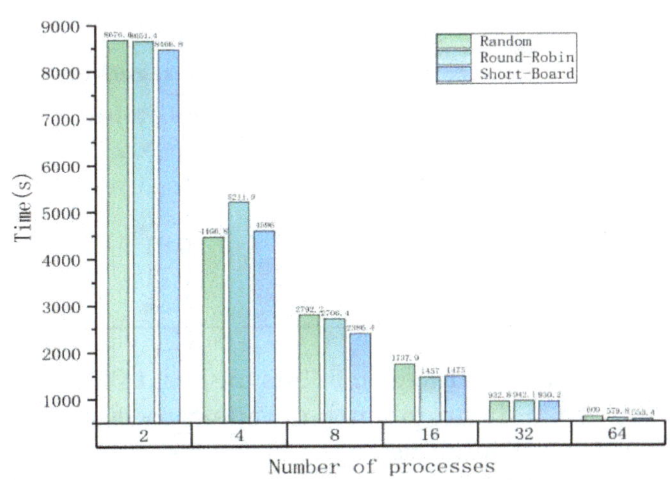

Figure 3. Effects of different load balancing strategies.

To profile different load balancing strategies, we summarize their effects under different parallel scales, as listed in Table 4. In all 6 test cases, the Short-Board strategy is the best in 4 cases and 2nd best in 2 remaining cases. We employ the load balancing efficiency (LBE) to quantify the effects of different strategies. Here, LBE is defined as:

$$LBE = AET/MET \tag{1}$$

Here, AET is the average execution time and MET is the maximum execution time. According to the above definition, the maxima of LBE is 1 (achieved if AET is equal to MET) and a greater LBE represents a better load balancing efficiency.

Table 4. Profiling of different load balancing strategies.

(a) Maximum times on different numbers of parallel processes with different strategies.

Number of Processes	Maximum Time among All Processes (s)		
	Random	Round-Robin	Short-Board
2	8676.60	8651.42	8466.89
4	4466.81	5211.96	4596.09
8	2792.29	2706.41	2386.45
16	1737.94	1457.01	1475.07
32	932.85	942.10	930.25
64	609.06	579.8	553.42

(b) Average times on different numbers of parallel processes with different strategies.

Number of Processes	Average Time of All Processes (s)		
	Random	Round-Robin	Short-Board
2	8513.10	8374.08	8392.46
4	4001.38	4535.42	4393.07
8	2287.19	2354.14	2242.42
16	1265.46	1174.12	1270.33
32	624.73	666.91	640.63
64	379.08	376.06	366.84

(c) Load balancing efficiencies on different numbers of parallel processes with different strategies.

Number of Processes	Efficiency (Average/Maximum)		
	Random	Round-Robin	Short-Board
2	0.98	0.97	0.99
4	0.90	0.87	0.96
8	0.82	0.87	0.94
16	0.73	0.81	0.86
32	0.67	0.71	0.69
64	0.62	0.65	0.66

Figure 4 shows that the Short-Board strategy exhibits the best LBE in almost all test cases. However, LBE values drop significantly when the number of parallel processes is greater than 16 in the 16M test set. The reason is that this test set contains only 175 articles, which means each process will only process two articles on average. If the input data set is big enough, the LBE will be maintained at a satisfactory level.

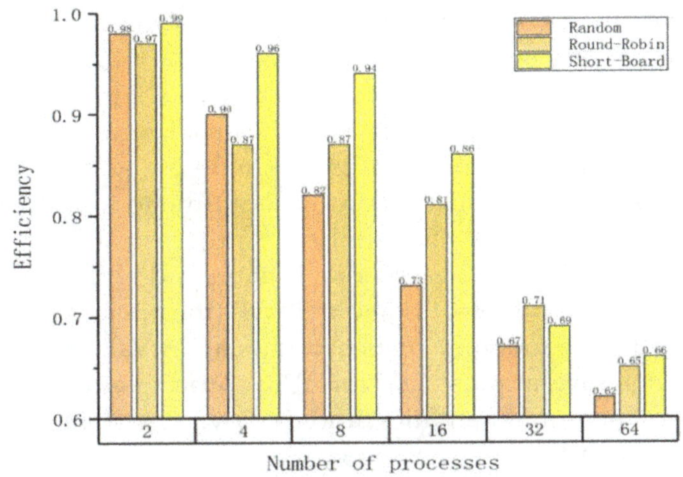

Figure 4. Load balancing efficiencies.

We also conducted an experiment on the whole 60 K corpus (61,078 articles). Table 5 shows that it took over 12 h to process 61,078 papers through three NER plugins (128 nodes under the Short-Board strategy, each node runs 5 processes). According to the results, we can see that parallelization greatly enhances the processing efficiency. To note, the speed-ups of different plugins differ as each plugin has its own characteristic computation and memory access patterns. To carry out a full-scale processing on the whole PMC-OA full-text dataset (over 1 million), it will take about 200 h if we only use 128 nodes. Fortunately, the computation capacity of Tianhe-2 is enormous, and we can reduce the total time down to several hours by harnessing the power of a few thousand nodes (over 16,000 available on Tianhe-2). We plan to carry out a full analysis on the whole PubMed dataset (the real large-scale biomedical texts) in the future. However, the cost of such a full run is currently beyond our funding support. We are currently in the application process of a bigger grant for this large-scale analysis. In our previous study, we have demonstrated that using text mining on a larger dataset does provide more comprehensive and insightful results compared with using a small dataset (say, can be handled by a few people) using thyroid cancer as a case study [19].

Table 5. The processing time of 61,078 papers running on 128 processes.

Number of Processes	Time (s)				
	Distribution	GNormPlus	tmVar2.0	Dnorm	Total
1	18,934.18	5,874,482.04	654,145.38	82,455.3	6,630,016.9
128	3643	23,733	16,214	233	43,823
Speed-up (x)	5.20	247.52	40.34	353.89	151.29

3. Materials and Methods

3.1. Data Sources and Storage

The biomedical literature has typical characteristics of large quantity, professional content, public resources, easy-accessibility, etc. Because of these characteristics, biomedical literature data has become one of the most noticeable data in biomedical field. For example, PubMed Central (PMC) is a free digital repository that archives publicly accessible articles. Until now, PMC has contained over 4.1 million references to full-text journal papers, covering a wide range of biomedical fields, and the literature data is stored in NXML format, from which we can extract some parts according to our interest.

However, most of the state-of-art NER tools do not support parallel processing, and it would take an enormous amount of time if we want to process the massive set of biomedical literature. One feasible solution is to harness the computing power provided by HPC systems by implementing a parallel NER processing framework. With this framework, text mining tools can be easily integrated into the framework and developers will not need to consider the details of parallel processing.

There are different levels of parallelism in text mining tasks. First, each input article is relatively independent; secondly, multiple sentences in each of the articles can be approximately regarded as independent. However, in practice, we usually use a single file as a processing unit, the reason is that many text mining tools spend a substantial amount of time to initialize on each processing pass. In addition, the memory size also limits the number of processes that can run in parallel on each computing node. For instance, on Tianhe-2 each node is equipped with 24 cores and 64 GB of memory, and the stable memory that users can control is about 50 GB (the operating system and other necessary tools need to use about 10 GB). The memory costs of a typical TmVar and gnormplus run for NER can be up to 5 GB and 10 GB. Therefore, at most 5 GNormPlus processes and 10 TmVar processes can run on one node. Figure 5 shows the implementation and deployment of a text mining system (paraBTM) in large-scale parallel environment.

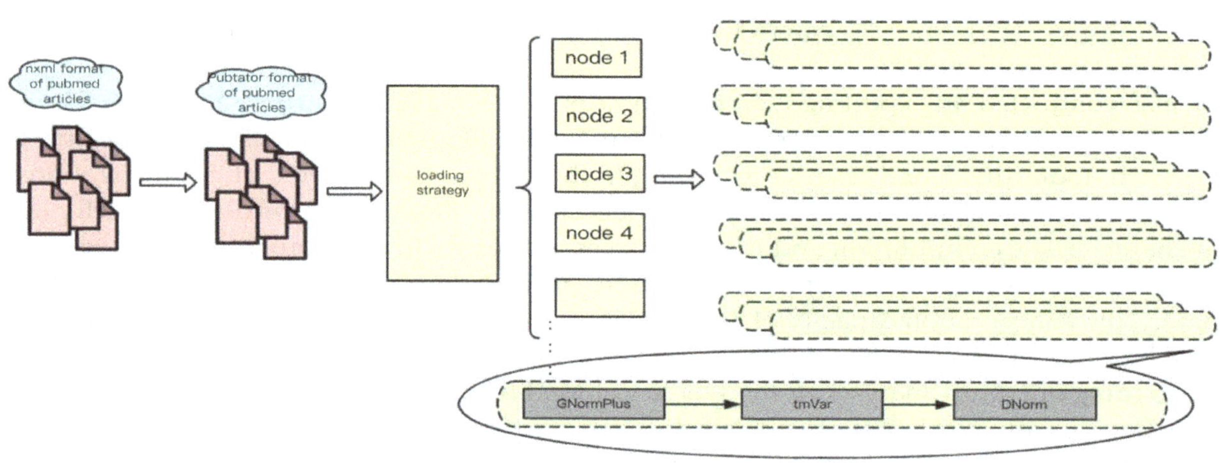

Figure 5. The implementation and deployment of paraBTM in large-scale parallel environment.

3.2. Parallel Processing

3.2.1. MPI-Based Multi-Node Computation

The message passing interface (MPI) is a standard model for parallel programming on HPCs. It is well established over 20 years, and has been implemented in different sorts of programming languages including C/C++ and Python. Our method can run on any supercomputer or cluster configured with MPI support. To note, different supercomputers might have different node configurations. When running on other platforms, the configuration (RAM, number of concurrent processes) might have to change accordingly.

In this work, we use MPI4PY (http://mpi4py.scipy.org/docs/) to implement the parallel processing. MPI4PY is a well-regarded, clean, and efficient implementation of MPI for Python. Our framework can simultaneously submit many jobs to cores distributed across computing nodes in Tianhe-2.

3.2.2. Load Balancing Strategy

A typical challenge in parallel computation is load unbalance, that is, workload is unevenly distributed among nodes, making some nodes very busy for a long time and others idle [20]. In this paper, we address this problem by designing an effective load balancing strategy.

Given a set F of files to be processed $F_0, F_1, F_2 \ldots F_{N-1}, F_n$, we initialize processes $P_0, P_1, P_2, \ldots\ldots, P_{n-1}, P_n$, and the number of processes is N, the problem is to allocate each file F_x to an appropriate process P_{rank}.

A naive solution is to randomly distribute target files into nodes. We can simply distribute files to by modulo operation rank $= \text{mod}(P, \text{size})$, P and size represent the position of the target file in the file list and number of processes respectively, and file F_p finally should be sent to P_{rank}. According to the formula, each file F to be processed is distributed to process $rank = P \text{ \% } size$ in turn. As the files are arranged in a random order, this process is actually a simulation of random distribution. This is a naïve strategy and easy to implement. However, this strategy does not take into consideration the length of each file, and will very likely cause an unbalanced load distribution, which would detriment the overall parallel efficiency. For instance, if the total length of files assigned to one specific node is far larger than others, then the overall running time will be prolonged until this slowest node finishes. Figure 6 shows an example of the naïve random load balancing strategy.

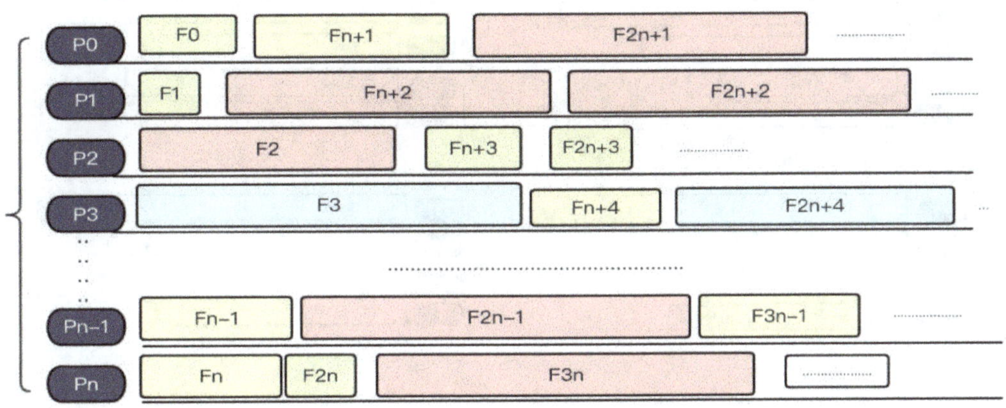

Figure 6. Random load balancing strategy.

A slightly more complex load balancing method is the round-robin (RR) method. Round-robin algorithm is a term that originally comes from the field of operating systems. Here, the general idea inspires us to mix small files with large files together into one process. After sorting files by size (see Figure 7), the system will assign files into processes in a snakelike way, making the size of files loaded in every process remains relatively balanced. Figure 8 shows an example of RR algorithm.

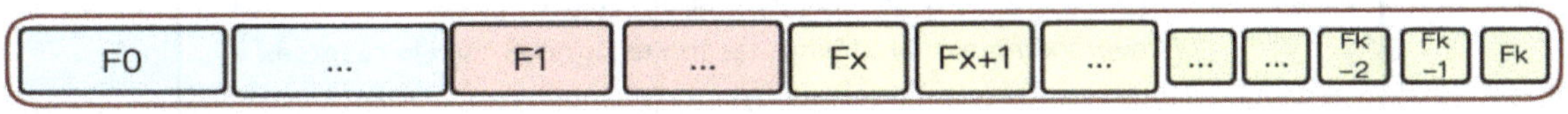

Figure 7. A sorted file list.

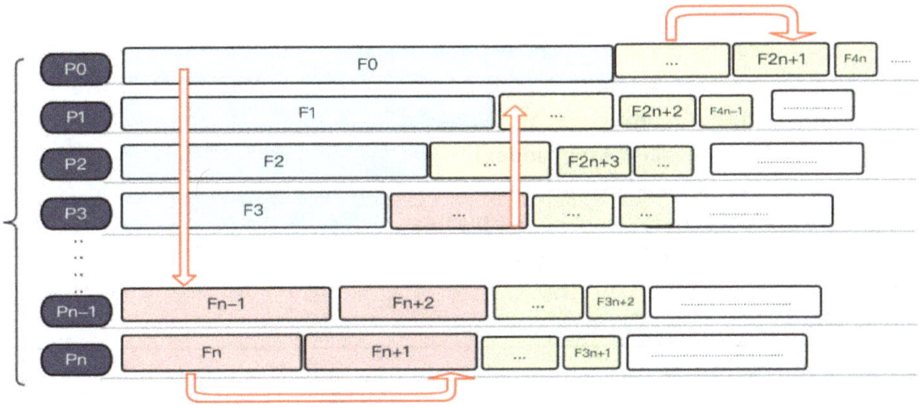

Figure 8. Round-robin algorithm.

The round-robin method also allocates the same number of files, and its serpentine way of load assignment ensures that the total size of the files in each process remains relatively balanced, since files were sorted by size in advance. However, in some circumstances, the lengths of input articles can be very biased, say, some files are extremely long while many others are short. In such cases, the RR method fails.

Instead of assignments based on the number of files, we proposed our "Short-Board" method. Firstly, the files that need to be processed are sorted in descending order according to the length of each file, and then files that need to be processed in the file list are sequentially fetched out and dispatched to the process whose current load is the smallest. Figure 9a–d shows an example of Short-Board algorithm. The pseudo code of Short-Board is shown in Figure 10.

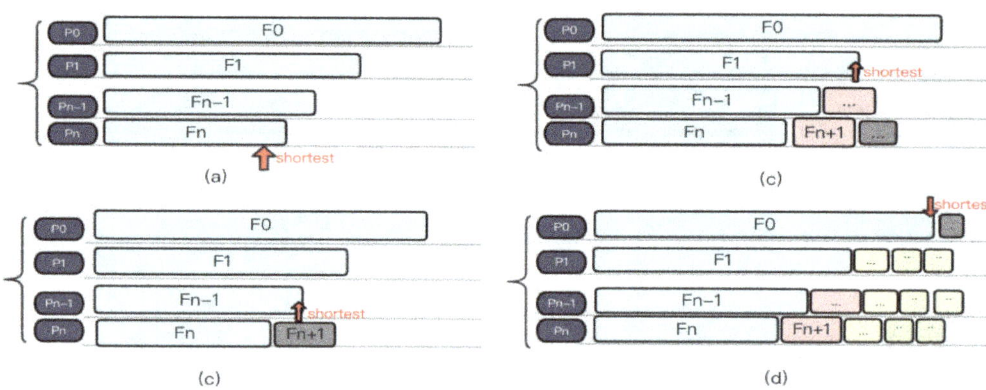

Figure 9. A demonstration of the Short-Board load balancing algorithm.

```
Loading Strategy: Short-Board

Input: sorted file information list by size in a descending order Flist
 number of process N

Initialization:ProcessList(a list consist of processinfo)
              ProcessInfo[index, capacity, subfilelist]
              (information for each process, including the index in the
processlist, the capacity, and the assigned subfile list )
              ReverseFlag(signal of whether the assignment list is reverse )

/* start the Short-Board Strategy */
for each file ∈ Flist do

          /* find the process which has smallest capacity */
          AssignIndex = FindMinCap(ProcessList)
          Use the AssignIndex to find the corresponding ProcessInfo in the
     ProcessList P[index, capacity, subfilelist]]
          P=ProcessList[AssignIndex]
          P[index]=AssignIndex
          P[capacity]=P[capacity] + filesize
          P[subfilelist] =P[subfilelist] ∪ fileindex

end for

Output: ProcessList
```

Figure 10. The pseudo code of Short-Board.

4. Conclusions

In this paper, we present paraBTM, a parallel framework for biomedical text mining developed on the Tianhe-2 supercomputer. It supports different types of components as plugins and its usage is straightforward. The parallel efficiency is guaranteed by a carefully devised load balancing strategy. We evaluated the performance of paraBTM on both small- and large-scale datasets. Experimental results validate that paraBTM effectively improve the processing speed of biomedical named entity recognition. On large scale of datasets, ParaBTM managed to process 60178 PubMed full-text articles in about 12 h. paraBTM is open-source and available at https://github.com/biotm/paraBTM.

Author Contributions: Y.X. and C.W. developed the algorithms and drafted the manuscript. They developed the codes of paraBTM together with X.Y. and W.W.; E.Z. and J.Y. proposed the idea of the work and revised the whole manuscript. All the authors have read and approve the manuscript.

Acknowledgments: Special thanks to Zhuo Song and Gen Li (from GeneTalks Ltd., Beijing) for providing knowledge support on genomic variations.

References

1. Rebholz-Schuhmann, D.; Oellrich, A.; Hoehndorf, R. Text-mining solutions for biomedical research: Enabling integrative biology. *Nat. Rev. Genet.* **2012**, *13*, 1–11. [CrossRef] [PubMed]
2. Wei, C.-H.; Harris, B.R.; Kao, H.-Y.; Lu, Z. tmVar: A text mining approach for extracting sequence variants in biomedical literature. *Bioinformatics* **2013**, *29*, 1433–1439. [CrossRef] [PubMed]
3. Thomas, P.; Rocktäschel, T.; Hakenberg, J.; Lichtblau, Y.; Leser, U. SETH detects and normalizes genetic variants in text. *Bioinformatics* **2016**, *32*, 2883–2885. [CrossRef] [PubMed]
4. Wei, C.-H.; Kao, H.-Y. Cross-species gene normalization by species inference. *BMC Bioinform.* **2011**, *12* (Suppl. 8), S5. [CrossRef] [PubMed]
5. Pan, X.; Shen, H.B. OUGENE: A disease associated over-expressed and under-expressed gene database. *Sci. Bull.* **2016**, *61*, 752–754. [CrossRef]
6. Leaman, R.; Lu, Z. TaggerOne: Joint named entity recognition and normalization with semi-Markov Models. *Bioinformatics* **2016**, *32*, 2839–2846. [CrossRef] [PubMed]
7. Leaman, R.; Islamaj Dogan, R.; Lu, Z. DNorm: Disease name normalization with pairwise learning to rank. *Bioinformatics* **2013**, *29*, 2909–2917. [CrossRef] [PubMed]
8. Quan, C.; Wang, M.; Ren, F. An unsupervised text mining method for relation extraction from biomedical literature. *PLoS ONE* **2014**, *9*, e102039. [CrossRef] [PubMed]
9. Xu, D.; Zhang, M.; Xie, Y.; Wang, F.; Chen, M.; Zhu, K.Q.; Wei, J. DTMiner: Identification of potential disease targets through biomedical literature mining. *Bioinformatics* **2016**, *32*, 3619–3626. [CrossRef] [PubMed]
10. Xu, R.; Wang, Q. A knowledge-driven conditional approach to extract pharmacogenomics specific drug-gene relationships from free text. *J. Biomed. Inform.* **2012**, *45*, 827–834. [CrossRef] [PubMed]
11. Percha, B.; Garten, Y.; Altman, R.B. Discovery and explanation of drug-drug interactions via text mining. *Pac. Symp. Biocomput.* **2012**, 410–421. [CrossRef]
12. Segura-Bedmar, I.; Martínez, P.; de Pablo-Sánchez, C. Using a shallow linguistic kernel for drug-drug interaction extraction. *J. Biomed. Inform.* **2011**, *44*, 789–804. [CrossRef] [PubMed]
13. Bravo, À.; Piñero, J.; Queralt-Rosinach, N.; Rautschka, M.; Furlong, L.I. Extraction of relations between genes and diseases from text and large-scale data analysis: Implications for translational research. *BMC Bioinform.* **2015**, *16*, 55. [CrossRef] [PubMed]
14. Pletscher-Frankild, S.; Pallejà, A.; Tsafou, K.; Binder, J.X.; Jensen, L.J. DISEASES: Text mining and data integration of disease–gene associations. *Methods* **2015**, *74*, 83–89. [CrossRef] [PubMed]
15. Franceschini, A.; Szklarczyk, D.; Frankild, S.; Kuhn, M.; Simonovic, M.; Roth, A.; Lin, J.; Minguez, P.; Bork, P.; von Mering, C.; et al. STRING v9.1: Protein-protein interaction networks, with increased coverage and integration. *Nucleic Acids Res.* **2013**, *41*, 808–815. [CrossRef] [PubMed]
16. Gerner, M.; Sarafraz, F.; Nenadic, G.; Bergman, C.M. BioContext: An integrated text mining system for large-scale extraction and contextualisation of biomolecular events. *Bioinformatics* **2012**, *28*, 2154–2161. [CrossRef] [PubMed]
17. Wei, C.-H.; Kao, H.-Y.; Lu, Z. GNormPlus: An Integrative Approach for Tagging Genes, Gene Families, and Protein Domains. *BioMed Res. Int.* **2015**, *2015*, 918710–918717. [CrossRef] [PubMed]

18. Liao, X.; Xiao, L.; Yang, C.; Lu, Y. MilkyWay-2 supercomputer: System and application. *Front. Comput. Sci.* **2014**, *8*, 345–356. [CrossRef]
19. Wu, C.; Schwartz, J.M.; Brabant, G.; Nenadic, G. Molecular profiling of thyroid cancer subtypes using large-scale text mining. *BMC Med. Genom.* **2014**, *7* (Suppl. 3), S3. [CrossRef] [PubMed]
20. Kaur, S.; Kaur, G. A Review of Load Balancing Strategies for Distributed Systems. *IJCA* **2015**, *121*, 45–47. [CrossRef]

11

A Resolution-Free Parallel Algorithm for Image Edge Detection within the Framework of Enzymatic Numerical P Systems

Jianying Yuan [1,2,3], Dequan Guo [1,2,3], Gexiang Zhang [4,5,*], Prithwineel Paul [5], Ming Zhu [2] and Qiang Yang [4]

1 The Postdoctoral Station at Xihua University Based on Collaboration Innovation Center of Sichuan Automotive Key Parts, Xihua University, Chengdu 610039, China; yuanjy@cuit.edu.cn (J.Y.); guodq@cuit.edu.cn (D.G.)
2 School of Control Engineering, Chengdu University of Information Technology, Chengdu 610225, China; zhuming@126.com
3 School of Aeronautics and Astronautics, University of Electronic Science and Technology, Chengdu 610054, China
4 Robotics Research Center, Xihua University, Chengdu 610039, China; qiangychd@126.com
5 School of Electrical Engineering, Southwest Jiaotong University, Chengdu 610031, China; prithwineelpaul@gmail.com
* Correspondence: zhgxdylan@126.com.

Abstract: Image edge detection is a fundamental problem in image processing and computer vision, particularly in the area of feature extraction. However, the time complexity increases squarely with the increase of image resolution in conventional serial computing mode. This results in being unbearably time consuming when dealing with a large amount of image data. In this paper, a novel resolution free parallel implementation algorithm for gradient based edge detection, namely EDENP, is proposed. The key point of our method is the introduction of an enzymatic numerical P system (ENPS) to design the parallel computing algorithm for image processing for the first time. The proposed algorithm is based on a cell-like P system with a nested membrane structure containing four membranes. The start and stop of the system is controlled by the variables in the skin membrane. The calculation of edge detection is performed in the inner three membranes in a parallel way. The performance and efficiency of this algorithm are evaluated on the CUDA platform. The main advantage of EDENP is that the time complexity of $O(1)$ can be achieved regardless of image resolution theoretically.

Keywords: membrane computing; edge detection; enzymatic numerical P system; resolution free

---

1. Introduction

In recent decades, image processing technology has experienced dramatic growth and widespread applications. Nearly no area escapes impact in some way by digital image processing. Normally, digital image processing includes three main levels, i.e., low-level, mid-level and high-level processing [1]. As one of the most basic operators in low-level image processing, edge detection can preserve the important structural properties of an image while significantly reducing the amount of data. This excellent property makes it a basic tool for many high-level image processing algorithms and is extensively applied in target tracking [2], image compression [3], and object recognition [4]. An edge can be defined as points in a digital image at which the image brightness changes sharply or has discontinuities. This phenomenon may be caused by depth discontinuous, illumination changes, or intrinsic texture properties of objects. In various edge detection algorithms, the gradient based method is a type of classic edge detection approach with the merit of simple theory and good performance. However, as convolution calculation (i.e., a classic neighborhood computing in image processing) [5] is involved in this kind of algorithm, the time complexity increases squarely with the increase of image resolution. So it is difficult to deal with images with large resolution, such as remote sensing images, medical images, etc., in real time processing.

In order to achieve real-time calculation of high resolution images, many researchers have put much effort into this problem and several methods have been proposed. Generally, there are two main categories of resolutions. The first type of resolution concerns computational algorithms. In this kind of method, an elaboratively computational algorithm is usually designed to reduce the computational complexity. For the template matching problem, integral image [6] and dual-bound algorithm [7] are two classical approaches to speed up the computation. In [8], Fast LDA feature extraction is present, where steepest descent and conjugate direction methods are combined to optimize the step size in each iteration. In [9], common orthogonal basis extraction is proposed to extract a common basis of collection of matrices. The second category is based on hardware with parallel architecture, such as Graphics Processing Unit (GPU) [10–12] and Field Programmable Gate Array (FPGA) [13,14]. GPU uses hundreds of parallel processor cores executing tens of thousands of parallel threads to rapidly solve large problems having substantial inherent parallelism. However, with the shrinking volume of chips, semiconductor technology begins to reach its physical limits, which means the performance of conventional computing technique based on silicon chip integrated circuit microprocessors will be difficult to improve further [15]. Under this background, some scholars have turned their attention to non-traditional computing, such as quantum computing [16], DNA computing [17] and membrane computing (MC) [18]. MC is a new active branch of natural computing that simulates the function and structure of living cells and tissues, abstracting their biochemical reactions and material exchanges [19]. One of the most prominent features of MC is its capability of generating exponential growth space over a polynomial time, which makes it a promising method to resolve the conflict between the ever-increasing amount of data in the image processing field and the backward computing power of conventional computer [20]. In recent years, image edge detection and image segmentation [21–24], image smoothing [25], obtaining homology groups of 2D images [26,27], counting cells [28], Enzymatic numerical P systems image thinning [29] and corner detection [30] in MC framework have been vividly studied. In the previous literature about MC and image processing, much work is based on tissue-like P systems. However, when designing a parallel implementation program of an existing image processing algorithm, it is difficult to realize the mathematical formula in "tissue-like P systems language". The reasons for this are as follows. First, the data type of an image is an integer between 0 and 255. When design image processing algorithm uses tissue-like P systems, the image data should be coded to symbolic variables and those symbolic variables need to be decoded to integer for display as the algorithm finished. Second, most image processing algorithms are composed of several steps in determined logical order, which means variables in the membrane system need to be calculated in a deterministic way, rather than in a random manner. Since the rules in tissue-like P systems are implemented randomly, it is difficult to control the execution orders of different rules.

In order to overcome the above shortcomings when tissue-like P systems are combined to image processing, we make the first attempt to introduce enzymatic numerical P system (ENPS) to image processing. Concretely, a parallel algorithm for gradient based edge detection algorithm is designed and tested in the framework of ENPS. Besides the features described in [25], ENPS has another two good properties which make it particularly appropriate for image processing. One is that numerical variables and numerical expressions can be used directly in ENPS. Thus, image data can be directly operated without the additional encoding and decoding process. Another important characteristic is that enzymatic variables can control the execution orders of multiple rules in ENPS, i.e., the algorithms with complex logical steps can be designed easily.

The main contribution of this paper is that a parallel algorithm for image edge detection in the framework of ENPS, namely EDENP, is designed. The significant advantage of EDENP is that it can achieve the time complexity of $O(1)$ theoretically, no matter how large the image resolution is.

Moreover, the performance is equivalent to the performance run on the serial computing platform. This is very important for real projects, because most of the classical image processing algorithms have been widely proven to be effective in practical engineering, so the designed parallel implementation algorithm can be directly applied to the real image processing project without the need to perform large-scale testing. To the best of our knowledge, it is the first time to bridge problems from image processing with ENPS.

The rest of this paper is structured as follows. Section 2 introduces the definition, characteristics and applications of MC and ENPS. The problem statement is elaborated in Section 3. Section 4 discusses the EDENP algorithm in detail. The experiments and results are presented in Section 5. Conclusions are drawn in Section 6.

2. MC and ENPS

MC is a young biocomputing model proposed by Gh.Păun in 2000 [19]. The computational devices in MC are called P systems. Generally, a P system includes three ingredients: (i) the membrane structure; (ii) multisets of objects; (iii) rules of a bio-chemical inspiration. The multisets of objects are placed in the membrane, and evolved according to given rules which are usually applied in a synchronous non-deterministic maximally parallel manner. Since being proposed, MC has received great attention from scientists in many fields [31–33]. In the past 20 years, both the theory [32,34–37] and application [31,38–41] of MC have been greatly developed, and many different classes of P systems have been investigated. According to the way in which membranes are structured, there are three major types of P systems, i.e., cell-like [19], tissue-like [42] and spiking neural P systems [43,44]. Enzymatic numerical P system comes from numerical P system (NPS). NPS is a new special research branch of cell-like P systems, proposed by the founder of MC, Gh.Păun in 2006 [45]. In NPS, multisets of objects associated to membranes are sets of numerical variables, and the evolutionary rules are composed of a production function and a repartition protocol [46–48]. The most common widely application area of NPS is robot controller design [49–52]. Although NPS can deal with numerical variables, it can only execute one production function per membrane at a time. When there are multiple production functions per membrane, one is selected randomly. This limits its application in some situations where the rules should be executed deterministically. In order to solve this problem and expand the application of NPS, ENPS is put forward [24]. It is extended from NPS by introducing enzyme-like variables which can make rules run deterministically [53]. The standard form of ENPS is defined as follows:

$$\Pi = \Big(m, H, \mu, (Var_1, E_1, \mathrm{Pr}_1, Var_1(0)), \ldots, (Var_m, E_m, \mathrm{Pr}_m, Var_m(0))\Big).$$

where:

1. m is the number of membranes used.
2. H is an alphabet that contains m symbols, and $H = \{1, 2, \ldots, m\}$.
3. μ is the membrane structure.
4. Var_i is the set of variables from membrane i and $Var_i(0)$ are the initial values for these variables.
5. Pr_i is the set of rules in membrane i, composed of a production function and a repartition protocol. A typical rule is as follows.

$$\mathrm{F}_{l,i}(y_{1,i}, \ldots, y_{k_l,i})|_{e_{j,i}} \rightarrow c_{l,1}|v_1 + c_{l,2}|v_2 + \ldots + c_{l,n_i}|v_{n_i},$$

where $e_{j,i}$ is a variable from Var_i different from $y_{1,i}, \ldots, y_{k_l,i}$ and $v_1, v_2, \ldots, v_{n_i}$. The rule can be executed at a time t only if $e_{j,i} > \min\{y_{1,i}(t), y_{2,i}(t), \ldots y_{k_l,i}(t)\}$. From the definition of ENPS, it is clear that with enzymes-like variables, the system can control multiple production functions to run in parallel in the same membrane deterministically [54]. Hence, it can overcome the disadvantages of traditional NPS that only run one rule nondeterministically at a time in a membrane. The ENPS with deterministic, parallel execution model has already been proved to be Turing universal [55,56]. In [57], it is shown that any ENPS working in all-parallel mode or one parallel model can be simulated by an equivalent one-membrane ENPS working in the same mode. Since the proposal of ENPS, this model has been successfully applied in a wide range of domains, such as robot control [58], big data field [59], and sequential minimal optimization [60] fields. In this paper, ENPS is used to solve the problem of gradient based image edge detection.

3. Problem Statement

Edges generally occur in areas where the brightness of the image changes dramatically. These changes can be described by image gradients. Usually, a pair of convolution masks are used to estimate the gradients in the x and y directions, respectively, as shown in Equations (1)–(3), where $(Sobel_x, Sobel_y)$, $(Prew_x, Prew_y)$, (Rob_x, Rob_y) are three classic pairs of convolution masks. In this paper, we take Sobel operator as an example of gradient based edge detection (GBED). When the masks are sliding over the image, a square of pixels are operated. Then both directional gradients and absolute gradient magnitudes of image are computed, as shown in Equations (4) and (5), where I is the image, (g_x, g_y) are gradients in x and y direction respectively, $g_{i,j}$ is the absolute gradient magnitude of a pixel with coordinate $(i, j), 2 \leq i, j \leq n-1$ for image with resolution of $n \times n$.

$$Sobel_x = \begin{bmatrix} -1 & 0 & 1 \\ -2 & 0 & 2 \\ -1 & 0 & 1 \end{bmatrix}; \quad Sobel_y = \begin{bmatrix} 1 & 2 & 1 \\ 0 & 0 & 0 \\ -1 & -2 & -1 \end{bmatrix} \tag{1}$$

$$Prew_x = \begin{bmatrix} -1 & 0 & 1 \\ -1 & 0 & 1 \\ -1 & 0 & 1 \end{bmatrix}; \quad Prew_y = \begin{bmatrix} 1 & 1 & 1 \\ 0 & 0 & 0 \\ -1 & -1 & -1 \end{bmatrix} \tag{2}$$

$$Rob_x = \begin{bmatrix} -1 & 0 \\ 0 & 1 \end{bmatrix}; \quad Rob_y = \begin{bmatrix} 0 & -1 \\ 1 & 0 \end{bmatrix} \tag{3}$$

$$g_x = Sobel_x * I; \quad g_y = Sobel_y * I; \tag{4}$$

$$g_{i,j} = \sqrt{{g_{x_{i,j}}}^2 + {g_{y_{i,j}}}^2} \tag{5}$$

When the gradient magnitude $g_{i,j}$ is computed, the difference between it and a predefined threshold θ is used to judge whether this pixel is an edge pixel or not, as presented in Equation (6), where $d_{i,j}$ is the difference. More concretely, if $d_{i,j}$ is greater than or equal to 0, then the pixel is assumed as an edge point, otherwise, it is a background point, as shown in Equation (7). It is worth noting that in real application, before thresholding, the gradient image should be filtered by "non-maximum suppression" for getting more real edges. In this paper, in order to simplify the algorithm, this step is ignored.

$$d_{i,j} = g_{i,j} - \theta \tag{6}$$

$$edg_{i,j} = \{ \begin{array}{ll} 1 & \text{if } (d_{i,j} \geq 0) \\ 0 & \text{if } (d_{i,j} < 0) \end{array} \tag{7}$$

The program pseudo code of GBED run on conventional serial computer is illustrated in Algorithm 1, where the initial value of $edg_{i,j}$ is set to 0. From Algorithm 1, it can be deduced that the computational complexity is $O(n^2)$ because two loops are involved. When n becomes large, the calculations are very time-consuming under the serial computing platform.

Algorithm 1 The pseudo code of GBED

Input: $I(n * n)$
Output: $edg(n * n)$

for $i = 2 : n - 2$ **do**
  for $j = 2 : n - 1$ **do**
    Computing $g_{x_{i,j}}$
    Computing $g_{y_{i,j}}$
    Computing $g_{i,j}$
    Computing $d_{i,j}$
    Computing $edg_{i,j}$
  end for
end for

In order to reduce the calculation time complexity, we attempt to introduce an enzymatic numerical P system to design a high parallel computing algorithm for edge detection. The details of how to design the algorithm will be given in the next section.

4. The EDENP Algorithm

This section starts with the mathematical model of EDENP followed by the detailed description of EDENP. The execution process and resources needed are discussed lastly.

4.1. Mathematical Model of EDENP

From Section 3, we know that the GBED algorithm contains four steps for a certain pixel in an image. In EDENP, the four steps will be executed in a cell-like P system under the control of enzyme variables, as illustrated in Figure 1. The initialization of variables, start and stop of the system will be controlled in the skin membrane. The directional gradients estimation will be completed in membrane 1. The absolute gradient magnitude estimation will take place in membrane 2. Membrane 3 is responsible for computing the image edge. The corresponding membrane structure is illustrated in Figure 2.

The mathematical expression of EDENP is as follows, and

$$\Pi = \Big(m, H, \mu, (Var_1, E_1, \Pr_1, Var_1(0)), ..., (Var_4, E_4, \Pr_4, Var_4(0))\Big),$$

where

1. $m = 4$.
2. $H = \{1, 2, 3, 4\}$.
3. $u = [[[[\ ]_1]_2]_3]_4$.
4. $Var_1 = \{g_{x_{i,j}}, g_{y_{i,j}}\}$, $Var_2 = g_{i,j}$, $Var_3 = \{ed_1, ed_2, ed_3, E_{i,j}, E_{D_{i,j}}\}$, $Var_4 = \{x_{i,j}, edg_{i,j}, \theta, e_{1,1}, E_D\}$.

   $x_{i,j} (1 \leq i, j \leq n)$, are the gray value of pixel with coordinate of (i, j) on the source image plane.

   $edg_{i,j} (1 \leq i, j \leq n)$, are the corresponding edge points of the source image with initial value 0.

   $\theta[threshold]$, is a numerical variable which is used as the threshold value for edge detection, and the value of threshold should be predefined.

   $g_{x_{i,j}} (1 \leq i, j \leq n)$, are the horizontal derivative approximations at each pixel.

   $g_{y_{i,j}} (1 \leq i, j \leq n)$, are the vertical derivative approximations at each pixel.

   $g_{i,j} (1 \leq i, j \leq n)$, are the gradient magnitude approximations at each pixel.

$ed_1[0]$, is a numerical variable with initial value 0, which is used as the background value of the edge image.

$ed_2[1]$, is a numerical variable with initial value 1, which is used as the edge point value of the edge image.

$ed_3[-256]$, is a numerical variable with initial value -256, which is used as a intermediate variable.

5. E_k is a set of enzyme variables from membrane k, i.e., $E_1 = [\ ], E_2 = [\ ], E_3 = \{E_{i,j}, E_{D_{i,j}}\}, E_4 = \{e_{1,1}, E_D\}$.
6. Pr_k is the set of programs (rules) in membrane k, composed of a production function and a repartition protocol.

$\mathrm{Pr}_{1,CE_{i,j}}$:
$(|x_{i,j+2} + 2x_{i+1,j+2} + x_{i+2,j+2} - x_{i,j} - 2x_{i+1,j} - x_{i+2,j}|)|_{e_{1,1}} \rightarrow 1|g_{x_{i,j}}, (2 \leq i,j \leq n-2),$

$\mathrm{Pr}_{2,CE_{i,j}}$:
$(|x_{i,j} + 2x_{i,j+1} + x_{i,j+2} - x_{i+2,j} - 2x_{i+2,j+1} - x_{i+2,j+2}|)|_{e_{1,1}} \rightarrow 1|g_{y_{i,j}}, (2 \leq i,j \leq n-2),$

$\mathrm{Pr}_{3,CE_{1,i}} : 0|_{e_{1,1}} \rightarrow |g_{x_{1,i}}; (1 \leq i \leq n),$

$\mathrm{Pr}_{4,CE_{n,i}} : 0|_{e_{1,1}} \rightarrow |g_{x_{n,i}}; (1 \leq i \leq n),$

$\mathrm{Pr}_{5,CE_{i,1}} : 0|_{e_{1,1}} \rightarrow |g_{x_{i,1}}; (2 \leq i \leq n-1),$

$\mathrm{Pr}_{6,CE_{i,n}} : 0|_{e_{1,1}} \rightarrow |g_{x_{i,n}}; (2 \leq i \leq n-1),$

$\mathrm{Pr}_{7,CE_{1,i}} : 0|_{e_{1,1}} \rightarrow |g_{y_{1,i}}; (1 \leq i \leq n),$

$\mathrm{Pr}_{8,CE_{n,i}} : 0|_{e_{1,1}} \rightarrow |g_{y_{n,i}}; (1 \leq i \leq n),$

$\mathrm{Pr}_{9,CE_{i,1}} : 0|_{e_{1,1}} \rightarrow |g_{y_{i,1}}; (2 \leq i \leq n-1),$

$\mathrm{Pr}_{10,CE_{i,n}} : 0|_{e_{1,1}} \rightarrow |g_{y_{i,n}}; (2 \leq i \leq n-1).$
Those rules are used to execute Formula (1). The enzyme in $\mathrm{Pr}_{1,CE_{i,j}} \sim \mathrm{Pr}_{10,CE_{i,j}}$ must exist in enough amount so that the rules can be activated. Specifically, if the value of the enzyme $e_{1,1}$ is greater than variable $x_{i,j} (1 \leq i,j \leq n)$, then rules $\mathrm{Pr}_{1,CE_{i,j}} \sim \mathrm{Pr}_{10,CE_{i,j}}$ are effective. Since variable $x_{i,j}$ is the gray value of image, the maximum value is 255. So, the initial value of $e_{1,1}$ is set to 256, such that the condition modeled by rule $\mathrm{Pr}_{1,CE_{i,j}} \sim \mathrm{Pr}_{10,CE_{i,j}}$ are satisfied. It is important to note that the number of rules are $n \times n$, and all the rules are executed in parallel.

7. $\mathrm{Pr}_{21,CE_{i,j}} : (\sqrt{g^2_{x_{i,j}} + g^2_{y_{i,j}}})|e_{1,1} \rightarrow 1|g_{i,j}; 1 \leq i,j \leq n$
$\mathrm{Pr}_{21,CE_{i,j}}$ are the rules which are executed by Formula (5). Hence, after executing $\mathrm{Pr}_{1,CE_{i,j}} \sim \mathrm{Pr}_{10,CE_{i,j}}$, the value of the variables $g_{x_{i,j}}, g_{y_{i,j}}$ are obtained. The maximum value of $g_{x_{i,j}}$ and $g_{y_{i,j}}$ is 255, and the enzyme $e_{1,1}$ is 256. So the condition of execution for rules $\mathrm{Pr}_{21,CE_{i,j}}$ is satisfied. Hence, all $n \times n$ rules are executed concurrently.

8. $\mathrm{Pr}_{31,CE_{i,j}} : (2*(g_{i,j} - \theta))| \rightarrow 1|\mathrm{g}_{i,j}+1|\mathrm{E}_{i,j}; 1 \leq i,j \leq n$
$\mathrm{Pr}_{31,CE_{i,j}}$ are the rules which compute $d_{i,j}$ in Formula (6). After executing $\mathrm{Pr}_{31,CE_{i,j}}$, the value of $d_{i,j}$ are obtained, which is equal to variables $g_{i,j}$ and $E_{i,j}$ in rule $\mathrm{Pr}_{31,CE_{i,j}}$.

9. $\mathrm{Pr}_{32,CE_{i,j}} : (ed_1 + 2 * ed_2)|_{\mathrm{E}_{i,j}} \rightarrow 1|\mathrm{edg}_{i,j} + 1|E_{D_{i,j}}$; $\mathrm{Pr}_{33,CE_{i,j}} : (0 * ed_1 + 0 * ed_3)|_{\mathrm{E}_{i,j}} \rightarrow 1|\mathrm{edg}_{i,j} + 1|E_{D_{i,j}}; 1 \leq i,j \leq n.$
$\mathrm{Pr}_{32,CE_{i,j}}$ and $\mathrm{Pr}_{33,CE_{i,j}}$ are rules for computing edge value as Formula (7). If $\mathrm{E}_{i,j}$ is greater than or equal to 0, then $\mathrm{Pr}_{32,CE_{i,j}}$ and $\mathrm{Pr}_{33,CE_{i,j}}$ are executed. Because ed_1 is 0, and ed_3 is -256, so $\mathrm{E}_{i,j} \geq \min(ed_1, ed_2)$ and $\mathrm{E}_{i,j} \geq \min(ed_1, ed_3)$. The execution condition of $\mathrm{Pr}_{32,CE_{i,j}}$ and $\mathrm{Pr}_{33,CE_{i,j}}$ is satisfied. If $\mathrm{d}_{i,j} < 0$, only $\mathrm{Pr}_{33,CE_{i,j}}$ will be executed. Because $\mathrm{E}_{i,j} \geq \min(ed_1, ed_3)$ and $\mathrm{E}_{i,j} < \min(ed_1, ed_2)$, only the execution condition of $\mathrm{Pr}_{33,CE_{i,j}}$ can be satisfied. After executing $\mathrm{Pr}_{32,CE_{i,j}}$ and $\mathrm{Pr}_{33,CE_{i,j}}$, variables $edg_{i,j}$ will be set to 1 if $\mathrm{d}_{i,j} \geq 0$ and every variable $E_{D_{i,j}}$ will be assigned.

10. $Pr_{main} : (0 * E_{D_{1,1}} + 0 * E_{D_{1,2}} + \ldots + 0 * E_{D_{m,n}} + 1)| \rightarrow 1|E_D$
Pr_{main} is a rule contained in membrane 4, which controls the stop condition of the P system. For pixel (i, j), if all the enzyme variables $E_{D_{i,j}}$ are assigned, the condition for Pr_{main} is meet. Enzyme variable E_D is set to 1 by rule Pr_{main}, and the system stops running.

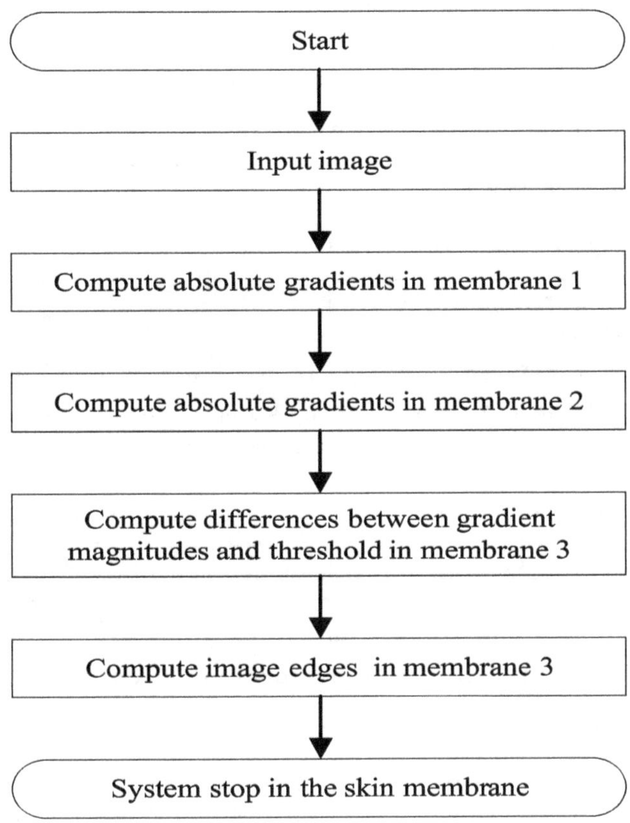

Figure 1. The flowchart of EDENP.

4.2. The Structure and Execution Processes of EDENP

As shown in Figure 2, the structure of EDENP includes four membranes. The system begins to start when the input variables $x_{i,j}$ representing the gray value of source image at location (i, j) appear in the skin membrane. The whole process includes five steps.

Step 1: Horizontal and vertical derivative approximations of every pixel are computed in membrane 1 by using rules of $Pr_{1,CE_{i,j}}$~$Pr_{10,CE_{i,j}}$ in a parallel manner. When the directional gradients are computed, membrane 2 will be activated.

Step 2: The gradient magnitude of all the pixels are obtained at the same time with rules of $Pr_{21,CE_{i,j}}$ in membrane 2.

Step 3: The comparisons between the gradient magnitudes of all pixels and the predefined threshold are executed by rules of $Pr_{31,CE_{i,j}}$ in membrane 3.

Step 4: The edge pixels are detected and marked with 1, while the background pixels are marked with 0 by rules of $Pr_{32,CE_{i,j}}$ and $Pr_{33,CE_{i,j}}$ in membrane 3.

Step 5: The system stop condition is satisfied and the system stops working by rules of Pr_{main} in membrane 4.

So as described above, only five steps are needed in the proposed algorithm for images with arbitrary resolution. Since we do not change the mathematical model of Sobel based edge detection, the detection result by our proposed method is the same as if run on a serial computing platform.

4 − *Main*

$$x_{ij}[\ ], edg_{ij}[0], e_{11}[256], E_D[0], \theta[thresh]; i,j = 1 \ldots n$$

$$\mathrm{Pr}_{main}: 0 * E_{D_{11}} + 0 * E_{D_{12}} + \ldots + 0 * E_{D_{mn}} + 1(\rightarrow) E_D$$

3-*ComputeEdge*

$$ed_1[0], ed_2[1], ed_3[-256], \mathrm{E}_{ij}[\ ], E_{D_{ij}}[\ ]$$

$$\mathrm{Pr}_{31,CE_{ij}}: 2*(g_{ij} - \theta)(\rightarrow) 1\mathrm{g}_{ij} + 1\mathrm{E}_{ij}; i,j = 1 \ldots n$$

$$\mathrm{Pr}_{32,CE_{ij}}: ed_1 + 2 * ed_2 (\mathrm{E}_{ij} \rightarrow) 1\mathrm{edg}_{ij} + 1E_{D_{ij}}; i,j = 1 \ldots n$$

$$\mathrm{Pr}_{33,CE_{ij}}: 0 * ed_1 + 0 * ed_3 (\mathrm{E}_{ij} \rightarrow) 1E_{D_{ij}}; i,j = 1 \ldots n$$

2-*ComputeGrdMag*

$$g_{ij}[]$$

$$\mathrm{Pr}_{21,CE_{ij}}: \sqrt{g_{x_{ij}}^2 + g_{y_{ij}}^2} (e_{11} \rightarrow) 1g_{ij}; i,j = 1 \ldots n$$

1-*ComputeDireGrad*

$$gx_{ij}[], gy_{ij}[]$$

$$\mathrm{Pr}_{1,CE_{ij}}: |x_{ij+2} + 2x_{i+1j+2} + x_{i+2j+2} - x_{ij} - 2x_{i+1j} - x_{i+2j}| (e_{11} \rightarrow) 1g_{x_{ij}}; i,j = 2 \ldots n-2$$

$$\mathrm{Pr}_{2,CE_{ij}}: |x_{ij} + 2x_{ij+1} + x_{ij+2} - x_{i+2j} - 2x_{i+2j+1} - x_{i+2j+2}| (e_{11} \rightarrow) 1g_{y_{ij}}; i,j = 2 \ldots n-2$$

$$\mathrm{Pr}_{3,CE_{1i}}: 0(e_{11} \rightarrow) 1g_{x_{1i}}; i = 1 \ldots n \qquad \mathrm{Pr}_{4,CE_{ni}}: 0(e_{11} \rightarrow) 1g_{x_{ni}}; i = 1 \ldots n$$

$$\mathrm{Pr}_{5,CE_{i1}}: 0(e_{11} \rightarrow) 1g_{x_{i1}}; i = 2 \ldots n-1 \qquad \mathrm{Pr}_{6,CE_{in}}: 0(e_{11} \rightarrow) 1g_{x_{in}}; i = 2 \ldots n-1$$

$$\mathrm{Pr}_{7,CE_{1i}}: 0(e_{11} \rightarrow) 1g_{y_{1i}}; i = 1 \ldots n \qquad \mathrm{Pr}_{8,CE_{ni}}: 0(e_{11} \rightarrow) 1g_{y_{ni}}; i = 1 \ldots n$$

$$\mathrm{Pr}_{9,CE_{i1}}: 0(e_{11} \rightarrow) 1g_{y_{i1}}; i = 2 \ldots n-1 \qquad \mathrm{Pr}_{10,CE_{in}}: 0(e_{11} \rightarrow) 1g_{y_{in}}; i = 2 \ldots n-1$$

Figure 2. The structure of EDENP.

4.3. Complexity and Resources Analysis

Taking into account that the size of the input data is $n \times n$, and the image is a gray image. The amount of resources needed is illustrated in Table 1. From Table 1, we can see that there are $(7n^2 + 6)$ variables, including $(2n^2 + 2)$ enzymatic variables and $(5n^2 + 4)$ numerical variables. $(6n^2 + 4)$ rules are involved in this system. The total storage space is 1 cell with $(13n^2 + 10)$ molecules. So the space complexity is $O(n^2)$ theoretically. The time complexity is $O(1)$ because the number of execution steps is 5, which implies the computational efficiency is constant for images under arbitrary resolutions.

From the above analysis, we can see that the core of the proposed algorithm is to use space to replace time to obtain high-performance parallel computing, which is exactly the prominent characteristic of MC. Since molecules are used as storage units in a real biological computer, huge storage space can be utilised when this algorithm is implemented on it. So we think the proposed parallel algorithm is effective for images with high resolutions, at least at a theoretical level.

Table 1. Complexity and resources needed for EDENP.

Term	Necessary Resources
Initial number of cells	1
Number of enzymatic variables	$2n^2 + 2$
Number of numerical variables	$5n^2 + 4$
Number of rules	$6n^2 + 4$
Execution steps	5

5. Experiments and Results

In this section, both the performance and efficiency of our proposed EDENP algorithm are evaluated. Since there is no hardware implementation of MC systems at present, the only way to test the behaviors of the designed P systems is to simulate them in conventional computers. In this paper, a parallel computing architecture, Compute Unified Device Architecture (CUDA), is used as the simulating platform, as it has been reported in literature [24,61]. The parameters of the platform on which our experiments are carried out are illustrated in Table 2. The threshold θ for all the experiments is set to 0.2.

Table 2. Parameters the computer used.

Term	Parameters
CPU model	Intel(R) Core(TM) i7-7700HQ
cache memory	8 MB, 16-Way, 64 byte lines
main memory	16 GB (2* DDR4 2400MHz)
hard disc	SSD, SK hynix SC308 SATA 128GB, 600 Mbps; MQ01ABD100, 1TB
GPU model	Nvidia GeForce GTX 1050 Ti (4 GB)
execution steps	5

5.1. *Performance Evaluation*

Two case studies are considered to evaluate the performance of the proposed method for different types of images. Since the proposed algorithm is in the framework of MC, edge detection methods based on tissue-like *P* systems [21,24] are chosen as comparison methods. Algorithms in the literature [21,24] are sketched and implemented on a CPU platform using the MATLAB program.

5.1.1. Qualitative Evaluation

Case study 1 is considered to evaluate the performance of the three algorithms for images with rich textures. Four images named *rice*, *cameraman*, *mri*, and *AT3_lm4_01* randomly collected from the MATLAB Image Tool Box are used as testing samples in this experiment, as shown in Figure 3a,e,i,m. Figure 3b–d,f–h,j–l,n–p show the detailed qualitative edge detection results of the three algorithms for the four images. It can be clearly observed from Figure 3b,f,j,n, that the contours of the objects can be detected, but meanwhile the noise in the background is also detected, which will make the following image processing, such as object recognition, more difficult to deal with. The results by reference [21] are shown in Figure 3c,g,k,o. It can be seen that there are too many small edges, and the main outlines of the targets can hardly be found even by human eyes. The results of EDENP are illustrated in Figure 3d,h,l,p, from which we can see that not only the main contours of objects can be detected successfully, but also the noise is well suppressed.

Case study 2 is used to test the performance of the three methods for images with less texture, in which images named *toyobjects*, *circbw*, *text*, *testpart1* randomly selected from MATLAB Image Tool Box are used as testing image samples. In image *toyobjects*, each object has a constant gray value, while the other three images are binary images. Like in Case 1, the detected edge results by the three approaches are shown in Figure 4. Figure 4b,f,j,n clearly show that there are many discontinuous edges when using algorithm in reference [24], while the other two methods can detect the edges completely. When comparing the thickness of the edges, it is obvious to see that the method in reference [21] can achieve the thinnest edges, then the EDENP method, and the edges detected by [24] is the thickest. Although the method in [21] can obtain the finest edges, those edges often have burrs, as shown in Figure 5. Figure 5a,e are the whole edge image of *toyobjects* and *circbw*. Figure 5b–d,f–h are the local enlargement of areas in pink rectangles in Figure 5a,e. Areas marked in green in Figure 5b,f are some examples of discontinuous edges by [24]. When comparing Figure 5c,g with Figure 5d,h, it is clear that edges by EDENP are much smoother than by algorithm [21].

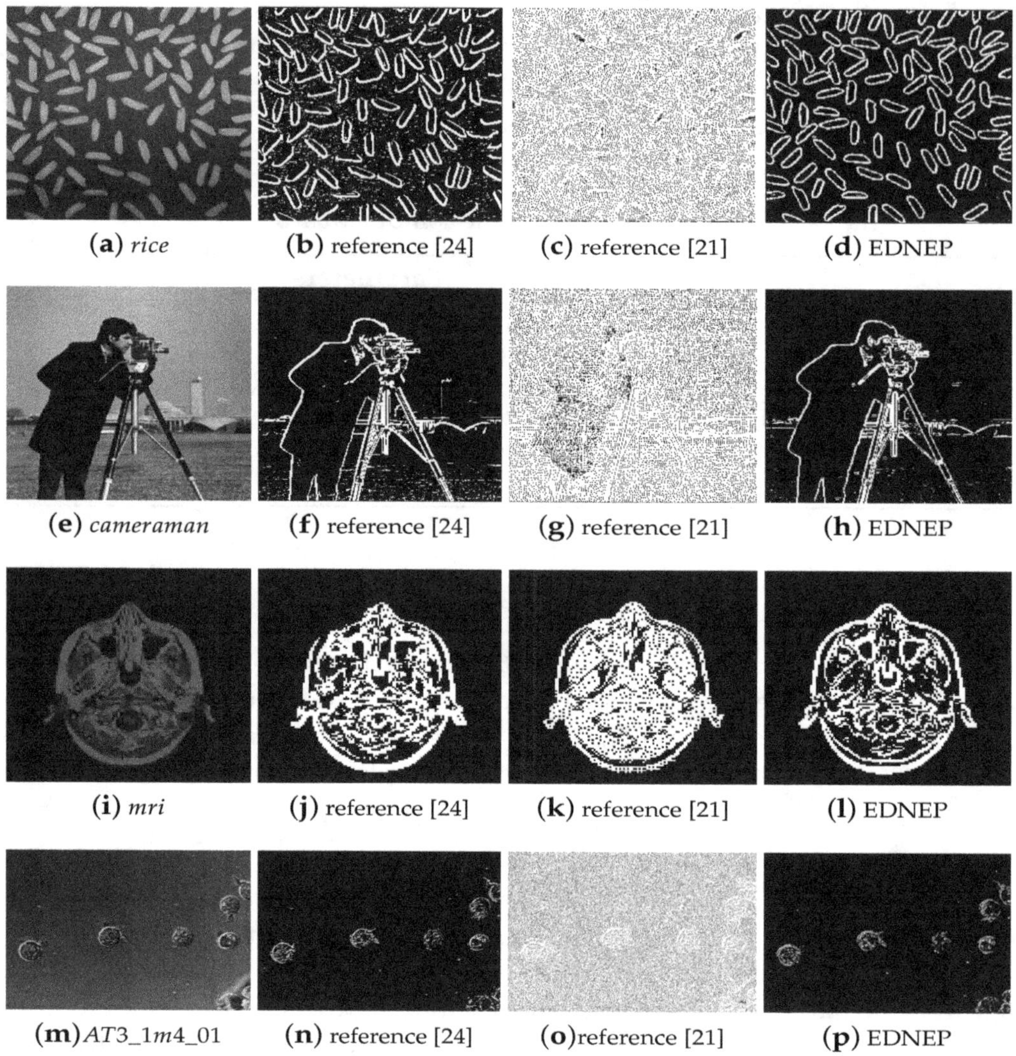

Figure 3. Edge detection results of images with rich texture (the first column: the source gray images; the second to the last column: results by using methods in [21,24] and EDENP respectively).

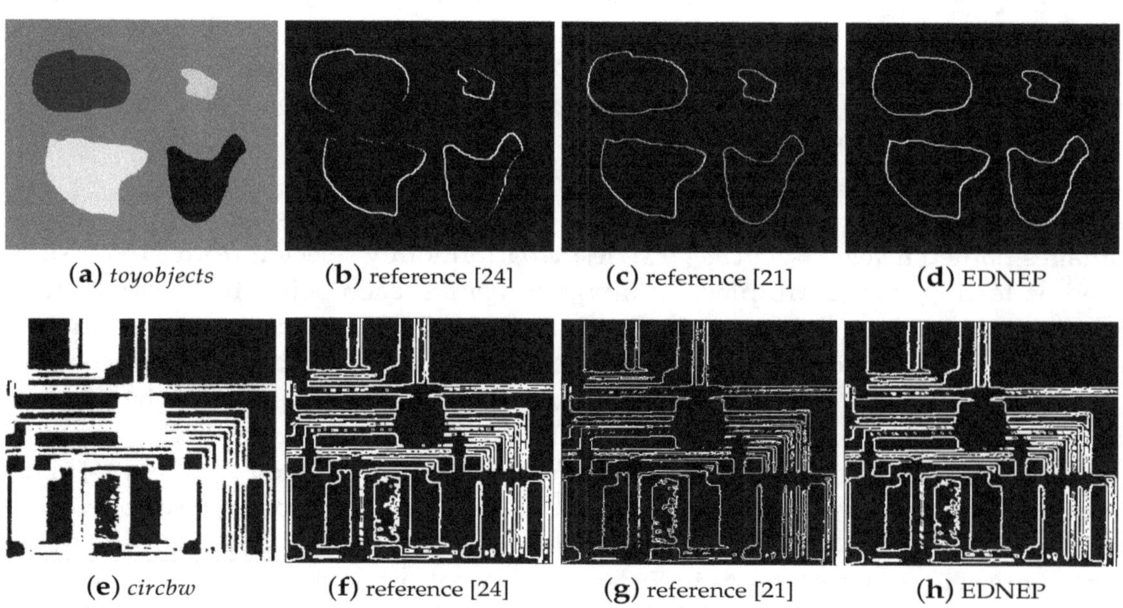

Figure 4. *Cont.*

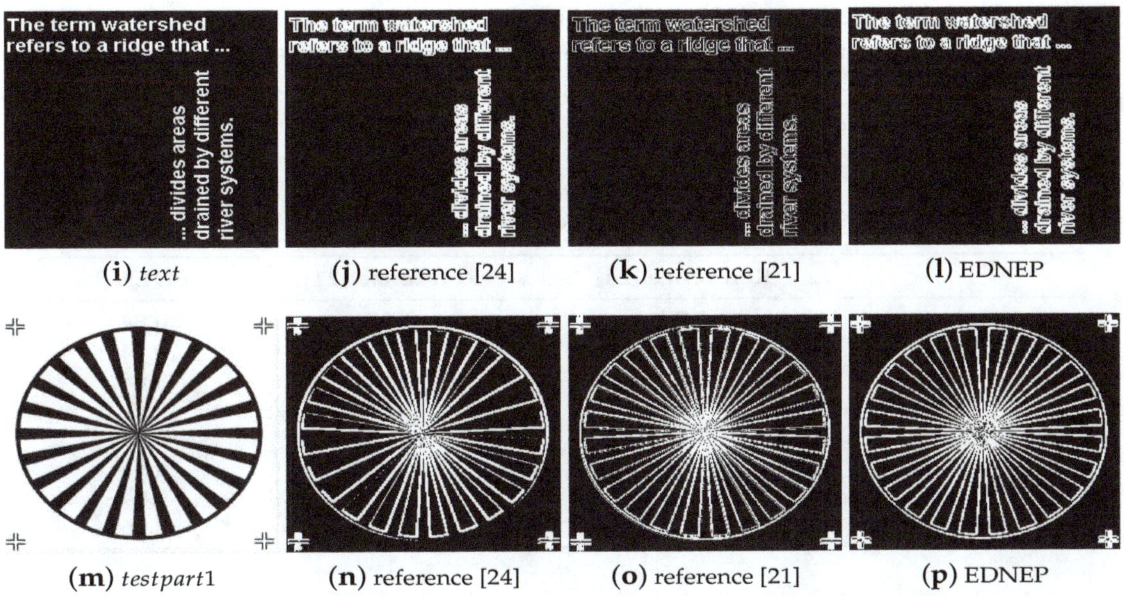

(i) *text* (j) reference [24] (k) reference [21] (l) EDNEP

(m) *testpart*1 (n) reference [24] (o) reference [21] (p) EDNEP

Figure 4. Edge detection results of images with less texture (the first column: the source gray images; the second to the last column: results by using methods in [21,24] and EDENP, respectively).

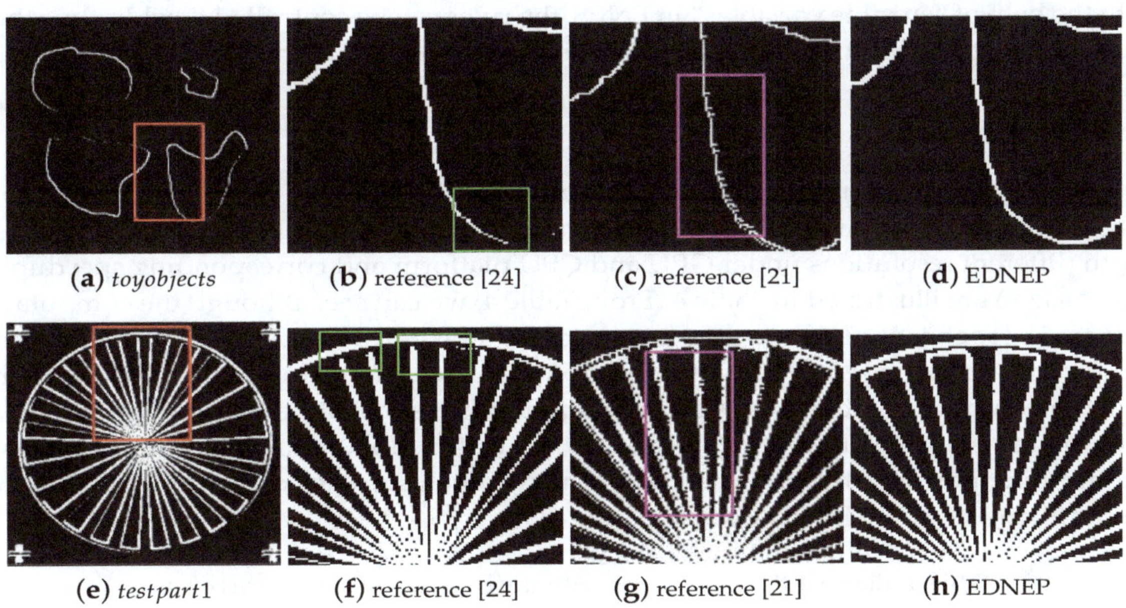

(a) *toyobjects* (b) reference [24] (c) reference [21] (d) EDNEP

(e) *testpart*1 (f) reference [24] (g) reference [21] (h) EDNEP

Figure 5. Edge detection results of *toyobjects* and *testpart*1 (the first column: the edge image; the second to the last columns: the local edge image enlarged by using methods in [21,24] and EDENP respectively).

5.1.2. Quantitative Evaluation

The confidence degree of the edge image is one of the most used indexes for evaluating the authenticity of the edge pixels. In general, the greater the edge confidence degree is, the more reliable the edges are. In this paper, we use this index to evaluate the performance of the edge detection algorithm quantitatively, whose mathematical definition is presented in reference [62].

Table 3 provides the comparison results of the three methods in terms of edge confidence degree. It can be seen from Table 3 that the EDENP method has the highest edge confidence degree for images with both high and low texture, which means edges detected by EDENP have less false edges.

Through the above quantitative and qualitative results, it can be deduced that the method in reference [21] is nearly invalid for grayscale images with rich texture. For images with less textures, this method can get the fine edges of the objects. However, the edges are not smooth in some cases because of the false burr edge points. The approach in [24] cannot get the whole contours of the objects due to the discontinous edges detected for images with both rich and less rich textures. The EDENP algorithm has the highest performance and can obtain clear, continuous, and authentic edges of images with both rich and less rich textures.

Table 3. The edge confidence degree.

	Reference [24]	Reference [21]	EDENP
rice	0.75	0.56	**0.84**
cameraman	0.66	0.32	**0.74**
mri	0.63	0.56	**0.68**
AT3_lm4_01	0.44	0.12	**0.5**
toyobjects	0.85	0.76	**0.86**
circbw	0.94	0.93	**0.95**
text	0.93	0.90	**0.94**
testpart1	0.81	0.79	**0.86**

In this paper, only edge detection methods in the framework of MC are chosen for a comparison. From the above experimental results, we can see that the proposed algorithm has better performance compared with the existing tissue-like based edge detection methods. The fundamental reason for this is that with the help of "enzyme variables" in ENPS, the rules can be controlled flexibly, thus the existing Sobel edge detection algorithm can be programmed in "membrane computing language" easily.

5.2. *Efficiency Evaluation*

To better describe the computation efficiency of EDENP, a speedup ratio is defined as the elapsed time of algorithm on CPU platform divided by running time on GPU platform. The running times of images with different resolutions under GPU and CPU platform and corresponding speedup ratios for one image (*camera*) are illustrated in Table 4. From Table 4, we can see, although the computation times of EDENP are independent of resolutions theoretically, it takes different times to execute the EDENP algorithm for the same image at different resolutions. The reason for this is that the programs do not run on real bio-computers. Table 5 gives the speedup ratios results of the other seven images. It can be found that the lowest speedup is 53, and the maximum speedup can reach up to 262. It is obvious that the computing power of the proposed algorithm is much superior compared with the traditional algorithm implemented on CPU platform.

Table 4. Elapsed time of images with different resolution (*cameraman*).

Image Resolution	256^2	384^2	512^2	768^2	1024^2	2048^2	Platform
Elapsed time(ms)	0.014	0.03	0.05	0.12	0.23	0.86	GPU
Elapsed time(ms)	3.5	9.1	4.4	9.6	41.9	72.8	CPU
Speedup ratio	250	303	88	80	182	130	

Table 5. The speedup ratio of seven images.

Image Resolution	256^2	384^2	512^2	768^2	1024^2	2048^2
rice	79	121	79	101	136	82
mri	60	80	77	62	73	66
AT3_lm4_01	80	90	102	172	71	75
toyobjects	187	162	163	81	182	62
circbw	193	213	262	210	176	66
text	167	180	194	118	57	65
testpart1	53	76	100	161	87	64

6. Conclusions

Membrane computing is a new branch of natural computing, and its amazing storage space and high parallel computing characteristics are very suitable for big data processing. Among various membrane systems, the ENPS can directly deal with numeric variables, and the enzyme variables can flexibly control the execution orders of different rules. In this paper, we attempt to apply ENPS to image processing, and take Sobel edge detection as an example. Compared with the previous works which are based on tissue-like P systems, the advantage of the proposed method is that it does not need to encode and decode the image data, and it is easy to write the program for algorithms with complex execution orders in "membrane computing language". The limitation of the proposed algorithm mainly has two aspects. One is that the execution of the algorithm is based on real biological computers. However, there are no universal biological computers at present, so it is difficult to evaluate the real computing efficiency of the proposed algorithm. The other shortage is that the space complexity is $O(n^2)$, which means large storage space is needed for the proposed algorithm. In future research, we will simulate the algorithm on FPGA hardware and try to combine the ENPS with other, more complex image processing algorithms.

Author Contributions: The research structure was conceived and designed by J.Y. and G.Z.; D.G., M.Z. and Q.Y. wrote the program and performed the experiments; J.Y. and P.P. wrote the paper and analyzed the data; G.Z. made revisions to the final manuscript. The final manuscript was read and corrected by all authors.

Acknowledgments: The authors would like to thank the anonymous reviewers for their valuable suggestions on improving this paper.

References

1. Zheng, S.; Yuille, A.; Tu, Z. Detecting object boundaries using low-, mid-, and high-level information. *Comput. Vis. Image Underst.* **2010**, *114*, 1055–1067. [CrossRef]
2. Zhao, L.; Zhao, Q.; Liu, H.; Lv, P.; Gu, D. Structural sparse representation-based semi-supervised learning and edge detection proposal for visual tracking. *Vis. Comput.* **2017**, *33*, 1169–1184. [CrossRef]
3. Hua, S.; Wen, H. Moment-preserving edge detection and its application to image data compression. *Opt. Eng.* **2013**, *32*, 1596–1608. [CrossRef]
4. Satpathy, A.; Jiang, X.; Eng, H.L. LBP-based edge-texture features for object recognition. *IEEE Trans. Image Process.* **2014**, *23*, 1953–1964. [CrossRef]
5. Saif, J.; Hammad, M.; Alqubati, I. Gradient based image edge detection. *Int. J. Eng. Technol.* **2016**, *8*, 153–156. [CrossRef]
6. Jung, J.; Lee, H.; Lee, J.; Park, D. A novel template matching scheme for fast full-Search boosted by an integral image. *IEEE Signal Proc. Lett.* **2010**, *17*, 107–110. [CrossRef]
7. Schweitzer, H.; Deng, R.; Anderson, R.F. A dual-bound algorithm for very fast and exact template matching. *IEEE Trans. Pattern Anal. Mach. Intell.* **2011**, *33*, 459–470. [CrossRef] [PubMed]
8. Ghassabeh, Y.A.; Rudzicz, F.; Moghaddam, H.A. Fast incremental LDA feature extraction. *Pattern Recognit.* **2015**, *48*, 1999–2012. [CrossRef]
9. Zhou, G.; Cichocki, A.; Zhang, Y.; Mandic, D.P. Group component analysis for multiblock data: common and individual feature extraction. *IEEE Trans. Neural Netw.* **2016**, *27*, 2426–2439. [CrossRef]
10. Herout, A.; Josth, R.; Juranek, R.; Havel, J.; Hradis, M.; Zemcik, P. Real-time object detection on CUDA. *J. Real-Time Image Process.* **2011**, *6*, 159–170. [CrossRef]
11. Jiang B. Real-time multi-resolution edge detection with pattern analysis on graphics processing unit. *J. Real-Time Image Process.* **2018**, *14*, 293–321. [CrossRef]
12. Zuo H.; Zhang Q.; Yong, X.; Zhao, R. Fast sobel edge detection algorithm based on GPU. *Opto-Electron. Eng.* **2009**, *36*, 8–12. [CrossRef]
13. Jiang, J.; Liu, C.; Ling, S.R. An FPGA implementation for real-time edge detection. *J. Real-Time Image Process.* **2018**, *15*, 787–797. [CrossRef]
14. Nausheen, N.; Seal, A.; Khanna, P.; Halder, S. A FPGA based implementation of Sobel edge detection. *Microprocess Microsy* **2018**, *56*, 84–91. [CrossRef]

15. Paolo,B.; Sipp, D. Regulation: Sell help not hope. *Nature* **2014**, *510*, 336–337. [CrossRef]
16. Jiang, N.; Dang, Y.; Wang, J. Quantum image matching. *Quantum Inf. Process.* **2016**, *15*, 3543–3572. [CrossRef]
17. Tsaftaris, S.A.; Katsaggelos, A.K.; Pappas, T.N.; Papoutsakis E.T. How can DNA computing be applied to digital signal processing?. *IEEE Signal Process. Mag.* **2004**, *21*, 57–61. [CrossRef]
18. Díaz-Pernil, D.; Gutierrez-Naranjo, M; Peng, H. Membrane computing and image processing: A short survey. *J. Membr. Comput.* **2019**. [CrossRef]
19. Păun, Gh. Computing with membranes. *J. Comput. Syst. Sci.* **2000**, *61*, 108–143. [CrossRef]
20. Alsalibi, B.; Venkat, I.; Subramanian, K.; Lebai, L.; De, W. The impact of bio-inspired approaches toward the advancement of face recognition. *ACM Comput. Surv.* **2015**, *48*, 1–33. [CrossRef]
21. Christinalh, A.; Díaz-Pernil, D.; Real, P. Region-based segmentation of 2D and 3D images with tissue-like P systems. *Pattern Recognit. Lett.* **2011** *32*, 2206–2212. [CrossRef]
22. Díaz-Pernil, D.; Gutiérrez-Naranjo, M.; Molina-Abril, H.; Real, P. Designing a new software tool for digital imagery based on P systems. *Nat. Comput.* **2012**, *11*, 381–386. [CrossRef]
23. Carnero, J.; Díaz-Pernil, D.; Molina-Abril, H.; Real, P. Image segmentation inspired by cellular models using hardware programming. In Prceedings of the 3rd International Workshop onComputational Topology in Image Context, Chipiona, Spain, 10–12 November 2010; Volume 1, pp. 143–150.
24. Díaz-Pernil, D.; Berciano, A.; Peña-Cantillana, F.; Gutiérrez-Naranjo, M. Segmenting images with gradient-based edge detection using membrane computing. *Pattern Recognit. Lett.* **2013**, *34*. 846–855. [CrossRef]
25. Peña-Cantillana, F.; Díaz-Pernil, D.; Christinal, H.; Gutiérrez-Naranjo, M. Implementation on CUDA of the smoothing problem with tissue-like P systems. *Int. J. Nat. Comput. Res.* **2011**, *2*, 25–34. [CrossRef]
26. Alsalibi, B.; Venkat, I.; Subramanian, K.; Christinal, H. A bio-inspired software for homology groups of 2D digital images. In Proceedings of the Asian Conference on Membrane Computing (ACMC), Coimbatore, India, 18–19 September 2014; IEEE Computer Society: Washington, DC, USA, 2014. [CrossRef]
27. Díaz-Pernil, D.; Christinal, H.; Gutiérrez-Naranjo, M.; Real, P. Using membrane computing for effective homology. *Appl. Algebr. Eng. Commun.* **2012** *23*, 233–249. [CrossRef]
28. Ardelean, I.; Díaz-Pernil, D.; Gutiérrez-Naranjo, M.; Peña-Cantillana, F.; Reina-Molina, R.; Sarchizian, I. Counting cells with tissue-like P systems. In Proceedings of the Tenth Brainstorming Week on Membrane Computing, Seville, Spain, 30 January–3 February 2012; Fénix Editora: Seville, Spain, 2012.
29. Reina-Molina, R.; Díaz-Pernil, D.; Gutiérrez-Naranjo, M. Cell complexes and membrane computing for thinning 2D and 3D images. In Proceedings of the Tenth Brainstorming Week on Membrane Computing, Seville, Spain, 30 January–3 February 2012; Fénix Editora: Seville, Spain, 2012.
30. Berciano, A.; Díaz-Pernil, D.; Christinal, H.; Venkat, I. First steps for a corner detection using membrane computing. In Proceedings of the Asian Conference on Membrane Computing, Coimbatore, India, 18–19 September 2014; IEEE Computer Society: Washington, DC, USA, 2014.
31. Enguix, G. Preliminaries about some possible applications of P systems in linguistics. *Lect. Notes Comput. Sci.* **2002**, *2597*, 74–89. [CrossRef]
32. Cabarle, F.; de la Cruz, R.; Zhang, X.; Jiang, M.; Liu, X.; Zeng, X. On string languages generated by spiking neural P systems with structural plasticity. *IEEE Trans. Nanobiosci.* **2018**, *17*, 560–566. [CrossRef]
33. Song, T.; Zeng, X.; Zheng, P.; Jiang, M.; Rodríguez-Patøn, A. A parallel workflow pattern modelling using spiking neural P systems with colored spikes. *IEEE Trans. Nanobiosci.* **2018**, *17*, 474–484. [CrossRef] [PubMed]
34. Mayne, R.; Phillips, N.; Adamatzky, A. Towards experimental P-systems using multivesicular liposomes. *J. Membr. Comput.* **2019**. [CrossRef]
35. Mitrana, V. Polarization: a new communication protocol in networks of bio-inspired processors. *J. Membr. Comput.* **2019**, published online. [CrossRef]
36. Pan, L.; Păun, Gh.; Zhang, G.; Neri, F. Spiking Neural P Systems with Communication on Request. *Int. J. Neural Syst.* **2017**, *27*, 1750042. [CrossRef]
37. Orellana-Martín, D.; Valencia-Cabrera, L.; Riscos-Núñez, A.; Pérez-Jiménez, M.J. P systems with proteins: A new frontier when membrane division disappears. *J. Membr. Comput.* **2019**. [CrossRef]
38. Sánchez-Karhunen, E.; Valencia-Cabrera, L. Modelling complex market interactions using PDP systems. *J. Membr. Comput.* **2019**. [CrossRef]
39. Zhang, G.; Rong, H.; Neri, F.; Pérez-Jiménez, M.J. An optimization spiking neural P system for approximately solving combinatorial optimization problems. *Int. J. Neural Syst.* **2014**, *24*, 1440006. [CrossRef] [PubMed]

40. Zhang, G.; Cheng, J.; Gheorghe, M.; Meng, Q. A hybrid approach based on differential evolution and tissue membrane systems for solving constrained manufacturing parameter optimization problems. *Appl. Soft Comput.* **2013**, *13*, 1528–1542. [CrossRef]
41. Wang, T.; Zhang, G.; Zhao, J.; He, Z.; Wang, J.; Pérez-Jiménez, M.J. Fault diagnosis of electric power systems based on fuzzy reasoning spiking neural P systems. *IEEE Trans. Power Syst.* **2015**, *30*, 1182–1194. [CrossRef]
42. Martín-Vide, C.; Păun, Gh.; Pazos, J.; Rodríguez-Patón, A. Tissue P systems. *Theor. Comput. Sci.* **2003**, *296*, 295–326. [CrossRef]
43. Ionescu, M.; Păun, Gh.; Yokomori, T. Spiking neural P systems. *Fund. Inform.* **2006**, *71*, 279–308. [CrossRef]
44. Zeng, X.; Pan, L.Q; Pérez-Jiménez, M.J. Small universal simple spiking neural P systems with weights. *Sci. China Inf. Sci.* **2014**, *57*, 1–11. [CrossRef]
45. Păun, G.; Păun, R. Membrane computing and economics: numerical P systems. *Fund. Inform.* **2006**, *73*, 213–227.
46. Zhang, Z.; Wu, T.; Paun, A.; Pan, L. Numerical P systems with migrating variables. *Theor. Comput. Sci.* **2016**, *641*, 85–108. [CrossRef]
47. Pan, L.; Zhang, Z.; Wu, T.; Xu, J. Numerical P systems with production thresholds. *Theor. Comput. Sci.* **2017**, *673*, 30–41. [CrossRef]
48. Zhang, Z.; Pan, L. Numerical P systems with thresholds. *Int. J. Comput. Commun.* **2017**, *11*, 292–304. [CrossRef]
49. Buiu, C.; Vasile, C.; Arsene, O. Development of membrane controllers for mobile robots. *Inform. Sci.* **2012**, *187*, 33–51. [CrossRef]
50. Wang, X.; Zhang, G.; Neri, F.; Jiang T.; Zhao, J.; Gheorghe, M.; Ipate, F.; Lefticaru, R. Design and implementation of membrane controllers for trajectory tracking of nonholonomic wheeled mobile robots. *Integr. Comput.-Aid Eng.* **2016**, *23*, 15–30. [CrossRef]
51. Zhang, G.; Gheorghe, M.; Pérez-Jiminez, M.J. *Real-Life Applications with Membrane Computing*; Springer International Publishing AG: Cham, Switzerland, 2017; pp. 130–141.
52. Mahalingam, K.; Rama, R.; Sureshkumar, W. Robot motion planning inside a grid using membrane computing. *Int. J. Imaging Robot.* **2017**, *17*, 33–51.
53. Pavel, A.; Arsene, O.; Buiu, C. Enzymatic numerical P systems—A new class of membrane computing systems. In Proceedings of the Fifth International Conference on Bio-Inspired Computing: Theories and Applications, Changsha, China, 23–26 September 2010; IEEE Computer Society: Washington, DC, USA, 2010. [CrossRef]
54. Zhang, Z.; Wu, T.; Paun, A.; Pan, L. Universal enzymatic numerical P systems with small number of enzymatic variables, *Sci. China Inf. Sci.* **2018**, *61*, 38–49. [CrossRef]
55. Vasile, C.; Pavel, A.; Dumitrache, I.; Păun, Gh. On the power of enzymatic numerical P system. *ACTA Inform.* **2012**, *49*, 395–412. [CrossRef]
56. Vasile, C.; Pavel, A.; Dumitrache, I. Universality of enzymatic numerical P systems. *Int. J. Comput. Math.* **2013**, *90*, 869–879. [CrossRef]
57. Leporati, A.; Porreca, A.; Zandron, C.; Mauri, G. Improved universality results for parallel enzymatic numerical P systems. *Int. J. Unconv. Comput.* **2013**, *9*, 385–404.
58. Pavel, A.; Buiu, C. Using enzymatic numerical P systems for modeling mobile robot controllers. *Nat. Comput.* **2012**, *11*, 387–393. [CrossRef]
59. Li, W.; Yang, J.; Zhang, J. Handling big data field with enzymatic numerical P System. *J. Sichuan Univ. Nat. Sci. Ed.* **2013**, *45*, 96–104.
60. Pang S.C.; Ding, T.; Rodriguez-Paton, A.; Song, T.; Zheng, P. A parallel bioinspired framework for numerical calculations using enzymatic P system with an enzymatic environment. *IEEE Access* **2018**, *6*, 65548–65556. [CrossRef]
61. Cecilia, J.; García, J.; Guerrero, G.; Martínet-del-Amor, M.; Pérez-Hurtado, I.; Pérez-Jiménez, M. Simulation of P systems with active membranes on CUDA. *Brief Bioinform.* **2009**, *11*, 313–322. [CrossRef] [PubMed]
62. Wang, J.; Bi, J.; Wang, L.; Wang, X. A non-reference evaluation method for edge detection of wear particles in ferrograph images. *Mech. Syst. Signal Process.* **2018**, *100*, 863–876. [CrossRef]

12

Genome-Wide identification of the LAC Gene Family and its Expression Analysis under Stress in *Brassica napus*

Xiaoke Ping [1], Tengyue Wang [1], Na Lin [1], Feifei Di [1], Yangyang Li [1], Hongju Jian [1], Hao Wang [2], Kun Lu [1], Jiana Li [1], Xinfu Xu [1] and Liezhao Liu [1,*]

[1] College of Agronomy and Biotechnology, Chongqing Engineering Research Center for Rapeseed, Academy of Agricultural Sciences, State Cultivation Base of Crop Stress Biology for Southern Mountainous Land, Southwest University, Chongqing 400715, China; xiaokeping1995@163.com (X.P.); tengyue1992@126.com (T.W.); linna123@yeah.net (N.L.); sddifeifei@163.com (F.D.); liyangyangswu@163.com (Y.L.); jianhongju1989@126.com (H.J.); drlukun@swu.edu.cn (K.L.); ljn1950@swu.edu.cn (J.L.); xinfuxu@126.com (X.X.)

[2] Hybrid Rapeseed Research Center of Shanxi Province, Shanxi Rapeseed Branch of National Centre for Oil Crops Genetic Improvement, Yangling 712100, China; wangzy846@sohu.com

* Correspondence: liezhao@swu.edu.cn.

Abstract: Lignin is an important biological polymer in plants that is necessary for plant secondary cell wall ontogenesis. The laccase (*LAC*) gene family catalyzes lignification and has been suggested to play a vital role in the plant kingdom. In this study, we identified 45 *LAC* genes from the *Brassica napus* genome (*BnLACs*), 25 *LAC* genes from the *Brassica rapa* genome (*BrLACs*) and 8 *LAC* genes from the *Brassica oleracea* genome (*BoLACs*). These *LAC* genes could be divided into five groups in a cladogram and members in same group had similar structures and conserved motifs. All *BnLACs* contained hormone- and stress- related elements determined by cis-element analysis. The expression of *BnLACs* was relatively higher in the root, seed coat and stem than in other tissues. Furthermore, *BnLAC4* and its predicted downstream genes showed earlier expression in the silique pericarps of short silique lines than long silique lines. Three miRNAs (miR397a, miR397b and miR6034) target 11 *BnLACs* were also predicted. The expression changes of *BnLACs* under series of stresses were further investigated by RNA sequencing (RNA-seq) and quantitative real-time polymerase chain reaction (qRT-PCR). The study will give a deeper understanding of the *LAC* gene family evolution and functions in *B. napus.*

Keywords: laccase; *Brassica napus;* lignification; stress

---

1. Introduction

B. napus originated from either the Mediterranean or Northern Europe and was formed by chromosome doubling after an interspecific natural cross between *B. rapa* (AA, 2n = 20) and *B. oleracea* (CC, 2n = 18) [1]. Rapeseed oil was once considered as a bad food choice because the seeds contain erucic acid and cholesterol, but with breeding selection and industrial improvement, *B. napus* has nowadays become the third largest source of vegetable oil. Unfortunately, *B. napus* is susceptible to various biotic and abiotic stresses, such as drought, heat, low temperature and fungi infection.

Lignin widely existed and composed of three monomers: coniferyl (G), sinapyl (S), and *p*-coumaryl (H) alcohols. Lignin in plants is involved in the formation of cell walls and together with cellulose increases cellular hardness. Studies have proved that lignin is related to drought stress [2] and a high content can improve the resistance to lodging and *Sclerotinia sclerotiorum* (*S. sclerotiorum*) [3,4]. Laccases are widely distributed with obvious functional differences in plants and fungi [5,6]. It can degrade lignin in *Pleurotus ostreatus* [7] and is expressed in lignifying cells in many plant species [8,9]. LACs are also named multicopper enzymes and supposed to catalyze lignin formation by polymerizing monolignols in plants [10].

To date, *LACs* have been characterized in many species. Lacquer tree contains LAC in the resin ducts and secreted resin [11]. In cotton, an ex-planta phytoremediation system was built based on the overexpression of *LACs* [12]. Forty-four, 46 and 84 *LACs* were identified from *Gossypium arboretum (G. arboretum), Gossypium raimondii (G. raimondii)* and *Gossypium hirsutum (G. hirsutum)*, respectively [13]. A total of 27 laccase candidates (*SbLAC1-SbLAC27*) were identified in *Sorghum bicolor* [14]. *LACs* are continually being detected in other species. An acidic *LAC* gene was found through cDNA cloning in sycamore maple and tobacco [15,16]. Five different LAC-encoding cDNA sequences were identified from ryegrass, with four from the stem and one from the meristematic tissue [17]. *Acer pseudoplatanus* has been found to produce and excrete LAC under cell culture [18,19]. In *Populus euramericana*, five distinct *LACs* were found in xylem tissue [20].

Many studies have indicated the relationship between LAC and lignification. In loblolly pine, LAC was purified from different xylem and shown to coincide with lignin formation in time and place. In *A. thaliana*, gene structure and molecular analysis of the laccase-like multicopper oxidase *(LMCO)* gene family noted that LAC genes *(AtLACs), AtLAC4, AtLAC7, AtLAC8* and *AtLAC15* were mainly expressed in the seed coat, root, pollen grains and cell walls, respectively, and all these tissues present high lignification [21,22]. In maize (*Zea mays*), the *ZmLAC2, ZmLAC3, ZmLAC4*, and *ZmLAC5* coincided with the tissues undergoing lignification [23]. Northern blot analysis indicated that five *LACs* (*LAC1, LAC2, LAC3, LAC90* and *LAC110*) in poplar were highly expressed in stems, although their sequences vary greatly [20]. *SofLAC* was reported as a new *LAC* gene and proven to participate in lignification in sugarcane [24]. In *B. napus* and *Brachypodium distachyon* LAC has been shown to affect the accumulation of lignin [25,26]. In addition to oxidative lignin polymerization, LAC can also protect plants from biotic stresses and abiotic stresses such as the toxic phytoalexins and tannins in the host environment. In another study about maize, *ZmLAC3* was induced by wound, whereas *ZmLAC2* and *ZmLAC5* were repressed and *ZmLAC4* gene expression was unaffected [23]. The *OsChI1* gene encodes a putative *LAC* precursor protein in rice (*Oryza sativa*), overexpression of *OsChI1* in *A. thaliana* improved plants drought and salt tolerance [27]. Compared with the numerous reports about *LAC* gene in other species, few reports are available for *B. napus* especially at the genome level and the influence of stress on *BnLACs* [28,29].

In our study, we identified *LAC* gene family in *B. napus* and characterized them by gene structure, motif and cis-element analysis. The expression patterns of all *BnLACs* were identified by RNA -seq and some of them were analyzed under different stresses by RNA-seq or qRT-PCR. The research not only uncovers the evolutionary relationship of *LAC* gene family but also provides information about LAC respond to biotic and abiotic stresses.

2. Results

2.1. Characterization of the 45 BnLACs

Basic Local Alignment Search Tool Protein (BLASTp) was performed and confirmed 45 *BnLACs* in the *B. napus* genome by using 17 *AtLACs* protein sequences as queries (Table 1). Except for *AtLAC2, AtLAC8* and *AtLAC16*, the remaining *AtLACs* had more than one homologous gene in the *B. napu* genome. *AtLAC3, AtLAC4, AtLAC5, AtLAC11, AtLAC12* and *AtLAC17* had four homologs genes were the most. The genomic sequences lengths of *BnLACs* had a wide range from 1937 (*BnLAC13-1*) to 7114 bp (*BnLAC5-4*). The average MW is 63.02 kDa. The pI values of these proteins varied from 6.10 (*BnLAC9-1*) to 9.74 (*BnLAC14-2*). Subcellular localization predicted results showed all the 45 proteins are located in secretory except for *BnLAC11-4*, which is predicted in mitochondrion. Thirty-nine *BnLACs* are accurately, unevenly mapped on the 12 *B. napus* chromosomes and no tandem duplication. The remaining six *BnLACs* are located on the unmapped scaffolds in the Ann_random and Cnn_ random genome.

Table 1. Characterization of *BnLACs* identified in *B. napus* genome.

LAC	Gene ID	Predicted Subcellular Location	Chr	Position	Genomic Sequences Length (bp)	cDNA Length (bp)	Protein Sequences Length (aa)	MW (kDa)	pI
BnLAC1-1	*BnaC08g17780D*	Secretory	C08	21234694:21237623	2929	1737	578	65.20	9.09
BnLAC1-2	*BnaA08g22770D*	Secretory	A08	16433641:16436613	2972	1740	579	65.23	9.09
BnLAC2	*BnaC04g54790D*	Secretory	C04	2100118: 2102571	2453	1737	578	64.46	9.53
BnLAC3-1	*BnaA04g17380D*	Secretory	A04	14120708:14122939	2231	1713	570	64.07	9.48
BnLAC3-2	*BnaC04g41010D*	Secretory	C04	41756886:41759146	2260	1713	570	64.12	9.59
BnLAC3-3	*BnaA05g12170D*	Secretory	A05	6998148: 7000182	2034	1632	543	60.91	9.41
BnLAC3-4	*BnaC04g14580D*	Secretory	C04	11956954:11959010	2056	1551	516	57.83	9.63
BnLAC4-1	*BnaA05g06610D*	Secretory	A05	3604302: 3607497	3195	1677	558	61.54	9.36
BnLAC4-2	*BnaC04g07220D*	Secretory	C04	5406354: 5409465	3111	1677	558	61.54	9.36
BnLAC4-3	*BnaA04g21810D*	Secretory	A04	16546132:16549220	3088	1680	559	61.79	9.41
BnLAC4-4	*BnaC04g45660D*	Secretory	C04	45245585:45248722	3137	1683	560	61.88	9.41
BnLAC5-1	*BnaC04g47080D*	Secretory	C04	46166799:46169078	2279	1710	569	63.00	8.57
BnLAC5-2	*BnaA05g05410D*	Secretory	A05	2791689: 2794263	2574	1725	574	63.65	8.92
BnLAC5-3	*BnaC04g04810D*	Secretory	C04	3510057: 3512615	2558	1725	574	63.52	8.85
BnLAC5-4	*BnaA04g29320D*	Secretory	A04	1313319: 1320433	7114	1722	573	63.47	8.57
BnLAC6-1	*BnaA04g27180D*	Secretory	A04	19140832:19142936	2104	1710	569	63.62	8.19
BnLAC6-2	*BnaC04g50890D*	Secretory	C04	48319588:48321669	2081	1743	580	65.01	8.7
BnLAC7-1	*BnaA05g29170D*	Secretory	A05	20475705:20479975	4270	1668	555	61.16	9.11
BnLAC7-2	*BnaCnng24340D*	Secretory	Cnn	22790373:22794992	4619	1707	568	62.68	9.1
BnLAC8	*BnaC02g03650D*	Secretory	C02	1747091: 1749618	2527	1542	513	57.09	8.88
BnLAC9-1	*BnaC03g00490D*	Secretory	C03	235782: 239591	3809	1704	567	62.86	6.1
BnLAC9-2	*BnaC03g00480D*	Secretory	C03	229247: 232215	2968	1779	592	65.59	6.67
BnLAC9-3	*BnaA03g00500D*	Secretory	A03	177682: 180567	2885	1734	577	63.99	6.83
BnLAC10-1	*BnaC02g03710D*	Secretory	C02	1776151: 1778498	2347	1680	559	61.18	9.46
BnLAC10-2	*BnaAnng13970D*	Secretory	Ann	15018444:15020836	2392	1692	563	61.62	9.46
BnLAC11-1	*BnaA10g26680D*	Secretory	A10	16966690:16969252	2562	1686	561	62.54	8.99
BnLAC11-2	*BnaC02g03260D*	Secretory	C02	1549295: 1551691	2396	1683	560	62.25	8.49
BnLAC11-3	*BnaAnng18410D*	Secretory	Ann	19628923:19631267	2344	1683	560	62.14	8.75
BnLAC11-4	*BnaCnng02950D*	Mitochondrion	Cnn	2437726: 2440207	2481	1731	576	64.31	8.88
BnLAC12-1	*BnaAnng01310D*	Secretory	Ann	804829: 807062	2233	1698	565	62.58	9.52
BnLAC12-2	*BnaA10g25010D*	Secretory	A10	16204889:16207264	2375	1698	565	62.65	9.1
BnLAC12-3	*BnaC02g02320D*	Secretory	C02	1036921: 1039163	2242	1695	564	62.50	9.49
BnLAC12-4	*BnaC09g49940D*	Secretory	C09	48045040:48047342	2302	1698	565	62.58	9.23
BnLAC13-1	*BnaA10g23590D*	Secretory	A10	15550049:15551986	1937	1704	567	63.05	6.65
BnLAC13-2	*BnaC09g48310D*	Secretory	C09	47172850:47174803	1953	1701	566	62.86	6.65
BnLAC14-1	*BnaC09g47160D*	Secretory	C09	46664083:46666581	2498	1731	576	65.13	9.67
BnLAC14-2	*BnaA10g22590D*	Secretory	A10	15188457:15190992	2535	1743	580	65.74	9.74
BnLAC15-1	*BnaC02g38340D*	Secretory	C02	41316880:41322634	5754	1692	563	63.60	9.01
BnLAC15-2	*BnaAnng08030D*	Secretory	Ann	8068829: 8073341	4512	1680	559	63.42	9.04
BnLAC15-3	*BnaA06g30430D*	Secretory	A06	20553666:20556168	2502	1683	560	63.26	9.09
BnLAC16	*BnaC09g34170D*	Secretory	C09	37560731:37563386	2655	1713	570	63.07	9.18
BnLAC17-1	*BnaA03g09140D*	Secretory	A03	4114497: 4116527	2030	1722	573	63.71	9.28
BnLAC17-2	*BnaC03g11450D*	Secretory	C03	5566603: 5568690	2087	1722	573	63.82	9.32
BnLAC17-3	*BnaA02g06580D*	Secretory	A02	3141348: 3143631	2283	1719	572	63.47	9.27
BnLAC17-4	*BnaA10g12900D*	Secretory	A10	10509237:10513003	3766	1722	573	63.59	9.28

Chromosome C04 has the most *LACs* (eight) and A06, C08 only have one *LAC* gene (Figure 1). To further infer the phylogenetic mechanisms of *BnLACs*, a comparative syntenic map of *B. napus* associated with *A. thaliana* was constructed. Thirteen *BnLACs* show syntenic relationship with those in *A. thaliana* and focus on chromosome 02 and chromosome 05 (Figure 2).

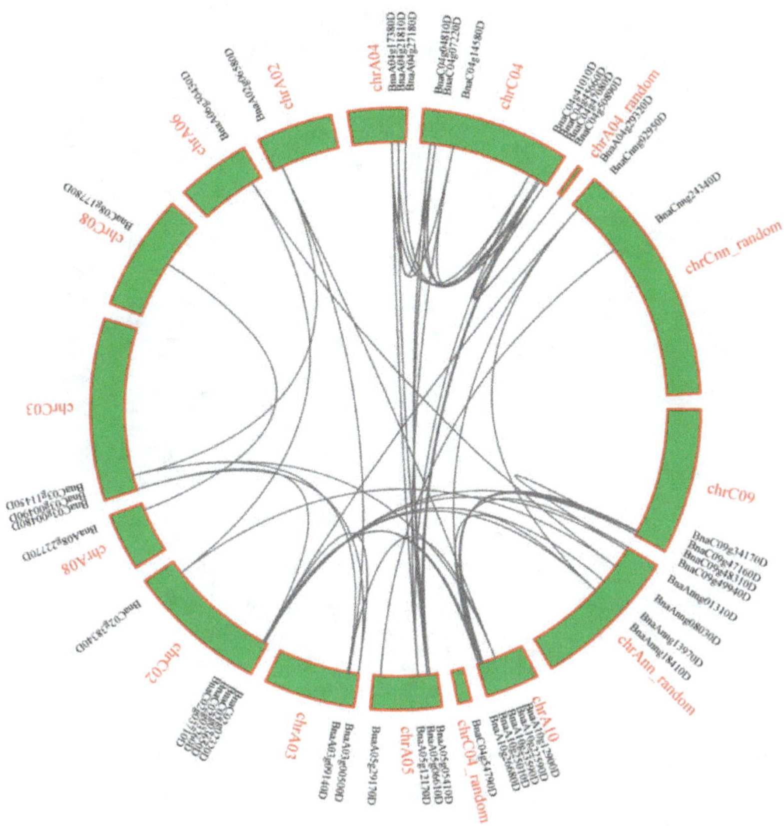

Figure 1. Interchromosomal relationships of *BnLACs* and gray lines indicate duplicated *LAC* gene pairs. The chromosome number and gene ID are indicated with red and black, respectively. The figure was generated by MCScanX with the default parameters [30].

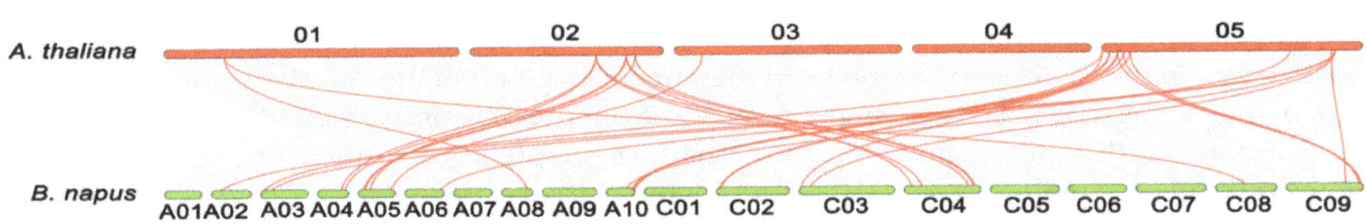

Figure 2. Synteny analysis of *LACs* between *A. thaliana* and *B. napus*. Gray lines in the background indicate all the collinear blocks, red lines highlight the syntenic *LAC* gene pairs.

2.2. Phylogenetic Analysis of LACs in A. thaliana, B. napus, B. rapa and B. oleracea

To study the evolutionary relationship of *LACs* in *A. thaliana*, *B. napus*, *B. rapa* and *B. oleracea*, a cladogram containing 45 *BnLACs*, 25 *BrLACs*, eight *BoLACs* and 17 *AtLACs* was constructed and divided into five groups with well-supported bootstrap values (Figure 3). Groups I, II, III, IV and V had 35, 26, 5, 15 and 14 members, respectively. Forty-five *BnLACs* were unevenly divided into five groups, the most being in Group I which contained 16 *BnLACs* the least being in Group III that only contained one. According to the bootstrap value in the tree, genes in same group were closely related but in different groups were far apart. Genes divided into the same group are thought to have similar functions and number of *LACs* in *B. oleracea* was far less than in *B. rapa* and *B. napus*, however, every group contained at least one *BoLAC*, which is essential to keep complete gene function of *LAC* in *B. oleracea*.

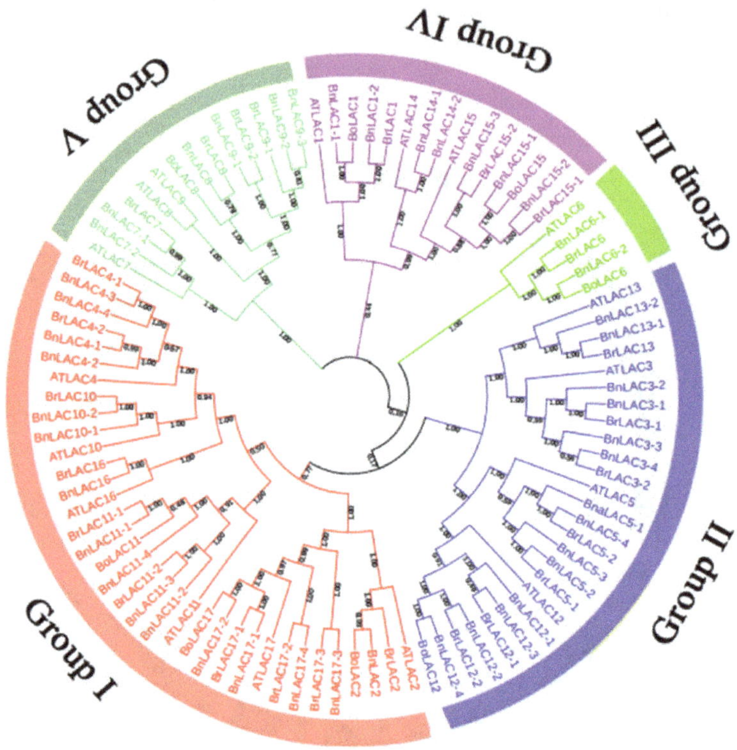

Figure 3. The cladogram of *LAC* proteins from *A. thaliana* (17), *B. napus* (45), *B. rapa* (25) and *B. oleracea* (8). The cladogram was constructed by the MEGA 7.0 software [31] using the neighbour-joining option with 1000 bootstrap replicates and pairwise deletion. Distinct colour segment represents different groups, bootstrap value are shown near nodes.

2.3. Gene Structure and Conservative Domain Analysis of BnLACs and AtLACs

As shown in the cladogram, members in the same group had highly similar gene structures, and the number of exons in the 45 *LACs* ranged from 4 to 7 (Figure 4). Compared with introns, exons were more stable in length. For example, *BnLAC5-4* had a separate longer intron but other members in the Group II didn't contain one. *BnLACs* homologous with *AtLAC4, AtLAC7, AtLAC8* and *AtLAC9* showed diversity with other *BnLACs* because of one or two long introns existed. For exploring more characteristics about *LACs*, introns number of *LACs* from *A. thaliana, S. bicolor, G. arboretum, G. raimondii* and *G. hirsutum* were compared. In *A. thaliana*, there was no *LAC* gene had more six introns. Most *LACs* in *S. bicolor* have one to three introns less than other species with five introns (Figure 5).

Full-length protein sequences of 17 *AtLACs* and 45 *BnLACs* were analyzed to identify their conserved motifs and further understand their functions (Figure 6). The length of the 20 motifs ranged from 6 to 50 amino acids and motif sequences are provided in Table S1. *AtLAC4, BnLAC8* had the fewest motifs (eight), *BnLAC11-4* had the most motifs (twenty). In contrast to others *LACs, AtLAC2* and *BnLAC11-4* had an extra motif 18, *BnLAC3-3; BnLAC3-4* and *BnLAC7-2* lose the motif 2. *LACs* divided into the same groups also had different motifs, *BnLAC5-4* and *AtLAC16* lost the motif 6 compared with other genes in their corresponding group. *AtLAC8, AtLAC9* and their homologous gene in *B. napus* genome lacked the motif 5 when compared with others in Group V. Motif 1 and motif 3 are highly conserved and can be found in all the protein sequences of 62 *LACs*.

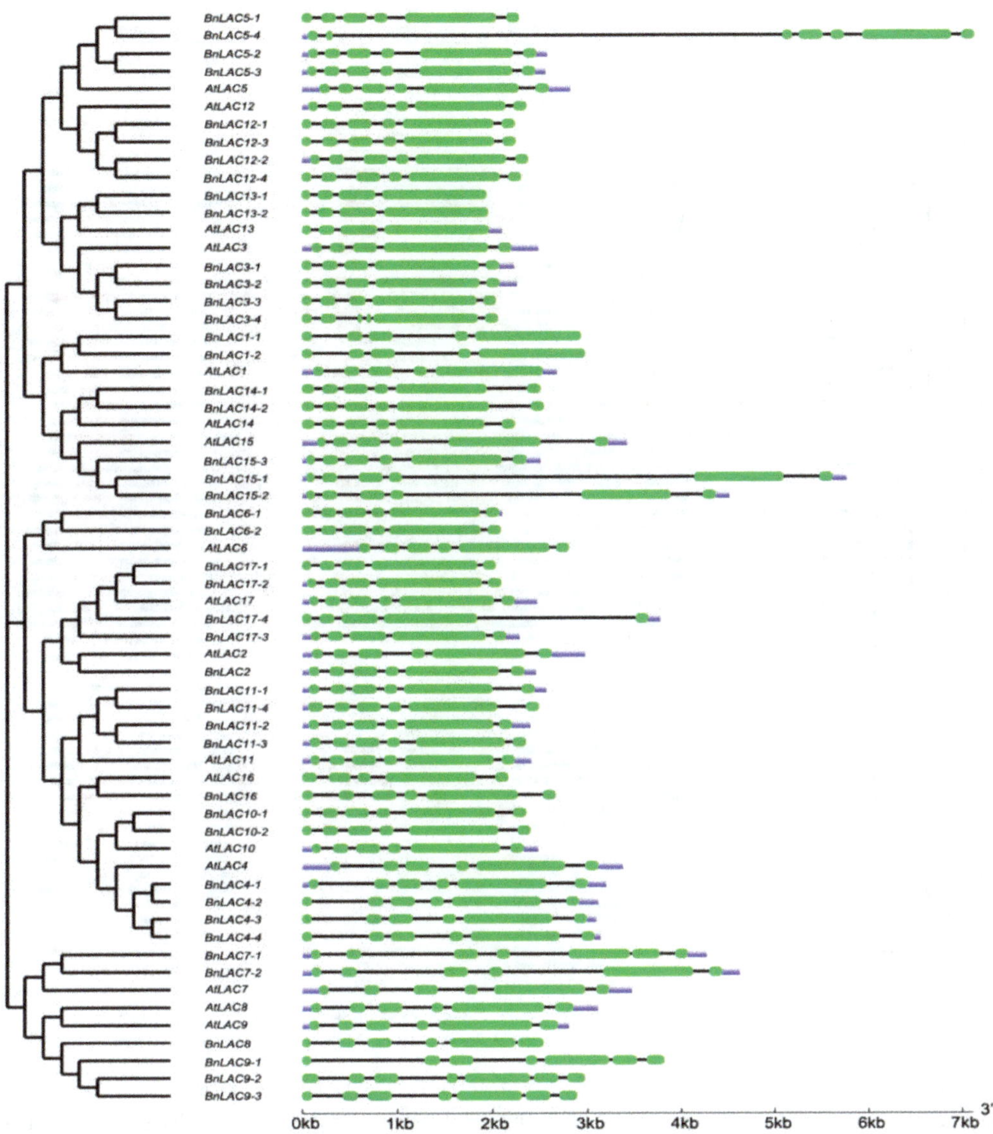

Figure 4. Exon-intron structures of *BnLACs* and *AtLACs* with a cladogram. The green boxes, black lines and blue box indicate exons, introns, untranslated region, respectively.

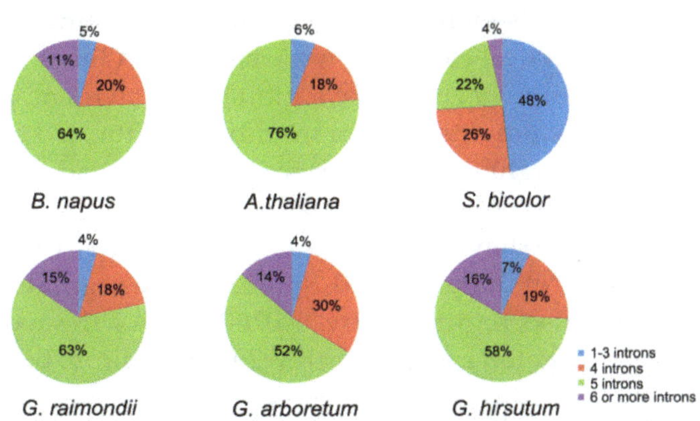

Figure 5. Number of introns in *A. thaliana, S. bicolor, G. arboretum, G. raimondii* and *G. hirsutum*. The figure was constructed by GraphPad Prism 6.

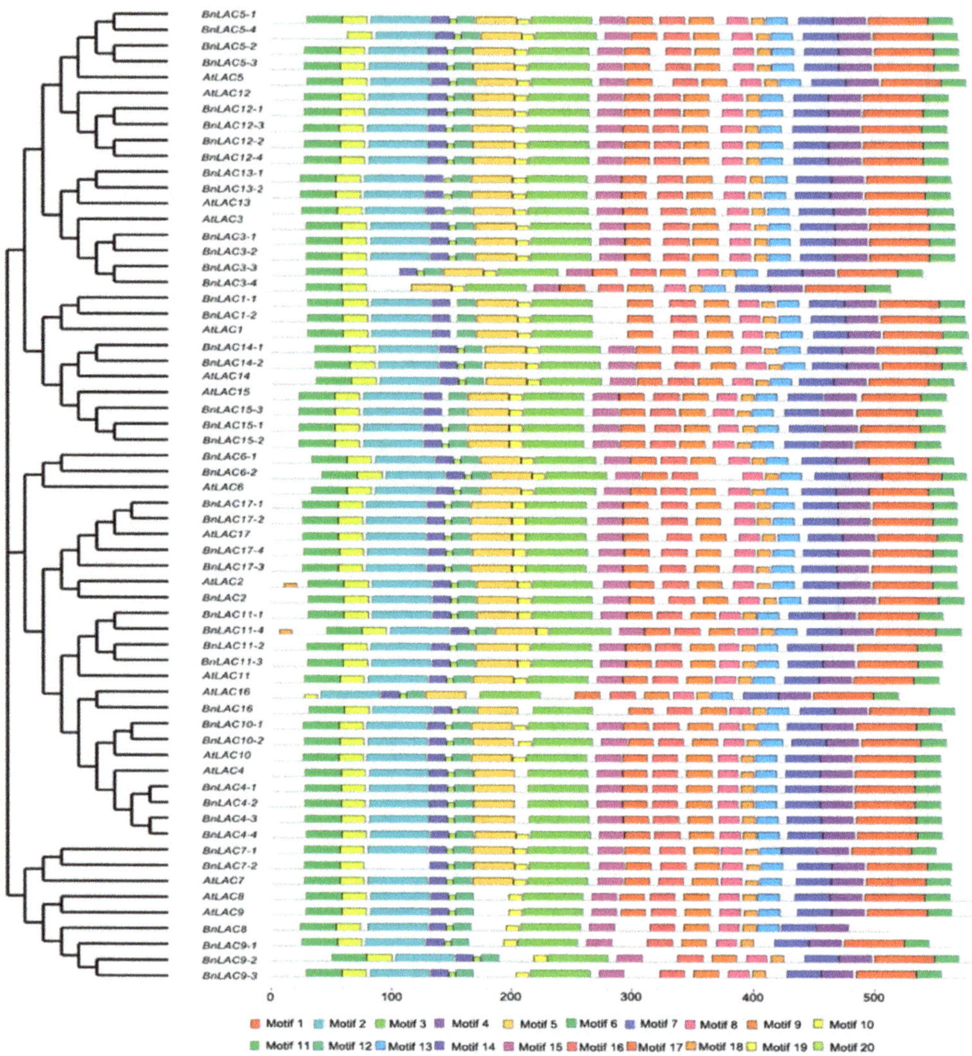

Figure 6. Conserved motif analysis of *BnLACs and AtLACs* proteins presented with a cladogram. Conserved motifs were generated by MEME 5.0.4 and boxes of different colors represent motifs 1-20. Motif sequences are provided in Table S1.

2.4. *Diverse cis Regulatory Elements and miRNAs are Predicted*

To investigate what condition could influence *BnLACs* expression, 1500 bp upstream of the initiation codons were analysed for cis-elements. Eleven kinds of elements including light responsive, six hormone-related, four stress-related elements were searched (Table S2). All the *BnLACs* are light and hormone responsive. Twenty-six *BnLACs* have a gibberellin element, 25 *BnLACs* have an ethylene element, which are the two most elements. Seven members contained a wound element and thirty *BnLACs* were influenced under drought stress. Fourteen members had low temperature elements and 34 members may have been influenced by heat stress.

Eleven *BnLACs* were predicted to have their expression regulated by miRNAs (Table S3, expectation number under 3.0 were selected). All the predicted *BnLACs* were regulated by miR397a and miR397b; *BnLAC4-1* and *BnLAC4-2* were also regulated by miR6034. As the expectation number showed, *BnLAC17-3* was the most likely targeted gene by miR397a and miR397b.

2.5. *Expression Pattern Analysis of BnLACs*

To investigate the expression patterns of 45 *BnLACs*, a heatmap was built based on the RNA-seq (BioProject ID PRJNA358784) using 32 different tissues and stages of *B. napus* as samples (Table S4, Figure 7). Results suggested that *BnLACs* have different expression patterns across tissues and stages. Strong expression occurred in highly lignified tissues such as roots and stems. *BnLAC15-1* and *BnLAC15-2* had the highest expression in the seeds and seed coats. *BnLAC4-1, BnLAC4-2, BnLAC4-3* and *BnALC4-4* had high expression in silique pericarps. *BnLAC5-2, BnLAC5-3, BnLAC15-1, BnLAC15-2* and *BnLAC15-3* highly expressed in seed coats. No *BnLACs* highly expressed in leaves. *BnLAC9-1* and *BnLAC14-2* rarely expressed in any tissue and stage. These results showed *BnLACs* functioned differently and some members are redundant.

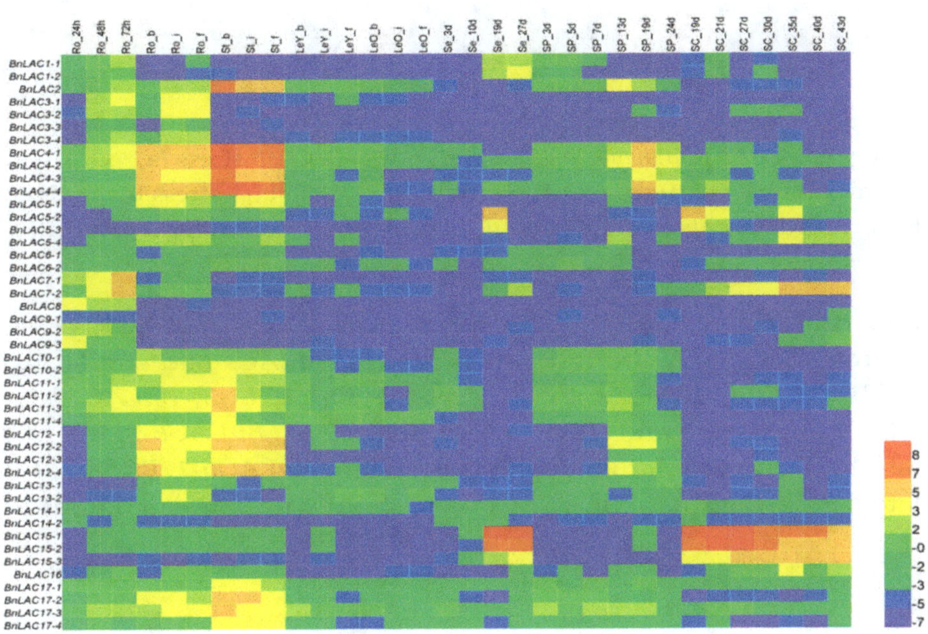

Figure 7. RNA-seq of *BnLACs* in *B. napus* 32 different tissues and stages. Ro, root; St, stem; LeY, young leaf; LeO, old leaf; Se, seed; SP, silique pericarp; SC, seed coat; s, seedling stage; b, bud stage; i, initial flowering stage; and, f, full-bloom stage. The 24, 48, and 72 h means the time that had passed after seed germination. The 3, 10, 27 d and other indicate the number of DAF. The bar on the lower right corner represents the Fragments Per Kilobase of Transcript Per Million Fragments Mapped (FPKM) values and different colors represent different expression levels.

2.6. *Responses of BnLACs upon Abiotic Stress*

As many studies described lignification can response to stresses, RNA-seq and qRT-RCR were used to analyze the expression patterns of several *BnLACs* under different stresses. RNA-seq showed that under Cd^{2+} stress, *BnLAC14-2, BnLAC15-1* and *BnLAC15-2* were up-regulated, *BnLAC12-1* and *BnLAC12-4* were down-regulated at 24 hours after stress then up-regulated at 72 hours after stressed. *BnLAC6-1, BnLAC9-1* and *BnLAC15-3* had no change and other *BnLACs* were down-regulated at the analysed time points. Most *BnLACs* had low expression after 72 hours of Cd^{2+} stress and only 7 *BnLACs* (*BnLAC4-1, BnLAC4-2, BnLAC7-2, BnLAC11-2, BnLAC11-3, BnLAC11-4* and *BnLAC17-3*) still highly expressed (Table S5, Figure 8a). Under NH_4^+ toxicity, Expression levels of *BnLAC6-2, BnLAC13-1* and *BnLAC13-2* were low and not influenced by NH_4^+ toxicity. *BnLAC4-1, BnLAC10-2, BnLAC11-1* and *BnLAC17-2* were up-regulated at 3 or 12 hours after stress and changed to normal level at 48 hours. Some members responsed NH_4^+ toxicity until 48 hours after stressed, *BnLAC2, BnLAC4-2* and *BnLAC14-2* were down-regulated and members such as *BnLAC3-2, BnLAC3-3, BnLAC3-4, BnLAC5-2, BnLAC7-2, BnLAC12-2, BnLAC12-4* and *BnLAC15-1* show significant upregulation (Table S5, Figure 8b). Under drought and wound stress, expression levels of 14 *BnLACs* in stems and leaves of three lines (ZS11, 7191 and D2) were analyzed by qRT-RCR.

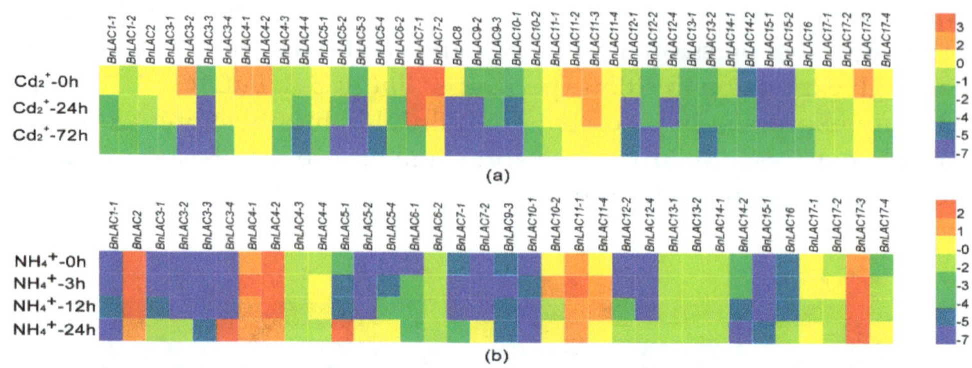

Figure 8. RNA-seq of *BnLACs* in *B. napus* tissues in response to Cd^{2+} stress (**a**), NH_4^+ stress (**b**). The 0, 3, 12 and other time points in (**a**,**b**) denotes the time after Cd^{2+}, NH_4^+ stress respectively. The bar on the lower right corner represents the FPKM values and different colors represent different expression levels.

High expression concentrated at stems, which were consistent with the RNA-seq analysis (Table S6). In leaves of three lines under drought stress, the 14 *BnLACs* had similar expression pattern (Figure 9). *BnLAC2, BnLAC4-1, BnLAC4-2, BnLAC4-4, BnLAC11-3, BnLAC12-1, BnLAC12-2* and *BnLAC17-3* have stable expression and close to zero. *BnLAC6-1, BnLAC6-2, BnLAC11-2, BnLAC11-4, BnLAC14-1* and *BnLAC17-1* have relatively high expression. *BnLAC6-1, BnLAC11-2* and *BnLAC11-4* were down-regulated, *BnLAC6-2* and *BnLAC17-1* were slightly up-regulated under drought stress.

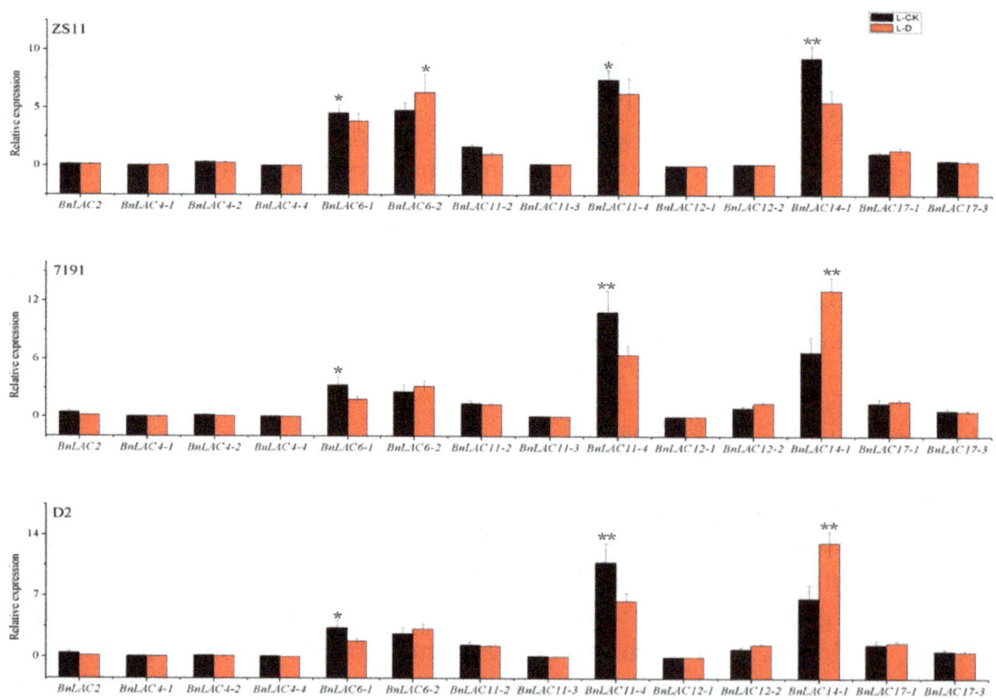

Figure 9. Quantitative RT-PCR of 14 *BnLACs* in *B. napus* leaves in response to drought stress. Expression of *Actin* and *UBC21* were used to normalize the expression level of each *LAC* gene. Bars represent means ± SEM of three biological replicates. Bars marked with asterisks indicate significant differences (Student's t-test) to corresponding control samples for the same time point, $*P < 0.05$, $**P < 0.01$. The results were presented by OriginPro 8 (OriginLab Corporation, Northampton, MA, USA). L-CK and L-D means no drought treated and drought treated, respectively.

In wounded leaves of line ZS11, the expression of *BnLAC2, BnLAC11-2, BnLAC11-3, BnLAC17-1* and *BnLAC17-3* rose firstly then returned to normal levels. *BnLAC6-1, BnLAC11-4, BnLAC12-2* and

BnLAC14-1 had similar expression patterns and showed high expression at 0.5 and 3 hours after wounding. *BnLAC4-1, BnLAC4-2, BnLAC4-4, BnLAC6-2* and *BnLAC12-1* showed no expression in any samples. In wounded leaves of line 7191, *BnLAC4-1, BnLAC4-2, BnLAC4-4, BnLAC11-4* and *BnLAC12-1* were up-regulated followed by down-regulation and showed the highest expression at different time points. *BnLAC4-1, BnLAC4-2, BnLAC4-4* and *BnLAC12-1* showed the highest expression at 1.5 hours after wounding and *BnLAC11-4* showed the highest expression at 0.5 hours after wounding. *BnLAC6-2, BnLAC12-2* and *BnLAC14-1* were down-regulated for all the analyzed time points and *BnLAC11-3* had no change after wounding. In wounded leaves of line D2, *BnLAC2, BnLAC6-2 BnLAC11-2, BnLAC12-2, BnLAC17-1* and *BnLAC17-3* had high expression at two time points, 0.5 and 1.5 hours after wounding, respectively. *BnLAC4-1, BnLAC4-2, BnLAC4-4, BnLAC11-3* and *BnLAC12-1* nearly have no expression in the control sample and kept stable under wounding stress. Expression levels of *BnLAC6-1, BnLAC11-4,* and *BnLAC14-1* responded to wounds slowly and showed high expression at 3 and 6 hours (Figure 10).

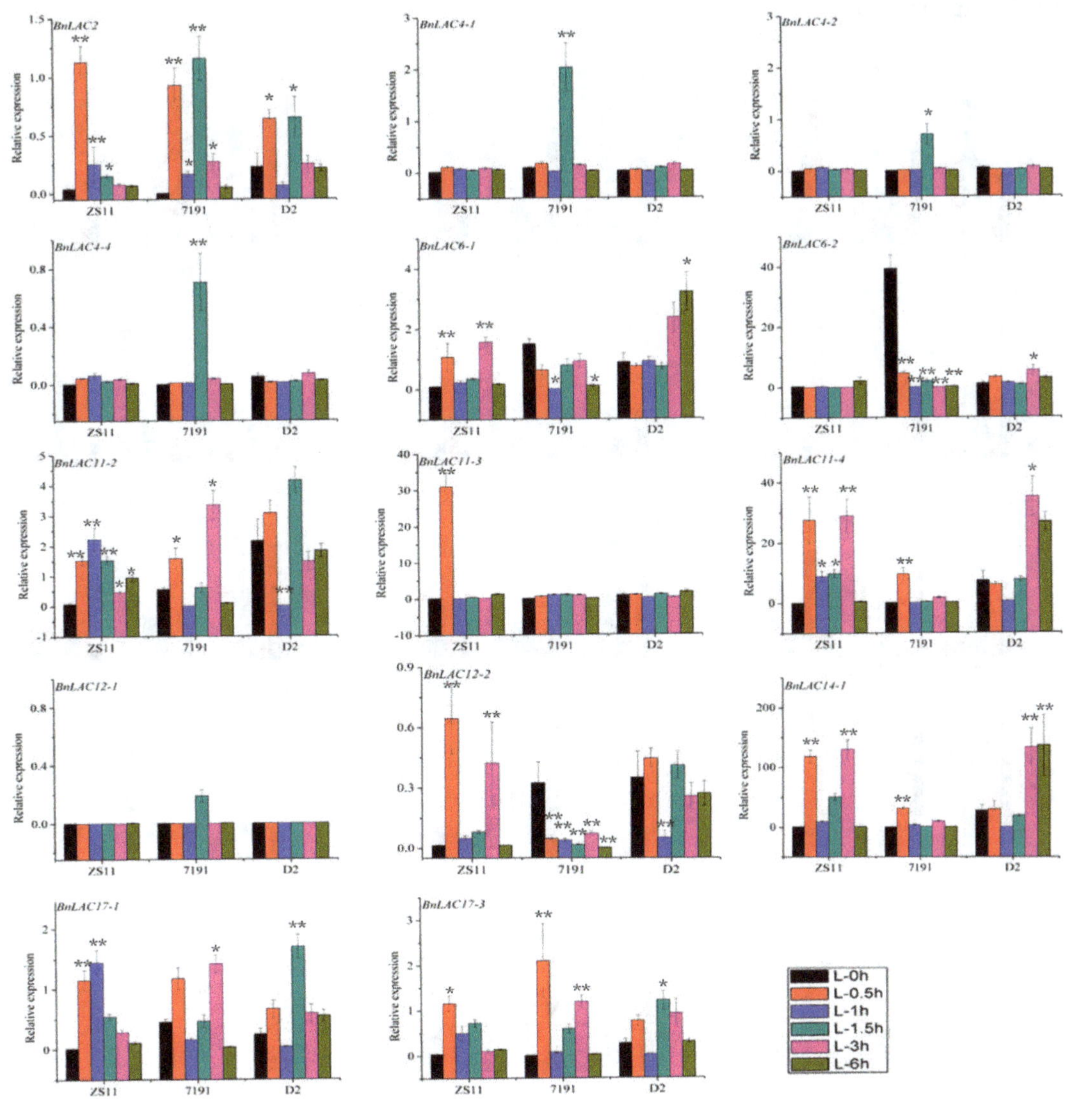

Figure 10. Quantitative RT-PCR of 14 *BnLACs* in *B. napus* leaves in response to wounding stress. Expression of *Actin* and *UBC21* were used to normalize the expression level of each *LAC* gene at each time point. Bars represent means ± SEM of three biological replicates. Bars marked with asterisks indicate significant differences (Student's t-test) to corresponding control samples for the same time point, $^{*}P < 0.05$, $^{**}P < 0.01$. The results were presented by OriginPro 8. Different color means different time after wound shown as bar at lower right corner.

In wounded stems of line ZS11, *BnLAC2, BnLAC4-1, BnLAC4-4, BnLAC11-3, BnLAC11-4* and *BnLAC12-1* had the highest expression at 1.5 hours after wounding and changed to zero at 6 hours. *BnLAC6-2, BnLAC11-2, BnLAC12-2, BnLAC17-1* and *BnLAC17-3* were up-regulated followed by down-regulation. *BnLAC6-1* and *BnLAC14-1* were down-regulated all the time after wounding. In wounded stems of 7191 line, *BnLAC4-1, BnLAC4-4, BnLAC11-2* and *BnLAC11-3* were down-regulated followed by up-regulation and showed high expression at 6 hours after wounding. *BnLAC12-1, BnLAC12-2* and *BnLAC17-1* were down-regulated at all the analyzed time points. *BnLAC6-1, BnLAC6-2, BnLAC11-4* and *BnLAC14-1* showed similar expression patterns and had their highest expression at 6 hours after wounding. In wounded stems of D2 line, *BnLAC2* and *BnLAC4-4* had the highest expression at 0.5 hour after wounding. *BnLAC4-1, BnLAC4-2, BnLAC11-3, BnLAC12-1, BnLAC17-1* and *BnLAC17-3* showed high expression at 0.5 and 1 hours after wounding. Expression levels of *BnLAC6-1, BnLAC12-2* and *BnLAC14-1* almost no change and were close to zero for the analyzed time points. Both *BnLAC11-2* and *BnLAC11-4* had the highest expression at 1 hour and lowest expression at 1.5 hours after wounding (Figure 11).

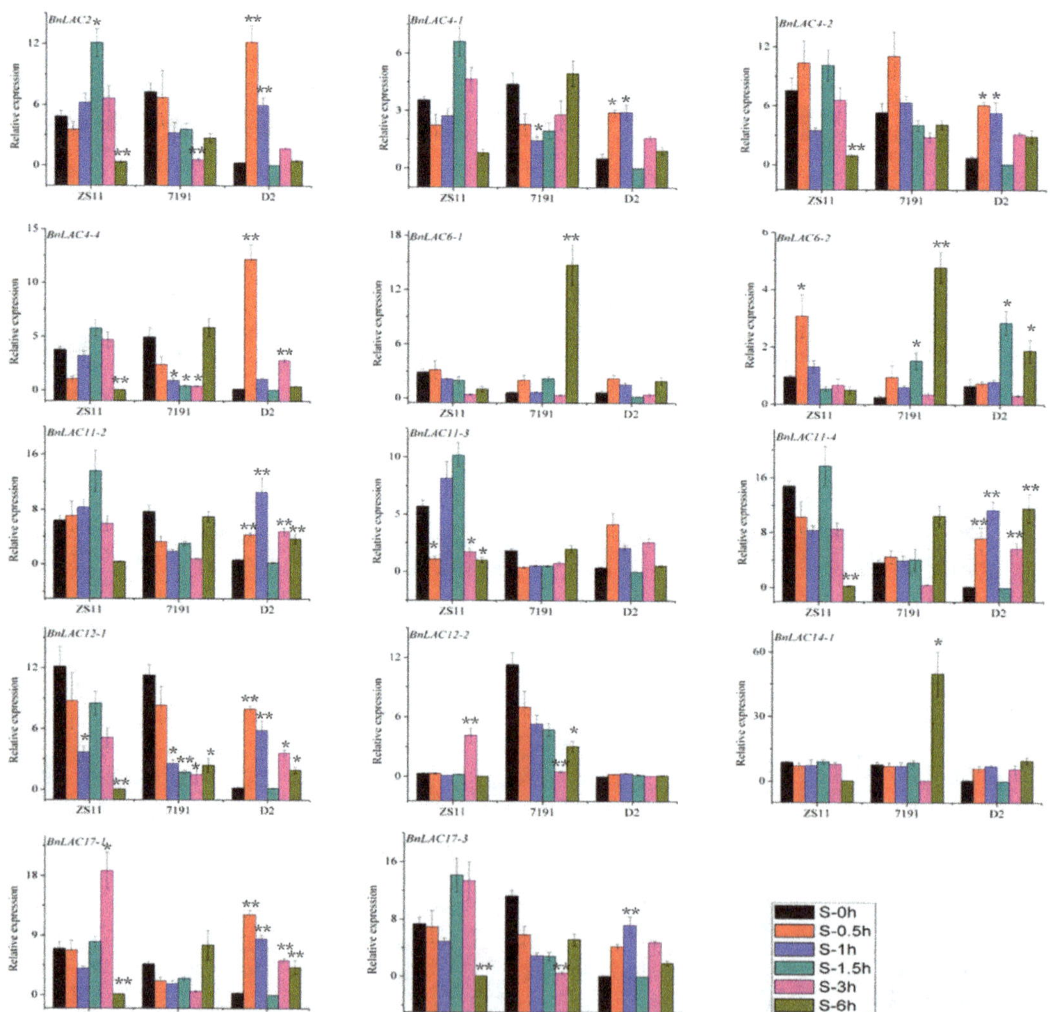

Figure 11. Quantitative RT-PCR of 14 *BnLACs* in *B. napus* stems in response to wounding stress. Expression of *Actin* and *UBC21* were used to normalize the expression level of each *LAC* gene at each time point. Bars represent means ± SEM of three biological replicates. Bars marked with asterisks indicate significant differences (Student's t-test) to corresponding control samples for the same time point, *$P < 0.05$, **$P < 0.01$. The results were presented by OriginPro 8. Different color means different time after wound shown as bar at lower right corner.

2.7. BnLAC4 and its Predicted Downstream Genes are Differentially Expressed in the Silique Pericarp between Long and Short Silique Lines

In STRING platform, CTL2, IRX3, CESA4, IRX1, LAC17, GAUT12, IRX6, PGSIP1, GLP10 and FLA11 were predicted to interact with protein AtLAC4. A total 70 homologous genes were identified in the B. napus genome and expression levels of them in silique pericarps of long and short siliques lines were showed by a heatmap (Table S7, Figure 12). Most of the identified genes showed higher expression in silique pericarps of short silique lines on the 16th Days After Flower (DAF) and almost equal expression in two kinds of silique pericarps on the 25th DAF. On the contrary, many genes showed a higher expression in long silique lines on the 35th DAF. Those results showed BnLAC4 and its' predicted downstream genes expressed earlier in silique pericarp of short siliques lines.

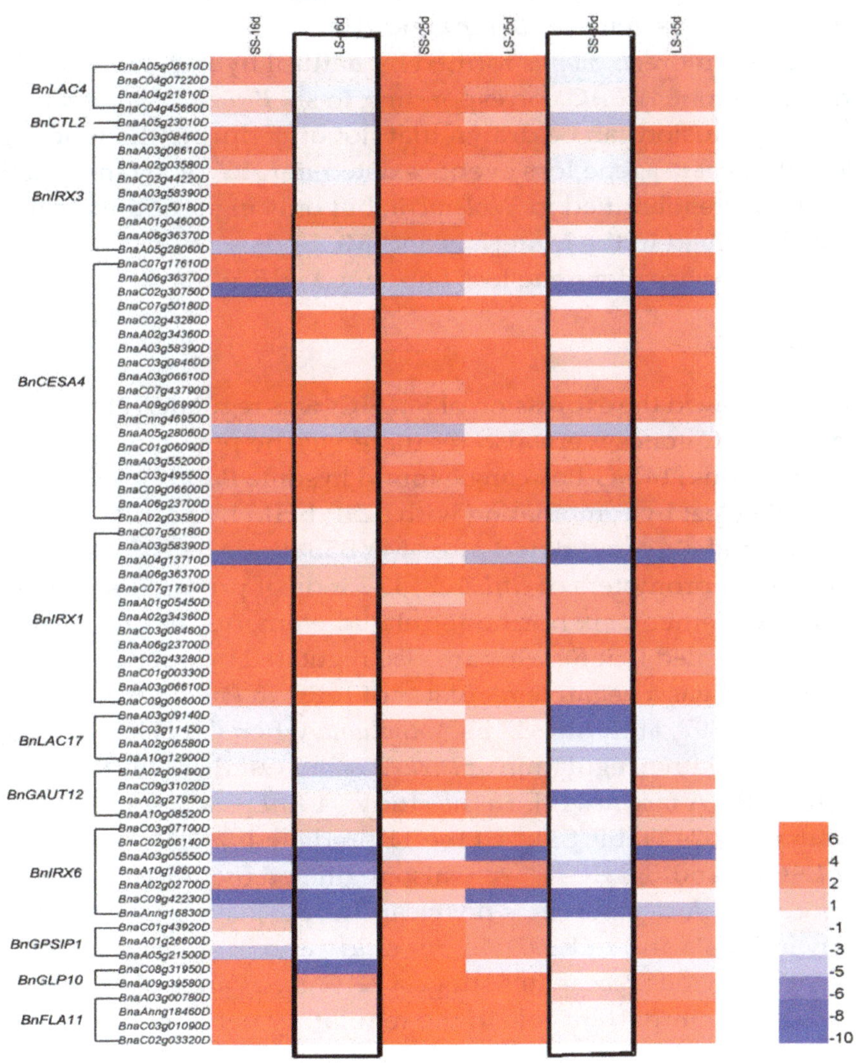

Figure 12. RNA-seq of BnLAC4 BnCTL2, BnIRX3, BnCESA4, BnIRX1, BnLAC17, BnGAUT12, BnIRX6, BnPGSIP1, BnGLP10 and BnFLA11 genes in six kinds of B. napus silique pericarps. LS, long silique lines; SS, short silique lines. The 16, 25, and 35 d means DAF. The bar on the lower right corner represents the FPKM values and different colors represent different expression levels. Heat map was generated by HemI 1.0 software. Lower expression between long and short silique lines at 16th and 35th DAF were indicated by black box.

3. Discussion

B. napus is the third largest source of vegetable oil worldwide and plays an important role in national economies and food industries. The rapeseed yield decreases frequently because of lodging and other biotic and abiotic stresses. Many studies have shown that lignin aids in the resistance to fungi, stress and the LAC acts as an enzyme related to lignification in plants. Previous studies on the *LAC* gene family have been performed, but no related reports on the analysis of this family in *B. napus* exist until now. We identified and analyzed the LAC gene family with the aim of providing a reference at the genome level and deeper understanding of lignification.

3.1. Loss Events Occurred in the LAC gene family Along with the Evolution

Loss events occur frequently during evolution because of hybridization and chromosome doubling [32]. As a result of whole-genome triplication (WGT), the genes in *A. thaliana* should have three homologs in *B. rapa* and *B. oleracea*. In the study, only *AtLAC1* had three orthologous genes in *B. rapa* and no *AtLACs* corresponding to three orthologous genes were found in *B. oleracea*. Some *AtLACs* even have no orthologous genes in *B. rapa* and *B. oleracea* like *AtLAC14* in *B. rapa*, *AtLAC3* and *AtLAC4* in *B. oleracea*. Furthermore, *B. napus* formed by natural hybridization and polyploidization of *B. rapa* and *B. oleracea*, but no one *AtLAC* corresponding to six *BnLACs* was seen in this study and the most is four. Hence, the conclusion can be drawn that not only during the whole-genome triplication but also the formation of *B. napus*, gene loss events existed in *LAC* gene family universally.

Twenty four *BoLACs* were identified in *B. oleracea,* but only eight of them contained four essential conserved domains as described in the Materials and Methods section. Compared with *B. rapa* that contained 25 *BrLACs*, faster or broader gene loss happened in *B. oleracea*.

3.2. Regulation of BnLAC Genes

Cis-element analysis showed the expression of *BnLACs* was regulated at transcriptional level (Table S2). Research has shown *LAC* genes were also regulated at the post-tanscriptional level or through post-translational modifications [14]. G-box can be found in some *BnLACs* and related to light-response and salt tolerance in rice flag leaf by combinating with bZIP, bHLH, and NAC TFs [33–35]. The *AtLAC4* gene has been confirmed to be up-regulated after MYB58 binds to AC elements [36]. Our results show diverse cis-elements in promoters of *BnLACs*, 11 kinds of promoters were selected for analysis including light responsive element, six hormone-related elements and four stress-related elements (Table S2). These findings revealed that *BnLACs* are also regulated by a series of factors. Related studies have reported that Ptr-miR397a is a negative regulator of *LACs* in *Populus trichocarpa* [37]. Several *LACs* are targeted by miR408, miR397, and miR857 in *A. thaliana* when Cu is absent [38]. Ptr-miR397a and Os-miR397 are involved in negative regulation of *PtrLACs* and *OsLACs* [38,39]. Seven *SbLACs* have also been predicted to be sbi-miRNA targets [14]. In the study, 11 *BnLACs* were predicted to be regulated by miR397a, miR397b and miR6034 (Table S3). All the 11 predicted *BnLACs* were regulated by miR397a and miR397b, and *BnLAC4-1* and *BnLAC4-2* were also regulated by miR6034. Research about miR6034 is very few and the process it participates in is not clear. Results have proved *BnLAC17-3* was the most likely targeted gene by miR397a and miR397b in our study and in *A. thaliana* [40]. The findings of the study and combined with previous researches suggest *LAC* genes are truly regulated by miRNAs, and miRNA397 likely plays a very important role in the regulatory network.

3.3. Expression Patterns and Response to Stress

Abundant expression focuses on the roots, stems, and seed coats, whereas the expression in leaves, petals, pistils, and stamens are very low. The expression coincides with lignification in different parts of the plant. *BnLAC13-1* and *BnLAC13-2* showed different expression pattern though contain same cis-elements (Table S2, Figure 7). The reason would be a network containing other factors exists and regulates expression patterns of *BnLACs* such as miRNA and epigenetic modifications [41]. Some members, such as *BnLAC9-1* and *BnLAC14-1,* were never highly expressed in any tissues or stages. It might be that their function was not required in biological processes or was only induced. by certain environmental factors, similar genes can also be found in cotton and Sorghum bicolor [13,14].

Many studies have illustrated that *LACs* are influenced by different kinds of stress and our study also proved that. *OsChL1* as a putative laccase precursor, its expression was increased under drought stress and overexpressed the gene in *A. thaliana* can increase drought and salt tolerance [27]. In the study, six out of 14 *BnLACs* were influenced by drought stress. *BnLAC6-2, BnLAC11-2* and *BnLAC11-4* were up-regulated, *BnLAC6-1* and *BnLAC11-4* were down-regulated. *BnLAC14-1* was up-regulated in line 7191 and D2 but down-regulated in line ZS11. Metal ions would influence the expression of *LACs* and miRNAs directly or indirectly. In *Citrus*, the expression of *LAC7* was up-regulated by boron toxicity [42]. miR397 has been confirmed as a regulatory factor of *LACs* and its expression were influenced by Cd^{2+} [43]. In our study, most *BnLACs* were down-regulated after Cd^{2+} treatment and only four *BnLACs* showed upregulation. Another study in our lab has indicated NH_4^+ enrichment treatment would increase the lignin content in stem and root. Consistent with phenotype, the expression of most *BnLACs* were up-regulated at different time after NH_4^+ enrichment treatment by RNA-seq. According to the results of the promoter analysis, some members like *BnLAC5-1, BnLAC6-1, BnLAC6-2* contain cis element about wound and a study has been reported *LACs* were influenced by wound [23]. qRT-RCR results showed the expression of the selected 14 *BnLACs* in leaves and sterms changed intricately after wounded. We also found gene in different lines responded to stress differently, *BnLAC11-4* was up-regulated in lines 7191 and D2 but down-regulated in line ZS11. Further works are needed to find out the relation between wound healing and *LACs*.

3.4. BnLAC4 and its Downstream Genes May Participate in Silique Elongation in B. napus

Studies have reported that miR397 (both miR397a and b) regulate lignin content and yield traits in *Rice* and *Populus trichocarpa* via modulating *LACs* [39]. In *A. thaliana,* overexpression of miR397b-resistant *AtLAC4* results in an increased silique length and decreased lignin content [40]. In our research, homologous genes of *AtLAC4, AtCTL2, AtIRX3, AtCESA4, AtIRX1, AtLAC17, AtGAUT12, AtIRX6, AtPGSIP1, AtGLP10* and *AtFLA11* in *B. napus* show earlier expression in silique pericarp of short silique lines than long silique lines. Some of those genes predicted like *AtIRX1* and *AtIRX6* are related to secondary cell wall biosynthesis. Combined with the reports, it could be that the period of *BnLAC4* and its downstream genes expression may regulate silique length in *B. napus.*

4. Materials and Methods

4.1. Plant Materials and Stress Treatment

Inbred line Zhongshuang11 (ZS11), 7191, and D2 were sown in humus and grown to the four-leaf-stage. Half of the plants in each line were transplanted for drought stress and the remaining were left for the control. After 25 days without irrigation, the lines in the drought stress treatment showed a wilted phenotype and young leaves from control and stressed lines were frozen immediately in liquid nitrogen and stored at −80 °C.

Leaves and stems were wounded at the bolt stage. Leaves were wounded by a plastic comb-like brush, which was 8.5 cm long and had 42 spikes with a diameter of 1 mm that were equally arrange. Every leaf received three rows of wounds on each side of the midrib and parallel with it; the total number of punctures in each leaf was 252. Stems were wounded to a centimeter depth by scalpel blades. Samples were harvested around the cut at 0.5, 1, 1.5, 3, and 6 hours after wound [44]. The collected samples were frozen immediately in liquid nitrogen and stored at −80 °C for RNA isolation.

The seeds were grown at 22 °C with a light intensity of 200 mol/m^2/s and a photoperiod of 16 hours for 7 days in hydroponic culture. Subsequently, the plants were collected after exposure to 1 mM Cd^{2+} ($CdCl_2$) at 0, 24, and 72 h immediately frozen in liquid nitrogen for RNA sequencing.

Seeds of the *B. napus* line ZS11 were surface-sterilized with 1.2% sodium hypochlorite and germinated in a chamber room (16 hours light 15000Lx/8 hours dark at temperature 25 °C) with Hoagland solution. At four-leaf-stage, a portion of the seedlings were cultivated with modified Hoagland solution (0 mM NO_3^-, 10 mM NH_4^+, pH 6.0; other ions were not changed) for NH_4^+ toxicity treatment. After 3, 12 and 48 hours, the third and fourth true leaves were immediately frozen in liquid nitrogen and stored at −80 °C for RNA sequencing.

Long and short silique lines were selected from a recombinant inbred line (RIL) population constructed from a cross between GH06 (female parent) and P174 (male parent). Lines were planted in open field and grew under normal condition, silique pericarps of two kinds lines were collected respectively at 16th, 25th and 35th DAF and frozen in liquid nitrogen for RNA sequencing.

4.2. Characterisation of the LAC Gene Family

To date, 17 *LACs* have been reported in *A. thaliana* [35]. BLASTp was performed in the *B. napus*, *B. rapa* and *B. oleracea* genome using the *AtLACs* protein sequences as queries and sequences with E-value less than 1×10^{-20} were selected [45]. Some repeated sequences were manually deleted according to the E-value. All the remaining genes were checked by InterProScan (http://www.ebi.ac.uk/interpro) [46], and the sequences with four essential conserved domains of multicopper oxidase type 1 (IPR001117), multicopper oxidase type 2 (IPR011706), multicopper oxidase type 3 (IPR011707) and laccase (IPR017761) were deemed as candidate *LACs* [13]. Another BLASTp was performed in *A. thaliana* genome using candidate *BnLACs* protein sequences as queries and hold those genes that corresponded to *AtLACs*. *LACs* identified from *B. napus*, *B. rapa* and *B. oleracea* were named according to the orthologous sequence in *A. thaliana*. Information about putative sequences were searched in the date bases of BRAD (http://brassicadb.org/) and *B. napus* Genome Browser (http://www.genoscope.cns.fr/brassicanapus/). The chromosomal locations were shown by MapChart software [47], and the number of amino acids, isoelectric point (pI) and molecular weight (MW) of the protein sequences were searched using the ExPASy website (http://web.expasy.org/). The subcellular localization pattern of *LAC* genes were predicted using the web-based tool TargetP1.1 server (http://www.cbs.dtu.dk/services/TargetP/) [48]. Multiple Collinearity Scan toolkit (MCScanX) was adopted to analyze the gene duplication events with the default parameters [30]. The synteny relationship of the *LACs* in *B. napus* and *A. thaliana* were constructed using the Dual Systeny Plotter software (https://github.com/CJ-Chen/TBtools).

4.3. Evolutionary relationship of the LAC Genes Family in A. thaliana, B. napus, B. rapa, and B. oleracea

A cladogram containing the sequences identified from the four species was built using MEGA 7.0 software [31], with 1000 bootstrap replicates performed, and it was then modified by iTOL (http://itol.embl.de/) and Photoshop CS 5 to further visualize evolutionary relationship.

4.4. Gene Structure and Conserved Motif Analysis

The cDNA sequences, genomic sequences and full-length protein sequences of *AtLACs* and *BnLACs* were obtained from the *A. thaliana* genome (http://www.arabidopsis.org/) and *B. napus* Genome Browser (http://www.genoscope.cns.fr/brassicanapus/) respectively. Gene structures were analysed by Gene Structure Display Server (GSDS2.0, http://gsds.cbi.pku.edu.cn/) [49], conserved motifs were tested by and Multiple EM for Motif Elicitation version 5.0.4 (MEME, http://meme-suite.org/tools/meme) with a limit of 20 motifs and any number of repetitions deemed as motif sites [50].

4.5. Cis-Elements Analysis and Prediction of miRNA Target BnLACs

One thousand and five hundred base pairs (bp) upstream of the initiation codons (ATG) were searched in the *B. napus* Genome Browser and cis-elements were analysed using the PlantCARE database (http://bioinformatics.psb.ugent.be/webtools/plantcare/html/). The genome sequences of the 45 *BnLACs* were submitted to the psRNATarget Server (http://plantgrn.noble.org/psRNATarget/) with default parameters to predicte the miRNAs with a target site on *BnLACs*. MiRNAs from the *B. napus* genome were selected and expectation number under 3 were selected [51].

4.6. Expression Patterns Analysis of B. napus LAC Genes

The expression patterns of *BnLACs* were based on RNA-seq, using data from the BioProject ID PRJNA358784. The data included the expression in different tissues in different stages of the *B. napus* cultivar ZS11. The clean reads were aligned to the *B. napus* reference genome and these sequence data and corresponding gene annotation files were downloaded from the genome website (http://www.genoscope.cns.fr/brassicanapus). The BWA and Bowtie softwares were used to map the reads to a reference genome and the reference genes, respectively [52]. The alignment results were visualized by IGV (Integrative Genomics Viewer) and genes expression levels were quantified on the basis of their FPKM values using Cufflinks with default parameters. For RNA-seq data about Cd^{2+} and NH_4^+ stresses, HISAT2 was used to map the reads to a reference genome and genes [53]. The alignment results were also visualized by IGV and the level of each gene expression was measured as FPKM by StringTie [54]. The heatmaps were built to represent the expression level of the *BnLACs* using HemI 1.0 software (http://hemi.biocuckoo.org/faq.php) [55].

4.7. RNA Extraction, Reverse Transcription and qRT-PCR

Total RNA was extracted using the EZ-10 DNAaway RNA Mini-prep Kit (Sangon Biotech, Shanghai, China). NanoDrop 2000 (Thermo Fisher Scientific, Worcester, MA, USA) and electrophoresis were used to measure concentrations and RNA integrity. Complementary DNA was obtained using the iScriptTM cDNA Synthesis Kit (Bio-Rad, Hercules, CA, USA) and diluted 15 times with distilled deionized water for qRT-PCR. The composition of qRT-PCR contained 2 μL of 15-fold diluted cDNA solution, 10 μL of SYBR®Green Supermix (Bio-Rad), 0.4 μL of 10 mM forward and reverse primers and 7.2 μL of distilled deionized water. Primers were designed on Primer Premier Software (version 5.0) (Table S8) [56] and qRT-PCR was performed on a CFX96 Real-time System (Bio-Rad) with the following conditions: 98 °C for 30 s, then 40 cycles of 98 °C for 15 s, 55 °C for 30 s, and an increase from 65–95 °C at increments of 0.5°C every 0.05 s. Three biological replicates and three technical replications were used for qRT-PCR. According to the $2^{-\Delta\Delta Ct}$ method using Actin7 and UBC21 as internal controls, the gene expression levels were determined and displayed by OriginPro 8 (OriginLab Corporation, Northampton, MA, USA).

4.8. Proteins Interaction with AtLAC4 and Identified their Homologous Genes in the B. napus genome

MiR397b regulated both lignin content and silique length via modulating *AtLAC4* has been identified [39]. To understand whether the similar interaction exit in *B. napus*, the protein sequences of *AtLAC4* was obtained from the *A. thaliana* genome (https://www.arabidopsis.org/) and used to predict the interacting proteins in STRING platform (https://string-db.org/?tdsourcetag=s_pctim_aiomsg). Homologous genes of predicted protein sequences were searched in the *B. napus* genome as the method of identifying *BnLACs*.

5. Conclusions

A total of 45 putative *BnLACs* were identified in the *B. napus* genome and unevenly mapped on the 12 *B. napus* chromosomes with no tandem duplication. *BnLACs* were divided into five groups in the cladogram and members in same group had similar structures and motifs. *BnLACs* had high expression in lignified tissues such as roots, stems and seed coats. After high concentration of NH_4^+ toxicity, most *BnLACs* were up-regulated and lignin more and faster deposited. Expression of many *BnLACs* were close to zero in leaves and uninfluenced by drought stress. Some *BnLACs* were down- regulated and individual gene showed different responses in different lines. Many members were intricately influenced by wounding stress, and significantly regulated members may take part in the healing process. By RNA-seq between long and short silique lines, we forecasted that earlier lignification may be a reason for the short siliques. The results in the study give a chance to further study the functions of *BnLACs* in lignification and the interaction with other biological processes.

Supplementary Materials:
Table S1: Motif sequences of conserved motif analysis, **Table S2:** Putative cis-elements in 45 *BnLACs* promoters, **Table S3:** List of *BnLACs* with putative miRNA target sites, **Table S4:** Expression levels of the *BnLACs* in different tissues and stages of *B. napus*. The values represent the FPKM values, **Table S5:** Expression levels of the *BnLACs* under Cd^{2+}, NH_4^+ enrichment treatment. The values represent the FPKM values, **Table S6:** Expression levels of 14 selected *BnLACs* in the L(leaf) and S(stem) of the three lines under drought and wound stress, **Table S7:** Expression levels of *BnLAC4, BnCTL2, BnIRX3, BnCESA4, BnIRX1, BnLAC17, BnGAUT12, BnIRX6, BnPGSIP1, BnGLP10* and *BnFLA11* genes in 6 kinds of silique pericarps. The values represent the FPKM values, **Table S8:** Primers used for qRT-PCR analysis.

Author Contributions: L.L., H.W. and J.L. designed the experiments. X.P., F.D., Y.L., T.W. and H.J. performed the experiments. X.P., T.W., N.L., X.X. and K.L. analyzed the data. X.P. wrote the paper.

References

1. Chalhoub, B.; Denoeud, F.; Liu, S.; Parkin, I.A.P.; Tang, H.; Wang, X.; Chiquet, J.; Belcram, H.; Tong, C.; Samans, B. Early Allopolyploid Evolution in the Post-Neolithic Brassica Napus Oilseed Genome. *Science* **2014**, *6199*, 950–953. [CrossRef] [PubMed]
2. Mourasobczak, J.; Souza, U.; Mazzafera, P. Drought Stress and Changes in the Lignin Content and Composition in Eucalyptus. *BMC Proc.* **2011**, *7*, 103.
3. Peng, D.; Chen, X.; Yin, Y.; Lu, K.; Yang, W.; Tang, Y.; Wang, Z. Lodging Resistance of Winter Wheat (*Triticum aestivum* L.): Lignin Accumulation and Its Related Enzymes Activities Due to the Application of Paclobutrazol or Gibberellin Acid. *Field Crops Res.* **2014**, *2*, 1–7. [CrossRef]
4. Cruickshank, A.W.; Cooper, M.; Ryley, M.J.; Cruickshank, A.W.; Cooper, M.; Ryley, M.J. Peanut Resistance to Sclerotinia Minor and S. Sclerotiorum. *Aust. J. Agric. Res.* **2002**, *10*, 1105–1110. [CrossRef]
5. Mayer, A.M.; Harel, E. Polyphenol Oxidase in Plants. *Phytochemistry* **1979**, *18*, 193–195. [CrossRef]
6. Solomon, E.I.; Sundaram, U.M.; Machonkin, T.E. Multicopper Oxidases and Oxygenases. *Chem. Rev.* **1996**, *7*, 2563. [CrossRef]
7. Cohen, R.; Persky, L.; Hadar, Y. Biotechnological Applications and Potential of Wood-Degrading Mushrooms of the Genus Pleurotus. *Appl. Microbiol. Biotechnol.* **2002**, *5*, 582–594. [CrossRef]
8. Bao, W.; O'malley, D.M.; Whetten, R.; Sederoff, R.R. A Laccase Associated with Lignification in Loblolly Pine Xylem. *Science* **1993**, *5108*, 672–674. [CrossRef] [PubMed]
9. Sterjiades, R.; Dean, J.F.; Eriksson, K.E. Laccase from Sycamore Maple (*Acer pseudoplatanus*) Polymerizes Monolignols. *Plant Physiol.* **1992**, *3*, 1162–1168.
10. Liang, M.; Haroldsen, V.; Cai, X.; Wu, Y. Expression of a Putative Laccase Gene, Zmlac1, in Maize Primary Roots under Stress. *Plant Cell Environ.* **2006**, *5*, 746. [CrossRef]
11. Hüttermann, A.; Mai, C.; Kharazipour, A. Modification of Lignin for the Production of New Compounded Materials. *Appl. Microbiol. Biotechnol.* **2001**, *4*, 387–394. [CrossRef]
12. Wang, G.D.; Li, Q.J.; Luo, B.; Chen, X.Y. Ex Planta Phytoremediation of Trichlorophenol and Phenolic Allelochemicals Via an Engineered Secretory Laccase. *Nat. Biotechnol.* **2004**, *7*, 893. [CrossRef]
13. Balasubramanian, V.K.; Rai, K.M.; Thu, S.W.; Mei, M.H.; Mendu, V. Genome-Wide Identification of Multifunctional Laccase Gene Family in Cotton (*Gossypium* spp.); Expression and Biochemical Analysis During Fiber Development. *Sci. Rep.* **2016**, *6*, 34309. [CrossRef]
14. Wang, J.; Feng, J.; Jia, W.; Fan, P.; Bao, H.; Li, S.; Li, Y. Genome-Wide Identification of Sorghum Bicolor Laccases Reveals Potential Targets for Lignin Modification. *Front. Plant Sci.* **2017**, *8*, 714. [CrossRef]
15. Lafayette, P.R.; Eriksson, K.E.; Dean, J.F. Nucleotide Sequence of a Cdna Clone Encoding an Acidic Laccase from Sycamore Maple (*Acer pseudoplatanus* L.). *Plant Physiol.* **1995**, *2*, 667–668. [CrossRef]
16. Kiefermeyer, M.C.; Gomord, V.; O'Connell, A.; Halpin, C.; Faye, L. Cloning and Sequence Analysis of Laccase-Encoding Cdna Clones from Tobacco. *Gene* **1996**, *1–2*, 205–207. [CrossRef]
17. Gavnholt, B.; Larsen, K.; Rasmussen, S.K. Isolation and Characterisation of Laccase Cdnas from Meristematic and Stem Tissues of Ryegrass (Lolium Perenne). *Plant Sci.* **2002**, *6*, 873–885. [CrossRef]

18. Bligny, R.; Gaillard, J.; Douce, R. Excretion of Laccase by Sycamore (*Acer pseudoplatanus* L.) Cells. Effects of a Copper Deficiency. *Biochemical Journal.* **1983**, *2*, 583–588.
19. Tezuka, K.; Hayashi, M.; Ishihara, H.; Onozaki, K.; Nishimura, M.; Takahashi, N. Occurrence of Heterogeneity of N-Linked Oligosaccharides Attached to Sycamore (*Acer pseudoplatanus* L.) Laccase after Excretion. *Biochem. Mol. Biol. Int.* **1993**, *3*, 395–402.
20. Ranocha, P.; Mcdougall, G.; Hawkins, S.; Sterjiades, R.; Borderies, G.; Stewart, D.; Cabanesmacheteau, M.; Boudet, A.M.; Goffner, D. Biochemical Characterization, Molecular Cloning and Expression of Laccases—A Divergent Gene Family—in Poplar. *FEBS J.* **1999**, *259*, 485–495. [CrossRef]
21. Turlapati, P.V.; Kim, K.W.; Davin, L.B.; Lewis, N.G. The Laccase Multigene Family in Arabidopsis Thaliana: Towards Addressing the Mystery of Their Gene Function(S). *Planta* **2011**, *3*, 439–470. [CrossRef] [PubMed]
22. Mccaig, B.C.; Meagher, R.B.; Dean, J.F.D. Gene Structure and Molecular Analysis of the Laccase-Like Multicopper Oxidase (Lmco) Gene Family in Arabidopsis Thaliana. *Planta* **2005**, *5*, 619–636. [CrossRef] [PubMed]
23. Caparrós-Ruiz, D.; Fornalé, S.; Civardi, L.; Puigdomènech, P.; Rigau, J. Isolation and Characterisation of a Family of Laccases in Maize. *Plant Sci.* **2006**, *2*, 217–225. [CrossRef]
24. Cesarino, I.; Araújo, P.; Mayer, J.L.S.; Vicentini, R.; Berthet, S.; Demedts, B.; Vanholme, B.; Boerjan, W.; Mazzafera, P. Expression of Soflac, a New Laccase in Sugarcane, Restores Lignin Content but Not S:G Ratio of Arabidopsis Lac17 Mutant. *J. Exp. Bot.* **2013**, *6*, 1769–1781. [CrossRef]
25. Zhang, K.; Lu, K.; Qu, C.; Liang, Y.; Wang, R.; Chai, Y.; Li, J. Gene Silencing of Bntt10 Family Genes Causes Retarded Pigmentation and Lignin Reduction in the Seed Coat of Brassica Napus. *PLoS ONE* **2013**, *4*, e61247. [CrossRef]
26. Wang, Y.; Le, B.P.; Antelme, S.; Soulhat, C.; Gineau, E.; Dalmais, M.; Bendahmane, A.; Morin, H.; Mouille, G.; Lapierre, C. Laccase 5 Is Required for Lignification of the Brachypodium Distachyon Culm. *Plant Physiol.* **2015**, *1*, 192–204. [CrossRef] [PubMed]
27. Cho, H.Y.; Lee, C.; Hwang, S.G.; Park, Y.C.; Lim, H.L.; Jang, C.S. Overexpression of the Oschi1 Gene, Encoding a Putative Laccase Precursor, Increases Tolerance to Drought and Salinity Stress in Transgenic Arabidopsis. *Gene* **2014**, *1*, 98–105. [CrossRef]
28. Sato, Y.; Bao, W.; Sederoff, R.; Whetten, R. Molecular Cloning and Expression of Eight Laccase Cdnas in Loblolly Pine (*Pinus taeda*). *J. Plant Res.* **2001**, *2*, 147–155. [CrossRef]
29. O'Malley, D.M.; Ross, W.; Bao, W.; Chen, C.; Sederoff, R.R. The Role of Laccase in Lignification. *Plant J.* **1993**, *5*, 751–757. [CrossRef]
30. Wang, Y.; Tang, H.; DeBarry, J.D.; Tan, X.; Li, J.; Wang, X.; Lee, T.-h.; Jin, H.; Marler, B.; Guo, H.; et al. Mcscanx: A Toolkit for Detection and Evolutionary Analysis of Gene Synteny and Collinearity. *Nucleic Acids Res.* **2012**, *7*, e49. [CrossRef]
31. Kumar, S.; Stecher, G.; Tamura, K. MEGA7: Molecular Evolutionary Genetics Analysis Version 7.0 for Bigger Datasets. *Mol. Biol. Evol.* **2016**, *33*, 1870–1874. [CrossRef]
32. Paterson, A.H.; Bowers, J.E.; Chapman, B.A. Ancient Polyploidization Predating Divergence of the Cereals, and Its Consequences for Comparative Genomics. *Proc. Natl. Acad. Sci. USA* **2004**, *26*, 9903–9908. [CrossRef]
33. Toledoortiz, G.; Huq, E.; Quail, P.H. The Arabidopsis Basic/Helix-Loop-Helix Transcription Factor Family. *Plant Cell* **2003**, *8*, 1749–1770. [CrossRef]
34. Guo, Y.; Gan, S. Atnap, a Nac Family Transcription Factor, Has an Important Role in Leaf Senescence. *Plant J.* **2006**, *4*, 601–612. [CrossRef]
35. Shen, H.; Cao, K.; Wang, X. Atbzip16 and Atbzip68, Two New Members of Gbfs, Can Interact with Other G Group Bzips in Arabidopsis Thaliana. *BMB Rep.* **2008**, *2*, 132–138. [CrossRef]
36. Zhou, J.; Lee, C.; Zhong, R.; Ye, Z.H. Myb58 and Myb63 Are Transcriptional Activators of the Lignin Biosynthetic Pathway During Secondary Cell Wall Formation in Arabidopsis. *Plant Cell* **2009**, *1*, 248–266. [CrossRef]
37. Lu, S.; Li, Q.; Wei, H.; Chang, M.J.; Tunlayaanukit, S.; Kim, H.; Liu, J.; Song, J.; Sun, Y.H.; Yuan, L. Ptr-Mir397a Is a Negative Regulator of Laccase Genes Affecting Lignin Content in Populus Trichocarpa. *Proc. Natl. Acad. Sci. USA* **2013**, *26*, 10848–10853. [CrossRef]
38. Abdelghany, S.E.; Pilon, M. Microrna-Mediated Systemic Down-Regulation of Copper Protein Expression in Response to Low Copper Availability in Arabidopsis. *J. Biol. Chem.* **2008**, *23*, 15932–15945. [CrossRef]

39. Zhang, Y.C.; Yu, Y.; Wang, C.Y.; Li, Z.Y.; Liu, Q.; Xu, J.; Liao, J.Y.; Wang, X.J.; Qu, L.H.; Chen, F. Overexpression of Microrna Osmir397 Improves Rice Yield by Increasing Grain Size and Promoting Panicle Branching. *Nat. Biotechnol.* **2013**, *9*, 848. [CrossRef]
40. Wang, C.-Y.; Zhang, S.; Yu, Y.; Luo, Y.-C.; Liu, Q.; Ju, C.; Zhang, Y.-C.; Qu, L.-H.; Lucas, W.J.; Wang, X. Mir397b Regulates Both Lignin Content and Seed Number in Arabidopsis Via Modulating a Laccase Involved in Lignin Biosynthesis. *Plant Biotechnol. J.* **2015**, *8*, 1132–1142. [CrossRef]
41. Bottcher, A.; Cesarino, I.; Santos, A.B.; Vicentini, R.; Mayer, J.L.; Vanholme, R.; Morreel, K.; Goeminne, G.; Moura, J.C.; Nobile, P.M. Lignification in Sugarcane: Biochemical Characterization, Gene Discovery, and Expression Analysis in Two Genotypes Contrasting for Lignin Content. *Plant Physiol.* **2013**, *4*, 1539. [CrossRef]
42. Jin, L.F.; Liu, Y.Z.; Yin, X.X.; Peng, S.A. Transcript Analysis of Citrus Mirna397 and Its Target Lac7 Reveals a Possible Role in Response to Boron Toxicity. *Acta Physiol. Plant.* **2016**, *1*, 18. [CrossRef]
43. Shen, C.; Huang, Y.Y.; He, C.T.; Zhou, Q.; Chen, J.X.; Tan, X.; Mubeen, S.; Yuan, J.G.; Yang, Z.Y. Comparative Analysis of Cadmium Responsive Micrornas in Roots of Two Ipomoea Aquatica Forsk. Cultivars with Different Cadmium Accumulation Capacities. *Plant Physiol. Biochem.* **2016**, *111*, 329–339. [CrossRef]
44. Bo, P.; Hopkins, R.; Rask, L.; Meijer, J. Differential Wound Induction of the Myrosinase System in Oilseed Rape (*Brassica napus*): Contrasting Insect Damage with Mechanical Damage. *Plant Sci.* **2005**, *3*, 715–722.
45. Altschul, S.F.; Madden, T.L.; Schäffer, A.A.; Zhang, J.; Zhang, Z.; Miller, W.; Lipman, D.J. Gapped Blast and Psi-Blast: A New Generation of Protein Database Search. *Nucleic Acids Res.* **1997**, *25*, 3389–3402. [CrossRef]
46. Mitchell, A.; Chang, H.Y.; Daugherty, L.; Fraser, M.; Hunter, S.; Lopez, R.; Mcanulla, C.; Mcmenamin, C.; Nuka, G.; Pesseat, S. The Interpro Protein Families Database: The Classification Resource after 15 Years. *Nucleic Acids Res.* **2015**, *43*, 213–221. [CrossRef]
47. Voorrips, R.E. Mapchart: Software for the Graphical Presentation of Linkage Maps and Qtls. *J. Hered.* **2002**, *1*, 77–78. [CrossRef]
48. Emanuelsson, O.; Nielsen, H.; Brunak, S.; Von Heijne, G. Predicting Subcellular Localization of Proteins Based on Their N-Terminal Amino Acid Sequence. *J. Mol. Biol.* **2000**, *4*, 1005–1016. [CrossRef]
49. Hu, B.; Jin, J.; Guo, A.Y.; Zhang, H.; Luo, J.; Gao, G. Gsds 2.0: An Upgraded Gene Feature Visualization Server. *Bioinformatics* **2015**, *8*, 1296. [CrossRef]
50. Bailey, T.L.; Boden, M.; Buske, F.A.; Frith, M.; Grant, C.E.; Clementi, L.; Ren, J.; Li, W.W.; Noble, W.S. Meme Suite: Tools for Motif Discovery and Searching. *Nucleic Acids Res.* **2009**, *37*, 202–208. [CrossRef]
51. Dai, X.; Zhao, P.X. Pssrnaminer: A Plant Short Small Rna Regulatory Cascade Analysis Server. *Nucleic Acids Res.* **2008**, *36*, W114–W118. [CrossRef]
52. Li, H.; Durbin, R. Fast and Accurate Short Read Alignment with Burrows–Wheeler Transform. *Bioinformatics* **2009**, *25*, 1754–1760. [CrossRef]
53. Daehwan, K.; Langmead, B.; Salzberg, S.L. Hisat: A Fast Spliced Aligner with Low Memory Requirements. *Nat. Methods* **2015**, *4*, 357–360.
54. Pertea, M.; Pertea, G.M.; Antonescu, C.M.; Chang, T.C.; Mendell, J.T.; Salzberg, S.L. Stringtie EnablesImproved Reconstruction of a Transcriptome from Rna-Seq Reads. *Nat. Biotechnol.* **2015**, *3*, 290–295. [CrossRef]
55. Deng, W.; Wang, Y.; Liu, Z.; Cheng, H.; Xue, Y. Hemi: A Toolkit for Illustrating Heatmaps. *PLoS ONE* **2014**, *11*, e111988. [CrossRef]
56. Lalitha, S. Primer Premier 5. *Biotech Softw. Internet Rep.* **2000**, *1*, 270–272. [CrossRef]

13

Detection of Protein Complexes based on Penalized Matrix Decomposition in a Sparse Protein–Protein Interaction Network

Buwen Cao [1,2,*], Shuguang Deng [1,*], Hua Qin [1], Pingjian Ding [2], Shaopeng Chen [3] and Guanghui Li [2,4]

1 College of Information and Electronic Engineering, Hunan City University, Yiyang 413000, China; qinhua_hcu@163.com

2 College of Computer Science and Electronic Engineering, Hunan University, Changsha 410082, China; dpj@hnu.edu.cn (P.D.); ghli16@163.com (G.L.)

3 College of Mathematics and Computer Science, Hunan Normal University, Changsha 410081, China; chenshaopeng2010@gmail.com

4 School of Information Engineering, East China Jiaotong University, Nanchang 330013, China

* Correspondence: cbwchj@126.com (B.C.); shuguangdeng@163.com (S.D.);

Academic Editors: Xiangxiang Zeng, Alfonso Rodríguez-Patón and Quan Zou

Abstract: High-throughput technology has generated large-scale protein interaction data, which is crucial in our understanding of biological organisms. Many complex identification algorithms have been developed to determine protein complexes. However, these methods are only suitable for dense protein interaction networks, because their capabilities decrease rapidly when applied to sparse protein–protein interaction (PPI) networks. In this study, based on penalized matrix decomposition (*PMD*), a novel method of penalized matrix decomposition for the identification of protein complexes (i.e., PMD_{pc}) was developed to detect protein complexes in the human protein interaction network. This method mainly consists of three steps. First, the adjacent matrix of the protein interaction network is normalized. Second, the normalized matrix is decomposed into three factor matrices. The PMD_{pc} method can detect protein complexes in sparse PPI networks by imposing appropriate constraints on factor matrices. Finally, the results of our method are compared with those of other methods in human PPI network. Experimental results show that our method can not only outperform classical algorithms, such as CFinder, ClusterONE, RRW, HC-PIN, and PCE-FR, but can also achieve an ideal overall performance in terms of a composite score consisting of F-measure, accuracy (ACC), and the maximum matching ratio (MMR).

Keywords: protein–protein interaction (PPI); clustering; protein complex; penalized matrix decomposition

---

1. Introduction

The identification of protein complexes is highly beneficial for the investigation of all kinds of organisms to understand biological processes and determine inherent organizational structures within cells [1]. The dramatic development of computational methods stimulates many protein complex identification algorithms for protein–protein interaction (PPI) networks, which are generally organized into three catalogs. The first catalog includes clustering methods that are also divided into three sub-catalogs. First, the local search approaches based on density are used to identify densely connected subgraphs in PPI networks, in which subgraphs with density above a pre-defined threshold, such as MCODE (Molecular Complex Detection) [2], CFinder (a software tool for network cluster detection) [3], DPCLus (a Density-Periphery based graph CLustering software) [4], and ICPM (Iterative

Clique Percolation Method) [5], are considered protein complexes. However, these approaches tend to neglect surrounding proteins that are connected to the kernel clusters with sparse links, which can show experimentally validated true interactions [6]. Another kind of method for detecting protein complexes uses classical hierarchy clustering techniques, which mainly depend on the distance between proteins to detect meaningful groups [6] and contain HC-PIN ((fast Hierarchical Clustering algorithm for Protein Interaction Network, agglomerative method) [7] and G-N algorithms (divisive method) [8]. Many hierarchical clustering methods employ similarities among the proteins that are calculated on the basis of network topology characteristics or biological meaning due to the further development of clustering technology. Such approaches mainly include NEMO (NEtwork MOdule identification) [9], ClusterONE (Clustering algorithm with Overlapping Neighborhood Expansion) [10], RFC (Rough Fuzzy Clustering) [11], MINE (Module Identification in Networks) [12], PageRankNibble [13], SPICi (Speed and Performance In Clustering,) [14], PCE-FR (Pseudo-Clique Extension based on Fuzzy Relation) [15], MTGO (Module detection via Topological information and GO knowledge) [16], WCOACH (Weighted COACH) [17], DCAFP (Density-based Clustering Approach for identifying overlapping protein complexes with Functional Preferences) [18], and cwMINE (Combined Weight of Module Identification in Networks) [19]. Experimental results show that these novel methods greatly outperform classical hierarchical clustering approaches. Except for the aforementioned clustering approaches, many other protein complex detection algorithms, such as RNSC (Restricted Neighborhood Search Clustering) [20], MCL [1], RRW (Repeated Random Walks algorithm) [21], CMC (Clustering-based on Maximal Cliques) [22], Coach [23], and AP (Affinity Propagation) with its variant [24] have achieved satisfactory results.

Another type of method used to detect protein complexes employs an intelligent optimization algorithm, which seeks the optimal solution of PPI based on a heuristic concept [25]. For large databases, the complexity of intelligent optimization algorithms is too high to run a correct consequence. The major weakness of the aforementioned methods is that their performance deteriorates when they are employed to sparse PPI networks [19,26]. To address this problem, matrix decomposition is proposed to improve the disadvantages of these methods. A co-clustering algorithm based on the adjacent matrix of PPI networks was proposed [6] and obtained overlapping and non-overlapping protein complexes successfully. The results show that the method reached a remarkable balance between network coverage and accuracy (ACC) and outperformed classical methods. Matrix factorization can be mainly organized into two main levels. The first level is the non-negative matrix factorization (NMF) (which integrates gene ontology (GO), gene expression data, and the PPI network to form the corresponding adjacency matrix and then decomposes it with common factors to achieve the overlapping functional modules with high ACC [27]). Zhang et al. [28] proposed sparse network-regularized multiple NMFs (SNMNMFs) to identify the microRNA regulatory modules and demonstrated the ideal performance of the proposed method in ovarian cancer dataset. The second level is the penalized matrix decomposition (PMD), which is widely applied in various datasets, such as microarray data [29], including gene expression data, and proteomic datasets [30].

Inspired by Ref. [24], PMD_{pc}, an approach used to identify the protein interaction network of protein complexes was originally proposed. First, the adjacent matrix of the protein interaction network was normalized. Second, the normalized matrix was decomposed into three factor matrices. Finally, the PMD_{pc} algorithm and several classical algorithms were executed from the well-investigated human PPI network. The experimental results show that our approach achieved satisfactory performance in terms of F-measure, ACC, and maximum matching ratio (MMR).

2. Results and Discussion

When PMD_{pc} is applied to identify the protein complexes in PPI network, the parameters of c_1, c_2, and k are crucial for the decomposition of the network. Considering that u should be sparse, we take $c_1 = 0.25 \times \sqrt{n}$ and $c_2 = 0.25 \times \sqrt{p}$ [31].

To study the parameter of k on the effect on the experimental results, we repeated the execution of algorithm and studied how the algorithm behaves in terms of F-measure and let $k \in (0, 2500]$ with a 100 increment. The detailed experimental results with different k values are presented in Figure 1. From Figure 1, we can clearly see that k is less than 1000; the experimental results fall short of satisfaction.

The value of the F-measure increases gradually until $k = 1600$ with the increase in k, such that the maximum value of 0.398, the F-measure, displays a steady state when it changes from 1600 to 2000. When k is greater than 2000, the value of F-measure shows a downward trend. Therefore, k is set to 2000.

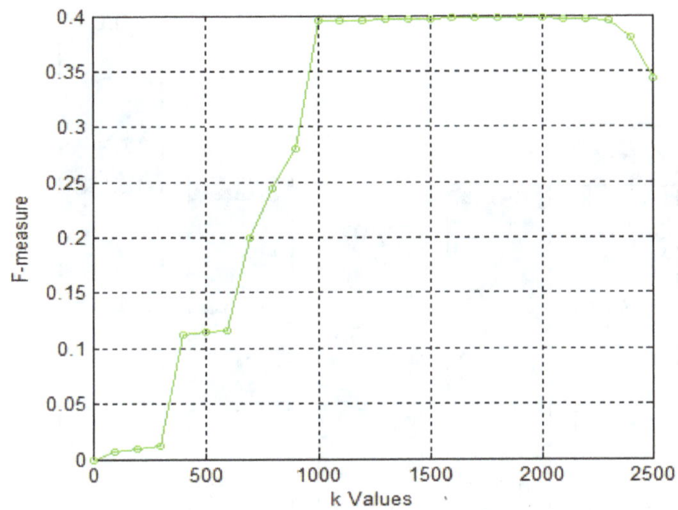

Figure 1. Values of F-measure for different values of $k \in (0, 2500]$ with a 100 increment in HPRD dataset.

Five classical protein complex algorithms, namely, CFinder [3], ClusterONE [10], RRW [21], HC-PIN [7], and PCE-FR [15], are applied on human PPI network of HPRD (Human Protein Reference Database, HPRD) to demonstrate the performance of PMD_{pc}. The complexes of the aforementioned algorithms with sizes less than 2 are filtered in our work. Moreover, the parameters of each method that is compared with our method are set using the default values recommended by the authors. The experimental result is shown in Table 1.

Table 1. Results of six protein complexes Algorithms in HPRD Dataset.

Algorithms	Number	Precision	Recall	F-Measure	ACC	Sep	MMR	MCC
CFinder	49	0.959	0.143	0.249	0.184	0.165	0.017	0.327
ClusterONE	755	0.295	0.186	0.229	0.333	0.209	0.084	0.391
RRW	167	0.671	0.190	0.296	0.236	0.231	0.034	0.209
HC-PIN	99	0.646	0.140	0.230	0.256	0.233	0.024	0.196
PCE-FR	274	0.534	0.178	0.267	0.279	0.169	0.029	0.035
PMD_{pc}	118	0.451	0.356	0.398	0.362	0.777	0.010	0.343

Table 1 shows that PMD_{pc} achieves a satisfactory performance on human PPI networks. Particularly, PMD_{pc} obtains the highest value of recall, F-measure, ACC, and Sep, which are 0.356, 0.398, 0.362, and 0.777, respectively. These results are significantly superior to the five other algorithms. Furthermore, CFinder achieves the highest precision of 0.959 and the lowest MMR of 0.017. ClusterONE identifies 755 protein complexes and achieves the highest MMR of 0.084. These values elaborate that our approach achieved an ideal result in identifying protein complexes from sparse PPI networks.

From Table 1, we can also clearly see that our method obtains the second highest value of MCC, which is 12.28% lower than that of ClusterONE. It demonstrates that our method achieved satisfactory performance in dealing imbalanced data.

To void the advantage of some evaluation metric, the composite score [24] is employed to wrap up the global performance. Interestingly, the composite comparison of our method shows absolute advantage in terms of F-measure, accuracy, and maximum matching ratio. Figure 2 presents the comparison results of the six algorithms on the HPRD dataset. The composite score of F-measure, accuracy, and maximum matching ratio is 0.770, which is 19.20% higher than the highest value of the five other methods. It further demonstrates the effectiveness of our method.

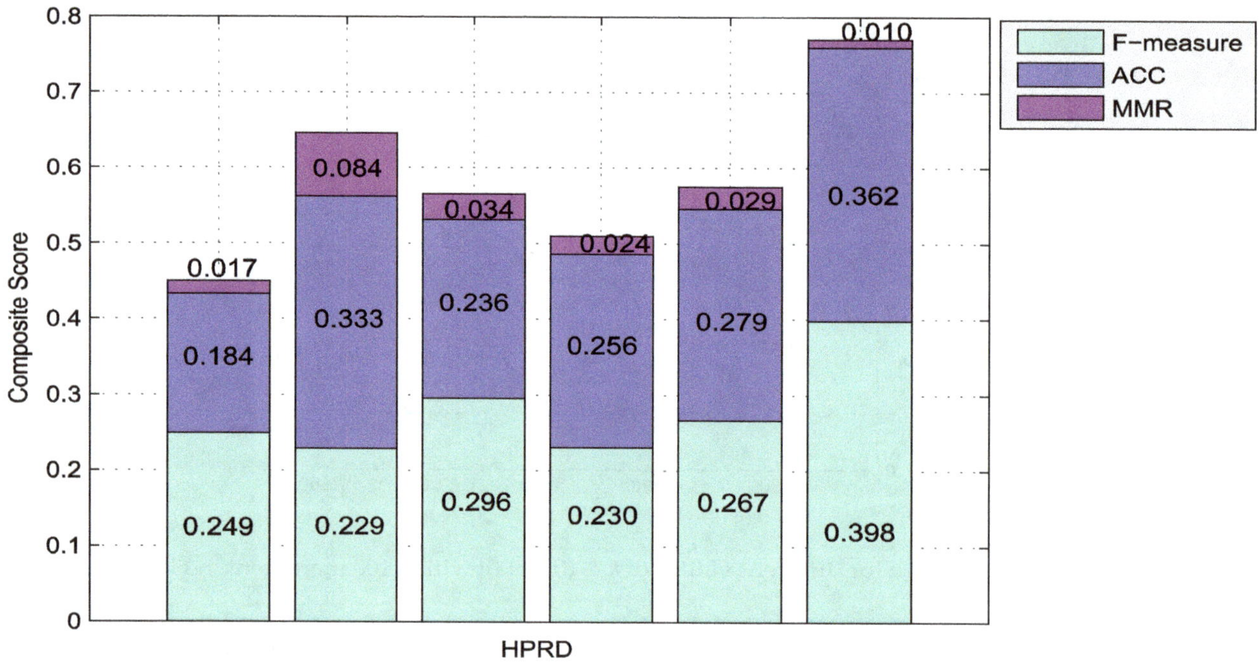

Figure 2. Results comparison of the six algorithms in HPRD dataset using CHPC2012 gold standard dataset. Columns correspond to the following algorithms, CFinder, ClusterONE, HC-PIN, PCE-FR, and PMD_{pc} from left to right. Various color of the same columns denotes the individual components of the composite score of the algorithm (cyan = F-measure, blue = ACC, and purple = MMR). The total height of each column is the value of the composite score for a special algorithm in a special dataset. Large score shows the clustering result is better.

3. Materials and Methods

3.1. Materials and Datasets

Our method is applied to detect the protein complexes in the human PPI dataset downloaded from Ref. [24], in which 9459 proteins and 36,935 interactions with the density of 0.0008 are included. The gold standard dataset is employed to evaluate the performance of the protein complexes identified in sparse PPI networks, which is CHPC2012 [32], integrating three databases, namely, CORUM [33], HPRD [34], and PINdb [35], and includes 1389 complexes and 3065 proteins.

3.2. Methods

Consider a sample dataset that consists of p eigenvectors in n samples, which is described by a matrix X with size $n \times p$ [30]. Without loss of generality, we assume that the means of column and row X are zero. The singular value decomposition of matrix X can be written as follows:

$$X = U\Delta V^T, U^T U = I_n, V^T V = I_p \tag{1}$$

The decomposition of sparse matrix is executed by imposing additional constraints on U and V. The single-factor PMD can be optimized using the following objective function, which is formulated as [30]

$$\begin{aligned} &\underset{\delta,u,v}{\operatorname{argmin}} \tfrac{1}{2}||\eta - \delta u v^T||_F^2, \\ &s.t.||u||_2^2 = 1, ||v||_2^2 = 1, \\ &P_1(u) \le c_1, P_2(v) \le c_2, \delta \ge 0. \end{aligned} \tag{2}$$

in which u is a column of U, v is a column of V, δ is a diagonal element of the matrix of η, $||\bullet||_F$ is the Frobenius norm, and P_1 and P_2 are penalty functions that have variety of forms [30].

Let U and V be $n \times R$ and $p \times R$ orthogonal matrices, respectively, and Δ a diagonal matrix with diagonal elements δ_r [30]

$$\frac{1}{2}\|\eta - U\Delta V^T\|_F^2 = \frac{1}{2}\|\eta\|_F^2 - \sum_{r=1}^{R} u_r^T \eta v_r \delta_r + \frac{1}{2}\sum_{r=1}^{R} \delta_r^2 \tag{3}$$

Therefore, when $R = 1$, we can infer that u and v satisfy Equation (7) and the following condition:

$$\begin{aligned} &\underset{u,v}{\operatorname{argmax}} u^T \eta v \\ &s.t.\ \|u\|_2^2 = 1,\ \|v\|_2^2 = 1,\ P_1(u) \le c_1,\ P_2(v) \le c_1 \end{aligned} \tag{4}$$

Moreover, δ satisfies Equation (2) when $\delta = u^T \eta v$.

The optimization problem in Equation (4) can be applied to the following biconvex optimization [30]:

$$\begin{aligned} &\underset{u,v}{\operatorname{argmax}} u^T \delta v \\ &s.t.\ \|u\|_2^2 \le 1,\ \|v\|_2^2 \le 1,\ P_1(u) \le c_1,\ P_2(v) \le c_2 \end{aligned} \tag{5}$$

Equation (5) satisfies Equation (4) based on the appropriate value of c [30]. Equation (5) is called the single factor PMD, and the iterative algorithm used to optimize it is described in Algorithm 1:

Algorithm 1. Calculating the single factor of PMD.

Step1. Initialize v and let unit $L_2 - norm$.
Step2. Interate until convergence:

(i) $u \leftarrow \arg\max_u u^T \delta v,\ s.t.\|u\|_2^2 \le 1, P_1(u) \le c_1$
(ii) $v \leftarrow \arg\max_v u^T \delta v,\ s.t.\|v\|_2^2 \le 1, P_2(v) \le c_2$

Step3. $d \leftarrow u^T \delta v$

Equation (2) is computed repeatedly to obtain other PMD factors. The corresponding algorithm is described in Algorithm 2.

Algorithm 2. Calculating the k factor of PMD.

Step1. $\eta^1 \leftarrow \eta$;
Step2. For $r \in 1, 2, \ldots, R$

(i) The single factor PMD (Algorithm 1) is executed on the matrix of η^r, computing u_r, v_r, δ_r, respectively;
(ii) $\eta^{r+1} \leftarrow \eta^r - \delta_r u_r v_r^T$

The constraint is imposed on u and v with $L_1 - norm$, i.e., $\|u\|_1 \le c_1, \|v\|_1 \le c_2$. By selecting parameters c_1 and c_2 appropriately, PMD can make factors u and v sparse. Generally, c_1 and c_2

should be restricted to ranges $1 \leq c_1 \leq \sqrt{n}$ and $1 \leq c_2 \leq \sqrt{p}$. Thus, the PMD method is shaped as $PMD(L_1, L_2)$, which is described as follows:

$$\begin{aligned} &\underset{u,v}{\text{argmax}} u^T \eta v \\ &s.t.\ \|u\|_2^2 \leq 1,\ \|v\|_2^2 \leq 1,\ \|u\|_1 \leq c_1, \|v\|_1 \leq c_2 \end{aligned} \tag{6}$$

Let S denote the operator of the soft threshold, i.e., $S(a,c) = \text{sgn}(a)(|a| - c)_+$, in which $c > 0$, $x_+ = \begin{cases} x & x > 0 \\ 0 & x \leq 0 \end{cases}$. The corresponding theorem is as follows:

Theorem 1. *Considering the optimization problem*

$$\begin{aligned} &\underset{u}{\text{argmax}} u^T a \\ &s.t.\ \|u\|_2^2 \leq 1,\ \|u\|_1 \leq c. \end{aligned} \tag{7}$$

The solution is $u = \frac{S(a,\Delta)}{\|S(a,\Delta)\|_2}$. If $\|u\|_1 \leq c$, then $\Delta = 0$; otherwise, $\|u\|_1 = c\ s.t.\ \Delta > 0$. The detailed proof regarding the theorem can be found in Ref. [30]. The analysis shows the solution of Equation (6) with Algorithm 1. According to Theorem 1, the single factor PMD can be optimized, as shown in Algorithm 3:

Algorithm 3. The optimization process of the single factor PMD.

Step1. Initialize v and let unit $L_2 - norm$.
Step2. Iterate until convergence:

(i) $u \leftarrow \frac{S(Xv,\Delta_1)}{\|S(Xv,\Delta_1)\|_2}$, if $\|u\|_1 \leq c_1$, then $\Delta_1 = 0$, else $\|u\|_1 = c_1, s.t., \Delta_1 > 0$

(ii) $u \leftarrow \frac{S(X^T u,\Delta_2)}{\|S(X^T u,\Delta_2)\|_2}$, if $\|v\|_1 \leq c_2$, then $\Delta_2 = 0$, else $\|v\|_1 = c_2, s.t., \Delta_2 > 0$

Step3. $d \leftarrow u^T \delta v$

To obtain the sparse factors of u and v, we let $c_1 = c\sqrt{n}, c_2 = c\sqrt{p}$, and the values of Δ_1 and Δ_2 are selected by the binary search.

For comprehensive discussion, discovered protein complexes and gold standard dataset are matched. The following evaluation measures are employed in this study.

F-measure. Two protein complexes, namely, p and g, are generated from the predicted protein complex and gold standard sets, respectively. The overlapping score $\text{os}(p,g)$ quantizes the closeness between the sets and is defined as follows [24]:

$$\text{os}(p,g) = \frac{|C_p \cap C_g|}{|C_p| \bullet |C_g|} \tag{8}$$

in which C_p, C_g denote protein complex sets p and g, respectively. If $\text{os}(p,g) \geq \theta$, then the two complexes are matched, in which θ is the threshold. θ is set as 0.2, which is consistent with many experiments for protein complex identification [24]. Let P and G represent the detected protein complex and gold standard sets, respectively; N_{cp} describes the number of identified protein complexes that match at least one gold standard set, i.e., $N_{cp} = |\{p|p \in P, \exists g \in G, os(p,g) \geq \theta\}|$; and N_{cp} presents the number of gold standard protein complexes that match at least one identified complex, that is $N_{cg} = |\{g|g \in G, \exists p \in P, os(p,g) \geq \theta\}|$. F-measure is mathematically defined as [24]

$$F - measure = \frac{2 \times Precision \times Recall}{Precision + Recall} \tag{9}$$

in which $Precision = N_{cp}/|P|$, $Recall = N_{cg}/|G|$. F-measure is defined as the harmonic mean of precision and recall, which can evaluate the overall performance of the detection methods.

ACC (Accuracy, ACC). ACC is used to quantify the quality of detected protein complexes, which is the geometric means of sensitivity and positive predictive value, PPV. The corresponding formulas are described as follows [24]:

$$ACC = \sqrt{S_n \times PPV} \tag{10}$$

in which $S_n = \frac{\sum_{i=1}^{n} \max_{j=1}^{m} t_{ij}}{\sum_{i=1}^{n} n_i}$, $PPV = \frac{\sum_{j=1}^{m} \max_{i=1}^{n} t_{ij}}{\sum_{j=1}^{m} \sum_{i=1}^{n} t_{ij}}$.

Sep (Separation, Sep). To void the case wherein proteins of a gold standard complex are matched with several identified protein complexes, Sep is used to measure the one-to-one correspondence between generated protein complexes and gold standard protein complexes. The formula is described as follows [24]:

$$\mathrm{Sep}_g = \frac{\sum_{i=1}^{n} \sum_{j=1}^{m} Sep_{ij}}{n},\ \mathrm{Sep}_p = \frac{\sum_{j=1}^{m} \sum_{i=1}^{n} Sep_{ij}}{m}, \mathrm{Sep} = \sqrt{\mathrm{Sep}_g \times Sep_p}, \tag{11}$$

in which $\mathrm{Sep}_{ij} = \frac{(t_{ij})^2}{\sum_{i=1}^{n} t_{ij} * \sum_{j=1}^{m} t_{ij}}$. In Formulas (10) and (11), n is the number of protein complexes in the gold standard dataset, m is the number of identified protein complexes, t_{ij} denotes the size of intersection between the ith gold standard complex and the jth detected complex, and n_i denotes the number of proteins included in the ith gold standard complex.

MMR (Maximum Matching Ratio). MMR is used to describe the maximum one-to-one matching between the identified and gold standard protein complexes, which are defined as follows [24]:

$$MMR(g,p) = \frac{\sum_{i=1}^{n} \max_{j=1}^{m} os(g_i, p_j)}{N_i} \tag{12}$$

in which os represents the overlapping score between two protein complexes, g_i is the ith gold standard complex, and p_j represents the jth identified protein complex.

MCC (Matthews Correlation Coefficient). MCC is widely used in bioinformatics as a performance metric that can handle imbalanced data. The formula is described as follows [24]:

$$MCC = \frac{TP \times TN - FP \times FN}{\sqrt{(TP + FN)(TP + FP)(TN + FP)(TN + FN)}} \tag{13}$$

in which TP, TN, FP, and FN mean the true positive, true negative, false positive, and false negative, respectively.

3.3. Detection of Protein Complexes Using PMD_{pc}

A PPI network is usually modeled as an undirected weight graph $G = (V, E, \omega)$, in which V represents a set of nodes (proteins), E is a set of edges (protein pairs), and ω is a set of similarity value between each protein pairs. The similarity of GO (Gene Ontology, GO) terms is mathematically expressed as follows [36]:

$$Sim(i,j) = \frac{|N(i) \cap N(j)|}{\min(N(i), N(j))} \tag{14}$$

in which $Sim(i,j)$ indicates the GO similarity of the protein pair (i,j). $N(i)$ denotes the number of GO terms that annotate the protein i. The PPI network is stocked as the matrix X with a size of $n \times n$, which is transformed into the vertex–PCA matrix X of size $n \times p$ by the principal component analysis, in which each row of X represents a protein in all n samples (protein complexes), and each column of X represents the expression level of a sample in all p proteins.

According to Section 3.2, the matrix X is decomposed into three matrices, namely, U, V, and Δ by PMD. The graphical description of PMD_{pc} is shown in Figure 3, in which u_k is the kth principal component, v_k is the kth expression model of the principal component, and u_{ik} indicates that the kth protein is projected on the kth protein complex. Therefore, matrix U is decomposed into several clusters (protein complexes) due to matrix decomposition.

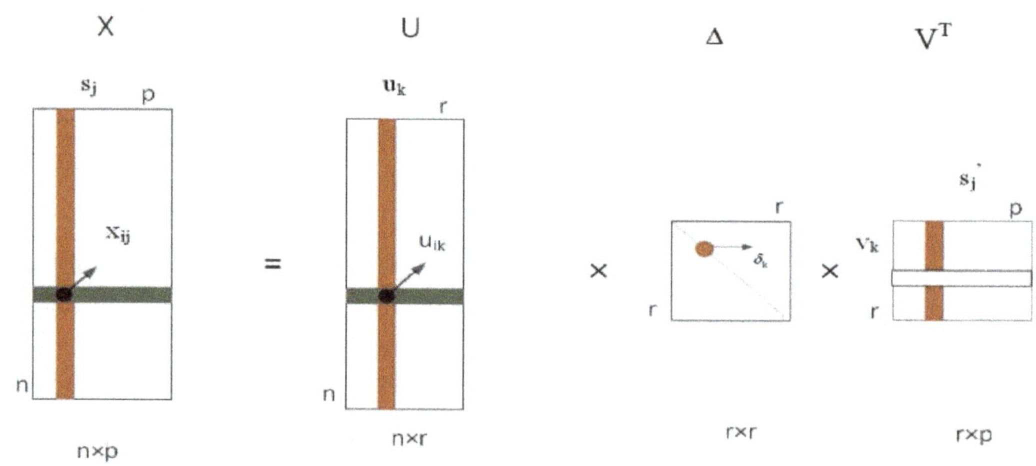

Figure 3. Graphical description of PMD_{pc}. Matrix X is decomposed into two base matrices, namely, U, V, and a diagonal matrix Δ.

PMD_{pc} is implemented in Java, and all experiments are performed on an Intel(R) Core(TM) i7-5557U CPU with 2.2 GHz and 8 GB RAM running Windows 7.0. The elapsed time is 9533 s.

4. Conclusions

The identification of protein complex helps us to discover and understand the cellular organizations and biological functions in PPI networks. Previous computational approaches mainly identified protein complexes in dense PPI networks, which had inferior performances in sparse PPI networks. In this work, PMD_{pc} is proposed on the basis of the penalized matrix decomposition to detect protein complexes in the human protein interaction network with 0.0008 density.

The performance of our method, PMD_{pc}, is compared with the performances of CFinder, ClusterONE, RRW, HC-PIN, and PCE-FR on the human PPI dataset derived from HPRD to validate the utilization of our method. The experimental results show that our proposed algorithm is better than the five classical approaches based on F-measure, ACC, and MMR. However, only the human PPI network was taken as the experimental dataset. The new method should be suitable for substructure detection with other sparse networks. Therefore, our algorithm will be used in the future to investigate other species of complex networks, such as gene regulatory and disease networks.

Author Contributions: Conceptualization: B.C. and S.D.; data curation: S.C.; formal analysis: S.D.; investigation: S.C.; methodology: B.C.; project administration: S.D.; resources: P.D.; software: S.C.; supervision: S.D.; validation: B.C., H.Q., and G.L.; writing-original draft: B.C.; writing-review & editing: P.D.

References

1. Enright, A.J.; Van Dongen, S.; Ouzounis, C.A. An efficient algorithm for large-scale detection of protein families. *Nucleic Acids Res.* **2002**, *30*, 1575–1584. [CrossRef] [PubMed]
2. Bader, G.D.; Hogue, C.W. An automated method for finding molecular complexes in large protein interaction networks. *BMC Bioinform.* **2003**, *4*, 2. [CrossRef]
3. Adamcsek, B.; Palla, G.; Farkas, I.J.; Derenyi, I.; Vicsek, T. Cfinder: Locating cliques and overlapping modules in biological networks. *Bioinformatics* **2006**, *22*, 1021–1023. [CrossRef] [PubMed]
4. Altaf-Ul-Amin, M.; Shinbo, Y.; Mihara, K.; Kurokawa, K.; Kanaya, S. Development and implementation of an algorithm for detection of protein complexes in large interaction networks. *BMC Bioinform.* **2006**, *7*, 1–13. [CrossRef] [PubMed]
5. Gao, L.; Sun, P.G.; Song, J. Clustering algorithm for detecting functional modules in protein interaction networks. *J. Bioinform. Comput. Biol.* **2011**, *7*, 217–242. [CrossRef]
6. Pizzuti, C.; Rombo, S.E. A coclustering approach for mining large protein-protein interaction networks. *IEEE ACM Trans. Comput. Biol.* **2012**, *9*, 717–730. [CrossRef] [PubMed]
7. Wang, J.X.; Li, M.; Chen, J.E.; Pan, Y. A fast hierarchical clustering algorithm for functional modules discovery in protein interaction networks. *IEEE ACM Trans. Comput. Biol.* **2011**, *8*, 607–620. [CrossRef] [PubMed]
8. Girvan, M.; Newman, M.E.J. Community structure in social and biological networks. *Proc. Natl. Acad. Sci. USA* **2002**, *99*, 7821–7826. [CrossRef] [PubMed]
9. Rivera, C.G.; Vakil, R.; Bader, J.S. Nemo: Network module identification in cytoscape. *BMC Bioinform.* **2010**, *11* (Suppl. 1), S61. [CrossRef] [PubMed]
10. Nepusz, T.; Yu, H.Y.; Paccanaro, A. Detecting overlapping protein complexes in protein-protein interaction networks. *Nat. Methods* **2012**, *9*, 471–472. [CrossRef] [PubMed]
11. Wu, H.; Gao, L.; Dong, J.H.; Yang, X.F. Detecting overlapping protein complexes by rough-fuzzy clustering in protein-protein interaction networks. *PLoS ONE* **2014**, *9*, 1856. [CrossRef] [PubMed]
12. Rhrissorrakrai, K.; Gunsalus, K.C. Mine: Module identification in networks. *BMC Bioinform.* **2011**, *12*, 192. [CrossRef] [PubMed]
13. Voevodski, K.; Teng, S.H.; Xia, Y. Finding local communities in protein networks. *BMC Bioinform.* **2009**, *10*, 297. [CrossRef] [PubMed]
14. Jiang, P.; Singh, M. Spici: A fast clustering algorithm for large biological networks. *Bioinformatics* **2010**, *26*, 1105–1111. [CrossRef] [PubMed]
15. Cao, B.W.; Luo, J.W.; Liang, C.; Wang, S.L.; Ding, P.J. Pce-fr: A novel method for identifying overlapping protein complexes in weighted protein-protein interaction networks using pseudo-clique extension based on fuzzy relation. *IEEE Trans. Nanobiosci.* **2016**, *15*, 728–738. [CrossRef] [PubMed]
16. Vella, D.; Marini, S.; Vitali, F.; di Silvestre, D.; Mauri, G.; Bellazzi, R. Mtgo: Ppi network analysis via topological and functional module identification. *Sci. Rep.* **2018**, *8*, 5499. [CrossRef] [PubMed]
17. Kouhsar, M.; Zare-Mirakabad, F.; Jamali, Y. Wcoach: Protein complex prediction in weighted ppi networks. *Genes Genet. Syst.* **2015**, *90*, 317–324. [CrossRef] [PubMed]
18. Hu, L.; Chan, K.C.C. A density-based clustering approach for identifying overlapping protein complexes with functional preferences. *BMC Bioinform.* **2015**, *16*, 174. [CrossRef] [PubMed]
19. Cao, B.; Luo, J.; Liang, C.; Wang, S. Identifying protein complexes by combining network topology and biological characteristics. *J. Comput. Theor. Nanosci.* **2016**, *13*, 1546–1955. [CrossRef]
20. King, A.D.; Przulj, N.; Jurisica, I. Protein complex prediction via cost-based clustering. *Bioinformatics* **2004**, *20*, 3013–3020. [CrossRef] [PubMed]
21. Macropol, K.; Can, T.; Singh, A.K. Rrw: Repeated random walks on genome-scale protein networks for local cluster discovery. *BMC Bioinform.* **2009**, *10*, 283. [CrossRef] [PubMed]
22. Liu, G.M.; Wong, L.; Chua, H.N. Complex discovery from weighted ppi networks. *Bioinformatics* **2009**, *25*, 1891–1897. [CrossRef] [PubMed]
23. Wu, M.; Li, X.L.; Kwoh, C.K.; Ng, S.K. A core-attachment based method to detect protein complexes in ppi networks. *BMC Bioinform.* **2009**, *10*, 169. [CrossRef] [PubMed]
24. Maulik, U.; Mukhopadhyay, A.; Bhattacharyya, M.; Kaderali, L.; Brors, B.; Bandyopadhyay, S.; Eils, R. Mining quasi-bicliques from hiv-1-human protein interaction network: A multiobjective biclustering approach. *IEEE ACM Trans. Comput. Biol.* **2013**, *10*, 423–435. [CrossRef] [PubMed]

25. Cao, B.; Luo, J.; Liang, C.; Wang, S.; Song, D. Moepga: A novel method to detect protein complexes in yeast protein-protein interaction networks based on multiobjective evolutionary programming genetic algorithm. *Comput. Biol. Chem.* **2015**, *58*, 173–181. [CrossRef] [PubMed]
26. Zhu, L.; Deng, S.-P.; You, Z.-H.; Huang, D.-S. Identifying spurious interactions in the protein-protein interaction networks using local similarity preserving embedding. *IEEE/ACM Trans. Comput. Biol. Bioinform.* **2017**, *14*, 345–352. [CrossRef] [PubMed]
27. Zhang, Y.; Du, N.; Ge, L. A collective nmf method for detecting protein functional module from multiple data sources. In Proceedings of the ACM Conference on Bioinformatics, Computational Biology and Biomedicine, Orlando, FL, USA, 7–10 October 2012; pp. 655–660. [CrossRef]
28. Zhang, S.H.; Li, Q.J.; Liu, J.; Zhou, X.J. A novel computational framework for simultaneous integration of multiple types of genomic data to identify microrna-gene regulatory modules. *Bioinformatics* **2011**, *27*, I401–I409. [CrossRef] [PubMed]
29. Zheng, C.H.; Zhang, L.; Ng, V.T.Y.; Shiu, S.C.K.; Huang, D.S. Molecular pattern discovery based on penalized matrix decomposition. *IEEE ACM Trans. Comput. Biol.* **2011**, *8*, 1592–1603. [CrossRef] [PubMed]
30. Witten, D.M.; Tibshirani, R.; Hastie, T. A penalized matrix decomposition, with applications to sparse principal components and canonical correlation analysis. *Biostatistics* **2009**, *10*, 515–534. [CrossRef] [PubMed]
31. Liu, J.-X.; Liu, J.; Gao, Y.-L.; Mi, J.-X.; Ma, C.-X.; Wang, D. A class-information-based penalized matrix decomposition for identifying plants core genes responding to abiotic stresses. *PLoS ONE* **2014**, *9*, e106097. [CrossRef] [PubMed]
32. Wu, M.; Yu, Q.; Li, X.L.; Zheng, J.; Huang, J.F.; Kwoh, C.K. Benchmarking human protein complexes to investigate drug-related systems and evaluate predicted protein complexes. *PLoS ONE* **2013**, *8*. [CrossRef] [PubMed]
33. Yang, P.; Li, X.; Wu, M.; Kwoh, C.K.; Ng, S.K. Inferring gene-phenotype associations via global protein complex network propagation. *PLoS ONE* **2011**, *6*, e21502. [CrossRef] [PubMed]
34. Peri, S.; Navarro, J.D.; Kristiansen, T.Z.; Amanchy, R.; Surendranath, V.; Muthusamy, B.; Gandhi, T.K.; Chandrika, K.N.; Deshpande, N.; Suresh, S.; et al. Human protein reference database as a discovery resource for proteomics. *Nucleic Acids Res.* **2004**, *32*, D497. [CrossRef] [PubMed]
35. Luc, P.V.; Tempst, P. Pindb: A database of nuclear protein complexes from human and yeast. *Bioinformatics* **2004**, *20*, 1413–1415. [CrossRef] [PubMed]
36. Shalgi, R.; Lieber, D.; Oren, M.; Pilpel, Y. Global and local architecture of the mammalian microrna-transcription factor regulatory network. *PLoS Comput. Biol.* **2007**, *3*, e131. [CrossRef] [PubMed]

14

Small Universal Bacteria and Plasmid Computing Systems

Xun Wang [1]**, Pan Zheng** [2]**, Tongmao Ma** [1] **and Tao Song** [1,3,*]

1 College of Computer and Communication Engineering, China University of Petroleum, Qingdao 266580, China; wangsyun@upc.edu.cn (X.W.); matongmao@163.com (T.M.)
2 Department of Accounting and Information Systems, University of Canterbury, Christchurch 8041, New Zealand; panzheng@ieee.org
3 Departamento de Inteligencia Artificial, Universidad Politcnica de Madrid (UPM), Campus de Montegancedo, 28660 Boadilla del Monte, Spain
* Correspondence: tsong@upc.edu.cn or t.song@upm.es.

Abstract: Bacterial computing is a known candidate in natural computing, the aim being to construct "bacterial computers" for solving complex problems. In this paper, a new kind of bacterial computing system, named the bacteria and plasmid computing system (BP system), is proposed. We investigate the computational power of BP systems with finite numbers of bacteria and plasmids. Specifically, it is obtained in a constructive way that a BP system with 2 bacteria and 34 plasmids is Turing universal. The results provide a theoretical cornerstone to construct powerful bacterial computers and demonstrate a concept of paradigms using a "reasonable" number of bacteria and plasmids for such devices.

Keywords: bacterial computing; bacteria and plasmid system; Turing universality; recursively enumerable function

1. Introduction

In cell biology, bacteria, despite their simplicity, contain a well-developed cell structure that is responsible for some of their unique biological structures and pathogenicity. The bacterial DNA resides inside the bacterial cytoplasm, for which transfer of cellular information, transcription, and DNA replication occurs within the same compartment [1,2]. Along with chromosomal DNA, most bacteria also contain small independent pieces of DNA called plasmids, which can be conveniently obtained and released by a bacterium to act as a gene delivery vehicle between bacteria in the form of horizontal gene transfer [3].

Bacterial computing was coined with the purpose of building biological machines, which are developed to solve real-life engineering and science problems [4]. Practically, bacterial computing proves mechanisms and the possibility of using bacteria for solving problems in vivo. If an individual bacterium can perform computation work as a computer, this envisions a way to build millions of computers in vivo. These "computers", combined together, can perform complicated computing tasks with efficient communication via plasmids. Using such conjugation, DNA molecules, acting as information carriers, can be transmitted from one cell to another. On the basis of the communication, information in one bacteria can be moved to another and can be used for further information processing [5,6].

Bacterial computing models belong to the field of bio-computing models, such as DNA computing models [7–9] and membrane computing models [10–12]. Because of the computational intelligence and parallel information processing strategy in biological systems, most of the bio-computing models have been proven to have the desired computational power. Most of these can do what a Turing machine can do (see, e.g., [13–19]). The proposed bacterial computing models can provide powerful computing models at the theoretical level but a lack of practical results. Current bacterial computing models are designed for solving certain specific biological applications, such as bacteria signal pathway detecting, but give no result for computing power analysis.

In general bacterial computing models, information to be processed is encoded by DNA sequences, and conjugation is the tool for communicating among bacteria. The biological process is shown in Figure 1.

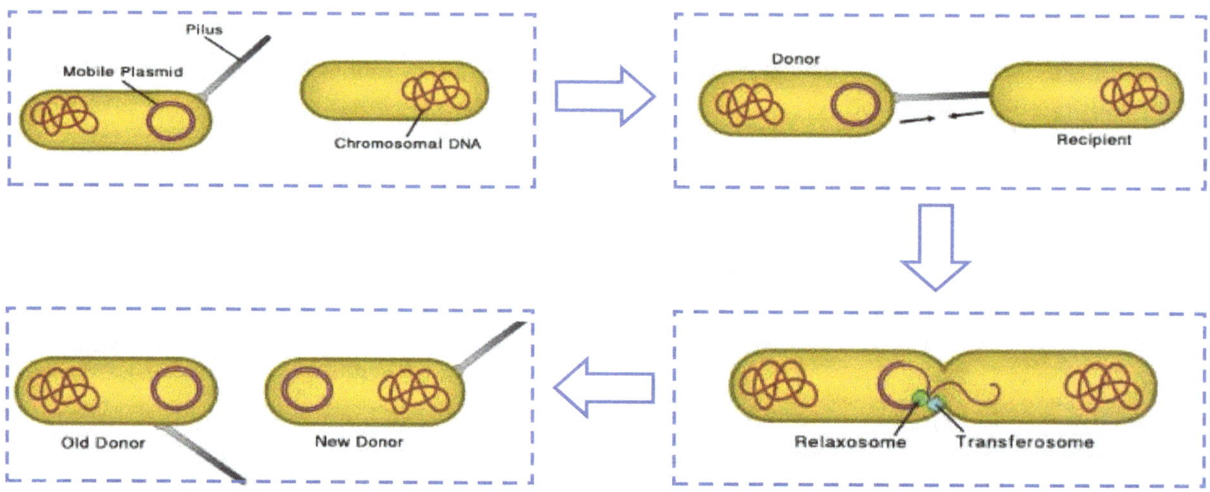

Figure 1. Bacteria conjugation from biological point of view.

Looking for small universal computing devices, such as small universal Turing machines [20,21], small universal register machines [22], small universal cellular automata [23], small universal circular Post machines [24], and so on, is a natural and well-investigated topic in computer science. Recently, this topic started to be considered also in the framework of bio-computing models [25–31].

In this work, we focus on designing small universal bacteria and plasmid computing systems (BP systems); that is, we construct Turing universal BP systems with finite numbers of bacteria and plasmids. Specifically, we demonstrate that a BP system with 2 bacteria and 34 plasmids is universal for computing recursively enumerable functions and families of sets of natural numbers. In the universality proofs, 2 bacteria are sufficient, as in [32], but the numbers of plasmids needed are reduced to about 10 from a possible infinite number. The results provide a theoretical cornerstone to construct powerful "bacterial computers" and demonstrate a concept of paradigms using a "reasonable" number of bacteria and plasmids for these devices.

2. The Bacteria and Plasmid System

In this work, as for automata in automata theory, the BP system is formally designed and defined. In general, the system is composed of three main components:

- a set of bacteria;
- a set of plasmids;
- a set of evolution rules in each bacterium, including conjugation rules and gene-editing (inserting/deleting) rules.

The evolution rules are in the form of productions in formal language theory, which are used to process and communicate information among bacteria. Such a system is proven to be powerful for a number of computing devices; that is, they can compute the sets of natural numbers that are Turing computable. However, in the universality proof, the number of plasmids involved is not limited. It is possible to use an infinite number of plasmids for information processing and exchanging. Such a feature is acceptable (as for the infinite tape in Turing machines) in mathematic theory but is not feasible with the biological facts.

A BP system of degree m is a construct of the following form:

$$\Pi = (O, b_1, b_2, \ldots, b_m, P, b_{out}), \text{where the following are true.}$$

- $O = \{g_1, g_2, \ldots, g_n\}$ is a set of genes in the chromosomal DNA of bacteria.
- $P = P_{crispr} \cup P_{temp} \cup \{p_{null}\}$ is a set of plasmids.
  - Plasmids in P_{crispr} are of the form $(cas9, gRNA^{\alpha}_{g_i})$ with $\alpha \in \{insert, delete\}$, which is used for cutting specific genes.
  - Plasmids in P_{temp} are of the form $(gRNA^{template}_{g_i})$, which takes templates of genes to be inserted.
  - Plasmid p_{null} is of the form (Pro^{Rel}_{Rap}) for bacteria conjugation.
- Variables $b_1, b_2, \ldots, b_m$ are m bacteria of the form $b_i = (w_i, R_i)$, where
  - w_i is a set of genes over O initially placed in bacterium b_i;
  - R_i is a set of rules in bacterium b_i of the following forms:
    (1) **Conjugation rule** is of the form (ATP-$P_c, b_i/b_j,$ ATP-P'_c), by which ATP in bacterium b_i is consumed and a set of plasmids $P'_c \subseteq P$ associated with ATP is transmitted into bacterium b_j.
    (2) **CRISPR/Cas9 gene inserting rule** is of the form $p_i p_{s_i} \times (g_j, g_k)$, where $p_i \in P_{crispr}$, $\alpha = insert$, $p_{s_i} \in P_{temp}$, and g_j and g_k are two neighboring genes. The insertion is operated if and only if g_j and g_k are neighboring genes and plasmids $p_i p_{s_i}$ are present in the bacterium.
    (3) **CRISPR/Cas9 gene deleting rule** is of the form $p_i \times (g_j, g_k)$ with $p_i p_{null} \in P_{crispr}$, $\alpha = detele$, and g_j and g_k being two neighboring genes. The rule can be used if and only there exists gene g_i placed between the two neighboring genes.
- Variable b_{out} is the output bacterium.

It is possible to have more than one enabled conjugation rule at a certain moment in a bacterium, but only one is non-deterministically chosen for use. This is due to the biological fact that ATP can support the transmission of one plasmid but not all of the plasmids. If a bacterium has more than one CRISPR/Cas9 operating rule associated with a certain common plasmid, only one of the rules is non-deterministically chosen for use; if the enabled CRISPR/Cas9 operating rules are associated with different plasmids, all of them will be used to edit the related genes.

The configuration of the system is described by chromosomal DNA encoding the information in each bacterium. Thus, the initial configuration is $\langle(w_1, w_2, \ldots, w_m\rangle$. Using the conjugation and CRISPR/Cas9 rules defined above, we can define the transitions among configurations. Any sequence of transitions starting from the initial configuration is called a computation. A computation is called successful if it reaches a halting configuration, that is, no rule can be used in any bacterium. The computational result is encoded by the chromosomal DNA in bacterium b_{out} when the system halts, where $b_{out} \in \{b_1, b_2, \ldots, b_m\}$ denotes the output bacterium. There are several ways to encode numbers by the chromosomal DNA. We use the number of genes in the chromosomal DNA to encode different numbers computed by the system.

The set of numbers computed by system Π is denoted by $N(\Pi)$. We denote by $NBP(bact_j, plas_k)$ the family of sets of numbers computed/generated by BP systems with m bacteria and k plasmids (if no limit is imposed on the values of parameters m and k, then the notation is replaced by $*$).

We need an input bacterium to receive genetic signals in the form of short DNA segments from the environment or certain bacteria, as well as an output bacteria, with which the system can compute functions. The input bacterium is denoted by b_{in} with $b_{in} \in \{b_1, b_2, \ldots, b_m\}$. Input bacterium b_{in} can read/receive information from the environment, where information is encoded by DNA segments or a string of genes. When a BP system has both input and output bacteria, it starts by reading/receiving information from the environment through input bacterium b_{in}. After reading the input information, the system starts its computation by using the conjugation and CRISPR/Cas9 gene inserting/deleting rules; it then finally halts. The computational result is stored in the output bacterium b_{out} encoded by a number of certain genes.

Mathematically, if the input information is x, which is encoded by DNA segments composed of x genes, when the system halts, bacterium b_{out} holds y genes. It is said that the BP system can compute the function $f(x) = y$. In general, if the inputs are $x_1, x_2, \ldots, x_n$ in the form of DNA strands containing x_i copies of gene g_i with $i = 1, 2, \ldots, n$, when the system halts, we obtain the computational result y in the output bacterium in the form of y copies of genes. The system is said to compute the function $f(x_1, x_2, \ldots, x_n) = y$.

3. Universality Results

In this section, we construct two small universal BP systems. Specifically, we construct a Turing universal BP system with 2 bacteria and 34 plasmids to compute recursively enumerable functions. As a natural-number computing device, a universal BP system with 2 bacteria and 34 plasmids is achieved.

In the following universality proofs, the notion of a register machine is used. A register machine is a construct of the form $M = (m, H, l_0, l_h, R)$, where m is the number of registers, H is the set of instruction labels, l_0 is the start label, l_h is the halt label (assigned to instruction HALT), and R is the set of instructions; each label from H labels only one instruction from R, thus precisely identifying it. The instructions are of the following forms:

- $l_i : (\text{ADD}(r), l_j, l_k)$ (add 1 to register r and then go to one of the instructions with labels l_j and l_k);
- $l_i : (\text{SUB}(r), l_j, l_k)$ (if register r is non-zero, then subtract 1 from it, and go to the instruction with label l_j; otherwise, go to the instruction with label l_k);
- l_h : HALT (the halt instruction).

A register machine M generates a set $N(M)$ of numbers in the following way: it starts with all registers being empty (i.e., storing the number zero) and then applies the instruction with label l_0; it continues to apply instructions as indicated by the labels (and made possible by the contents of registers). If the register machine finally reaches the halt instruction, then the number n present in specified register 0 at that time is said to be generated by M. If the computation does not halt, then no number is generated. It is known (e.g., see [33]) that register machines generate all sets of numbers that are Turing computable.

A register machine can also compute functions. In [22], register machines are proposed for computing functions, with the universality defined as follows: Let $\varphi_x(y)$ be a fixed admissible enumeration of the unary partial recursive functions. A register machine M is said to be universal if there is a recursive function g such that for all natural numbers x and y, it holds $\varphi_x(y) = M(g(x), y)$; that is, with input $g(x)$ and y introduced in registers 1 and 2, the result $\varphi_x(y)$ is obtained in register 0 when M halts.

A specific universal register machine M_u shown in Figure 2 is used here, which was modified by a universal register machine from [22]. Specifically, the universal register machine from [22] contains a separate check for zero of register 6 of the form l_8 : (SUB(6), l_0, l_{10}); this instruction was replaced in M_u by l_8 : (SUB(6), l_9, l_0), l_9 :(ADD(6), l_{10}) (see Figure 2). Therefore, in the modified universal register machine, there are 8 registers (numbered from 0 to 7) and 23 instructions (hence 23 labels), the last instruction being the halting instruction. The input numbers are introduced in registers 1 and 2, and the result is obtained in register 0.

l_0 : (SUB(1), l_1, l_2),	l_1 : (ADD(7), l_0),
l_2 : (ADD(6), l_3),	l_3 : (SUB(5), l_2, l_4),
l_4 : (SUB(6), l_5, l_3),	l_5 : (ADD(5), l_6),
l_6 : (SUB(7), l_7, l_8),	l_7 : (ADD(1), l_4),
l_8 : (SUB(6), l_9, l_0),	l_9 : (ADD(6), l_{10}),
l_{10} : (SUB(4), l_0, l_{11}),	l_{11} : (SUB(5), l_{12}, l_{13}),
l_{12} : (SUB(5), l_{14}, l_{15}),	l_{13} : (SUB(2), l_{18}, l_{19}),
l_{14} : (SUB(5), l_{16}, l_{17}),	l_{15} : (SUB(3), l_{18}, l_{20}),
l_{16} : (ADD(4), l_{11}),	l_{17} : (ADD(2), l_{21}),
l_{18}: (SUB(4), l_0, l_h),	l_{19} : (SUB(0), l_0, l_{18}),
l_{20} : (ADD(0), l_0),	l_{21} : (ADD(3), l_{18}),
l_h : HALT	

Figure 2. The universal register machine for computing Turing-computable functions [22].

3.1. A Small Universal BP System as Function Computing Device

Theorem 1. *There exists a Turing universal BP system with 2 bacteria and 34 plasmids that can compute Turing-computable recursively enumerable functions.*

Proof. To this aim, we construct a BP system Π with 2 bacteria and 34 plasmids to simulate the register machine M_u shown in Figure 2. The system Π is of the following form:

$$\Pi = (O, b_1, b_2, P, b_{in}, b_{out}), \text{where the following are true.}$$

- $O = \{g_0, g_1, \ldots, g_7, g_m\}$ is set of genes in chromosomal DNA of bacteria.
- $P = P_{crispr} \cup P_{temp} \cup \{p_{null}\}$ is a set of plasmids shown in Table 1, where
  - $P_{crispr} = \{p_1, p_2, \ldots, p_{22}, p_h\}$, whose elements associated with the labels of instructions are used for gene cutting;
  - $P_{temp} = \{p_{s_i} \mid i = 1,2,5,7,9,16,17,20,21\}$ are plasmids taking templates of genes to be inserted, which are used for simulating ADD instructions;
  - plasmid p_{null} for bacteria conjugation is used for simulating SUB instructions.
- $b_1 = (w_1, R_1)$, where $w_1 = \lambda$, meaning no initial chromosomal DNA is placed in bacteria b_1; the set of rules R_1 is shown in Table 2.
- $b_2 = (w_2, R_2)$, where $w_2 = g_0g_mg_1g_mg_2g_mg_3g_mg_4g_mg_5g_mg_6g_mg_7g_m$, indicating the initially placed chromosomal DNA in bacterium b_2; the set of rules R_2 is shown in Table 2.
- $b_{in} = b_{out} = b_2$, which means bacterium b_2 can read signals from the environment, and when the system halts, the computational result is stored in bacterium b_2.

In general, for each add instruction l_i acting on register $r \in \{0,1,2,3,4,5,6,7\}$, plasmids $p_i = (cas9, gRNA_{g_r}^{insert})$ and $p_{s_i} = (gRNA_{g_r}^{template})$ are associated; for any SUB instruction l_i acting on register $r \in \{0,1,2,3,4,5,6,7\}$, a plasmid $p_i = (cas9, gRNA_{g_r}^{delete})$ is associated in system Π. The numbers stored in register r are encoded by the number of copies of gene g_r with $r \in \{0,1,2,3,4,5,6,7\}$ in chromosomal DNA of bacterium b_2. Specifically, if the number stored in register r is $n \geq 0$, then bacterium b_2 contains $n+1$ copies of gene g_r.

During the simulation of register machine M_u by system Π, when bacterium b_1 holds a pair of plasmids $p_ip_{s_i}$ (respectively p_ip_{null}) and ATP, the system starts to simulate an ADD instruction (respectively a SUB instruction) l_i of M_u: plasmids $p_i p_{s_i}$ (respectively $p_i p_{null}$) are transmitted to bacterium b_2 by the conjugation rule; then one copy of gene g_r between neighboring genes g_r and g_m is inserted (respectively deleted) to simulate increasing (respectively decreasing) the number in register r by 1; after this, bacterium b_2 sends ATP and plasmids $p_j p_{null}$ to bacterium b_1 if the proceeding instruction l_j is a SUB instruction or plasmids $p_j p_{s_j}$ if the proceeding l_j is an ADD instruction.

Table 1. Plasmids in system Π.

Plasmid	Forms of Plasmids	Plasmid	Forms of Plasmids
p_0	$p_0 = (cas9, gRNA_{g_1}^{delete})$	p_{16}	$p_{16} = (cas9, gRNA_{g_4}^{insert})$
p_1	$p_1 = (cas9, gRNA_{g_7}^{insert})$	p_{17}	$p_{17} = (cas9, gRNA_{g_2}^{insert})$
p_2	$p_2 = (cas9, gRNA_{g_6}^{insert})$	p_{18}	$p_{18} = (cas9, gRNA_{g_4}^{delete})$
p_3	$p_3 = (cas9, gRNA_{g_5}^{delete})$	p_{19}	$p_{19} = (cas9, gRNA_{g_3}^{delete})$
p_4	$p_4 = (cas9, gRNA_{g_6}^{delete})$	p_{20}	$p_{20} = (cas9, gRNA_{g_0}^{insert})$
p_5	$p_5 = (cas9, gRNA_{g_5}^{insert})$	p_{21}	$p_{21} = (cas9, gRNA_{g_3}^{insert})$
p_6	$p_6 = (cas9, gRNA_{g_7}^{delete})$	p_h	$p_h = (cas9, gRNA_{g_h}^{delete})$
p_7	$p_7 = (cas9, gRNA_{g_1}^{insert})$	p_{s_1}	$p_{s_1} = (gRNA_{g_7}^{template})$
p_8	$p_8 = (cas9, gRNA_{g_6}^{delete})$	p_{s_2}	$p_{s_2} = (gRNA_{g_6}^{template})$
p_9	$p_9 = (cas9, gRNA_{g_6}^{insert})$	p_{s_5}	$p_{s_5} = (gRNA_{g_5}^{template})$
p_{10}	$p_{10} = (cas9, gRNA_{g_4}^{delete})$	p_{s_7}	$p_{s_7} = (gRNA_{g_1}^{template})$
p_{11}	$p_{11} = (cas9, gRNA_{g_5}^{delete})$	p_{s_9}	$p_{s_9} = (gRNA_{g_6}^{template})$
p_{12}	$p_{12} = (cas9, gRNA_{g_5}^{delete})$	$p_{s_{16}}$	$p_{s_{16}} = (gRNA_{g_4}^{template})$
p_{13}	$p_{13} = (cas9, gRNA_{g_2}^{delete})$	$p_{s_{17}}$	$p_{s_{17}} = (gRNA_{g_2}^{template})$
p_{14}	$p_{14} = (cas9, gRNA_{g_5}^{delete})$	$p_{s_{20}}$	$p_{s_{20}} = (gRNA_{g_0}^{template})$
p_{15}	$p_{15} = (cas9, gRNA_{g_3}^{delete})$	$p_{s_{21}}$	$p_{s_{21}} = (gRNA_{g_3}^{template})$
p_{null}	(Pro_{Rap}^{Rel})	p_{s_h}	$p_{s_h} = (Pro_{Rap}^{Rel})$

Initially, there is no chromosomal DNA initially placed in bacterium b_1, but bacterium b_2 has genes $w_2 = g_0 g_m g_1 g_m g_2 g_m g_3 g_m g_4 g_m g_5 g_m g_6 g_m g_7 g_m$. At the beginning, the system receives $g(x)$ copies of gene g_1 and y copies of gene g_2 from the environment through input bacterium b_2, which simulates the numbers $g(x)$ and y being introduced in registers 1 and 2 for register machine M_u. In this way, the chromosomal DNA of bacterium b_2 becomes

$$g_0 g_m g_1^{g(x)+1} g_m g_2^{y+1} g_m g_3 g_m g_4 g_m g_5 g_m g_6 g_m g_7 g_m.$$

Once completing the reading of information from the environment, a pair of plasmids $p_0 p_{s_0}$ and one unit of ATP is placed in bacterium b_1 to trigger the computation; meanwhile no plasmid or ATP is initially contained in bacterium b_2. The transition of system Π by reading input signals encoded by $g(x)$ copies of genes g_1 and y copies of gene g_2 through input bacterium b_2 is shown in Figure 3.

In what follows, we explain how system Π simulates ADD instructions and SUB instructions and outputs the computational result.

Simulating the ADD instruction: l_i : (DD(r), l_j).

We assume at a certain moment that system Π starts to simulate an ADD instruction l_i of M_u, acting on register $r \in \{0, 1, 2, \ldots, 7\}$. At that moment, bacterium b_1 holds two plasmids $p_i p_{s_i}$ and ATP, such that the conjugation rule (ATP-$p_i p_{s_i}$, b_1/b_2, ATP-$p_i p_{s_i}$) is used. By using the conjugation rule, plasmids $p_i p_{s_i}$ and ATP are transmitted to bacterium b_2. In system Π, plasmids p_i and p_{s_i} are associated with the ADD instruction l_i, where plasmid p_i is of the form $p_i = (cas9, gRNA_{g_r}^{insert})$ for cutting a certain site of chromosomal DNA, and p_{s_i} is of the form $p_i = (gRNA_{gr}^{template})$ carrying the gene to be inserted.

In bacterium b_2, the CRISPR/Cas9 inserting rule $p_i \times (g_r, g_m)$ is used to insert gene g_r between neighboring genes g_r and g_m. In this way, the number of gene g_r of bacterium b_2 is increased by 1, which simulates the number in register r being increased by 1. We note that there is a unique position at which gene g_r can be inserted with the context of neighboring g_r and g_m.

Table 2. Rules in each bacterium of system Π.

Sim.	Rules	Bac.
l_0	$(\text{ATP-}p_0p_{null}, b_1/b_2, \text{ATP-}p_0p_{null})$	b_1
	$p_0 \times (g_1, g_m)$, $(\text{ATP-}p_{null}, b_2/b_1, \text{ATP-}p_1p_{s_1})$, $(\text{ATP-}p_0p_{null}, b_2/b_1, \text{ATP-}p_2p_{s_2})$	b_2
l_1	$(\text{ATP-}p_1p_{s_1}, b_1/b_2, \text{ATP-}p_1p_{s_1})$	b_1
	$p_1 \times (g_7, g_m)$, $(\text{ATP-}p_{s_1}, b_2/b_1, \text{ATP-}p_0p_{null})$	b_2
l_2	$(\text{ATP-}p_2p_{s_2}, b_1/b_2, \text{ATP-}p_2p_{s_2})$	b_1
	$p_2 \times (g_6, g_m)$, $(\text{ATP-}p_{s_2}, b_2/b_1, \text{ATP-}p_3p_{null})$	b_2
l_3	$(\text{ATP-}p_3p_{null}, b_1/b_2, \text{ATP-}p_3p_{null})$	b_1
	$p_3 \times (g_5, g_m)$, $(\text{ATP-}p_{null}, b_2/b_1, \text{ATP-}p_2p_{s_2})$, $(\text{ATP-}p_3p_{null}, b_2/b_1, \text{ATP-}p_4p_{null})$	b_2
l_4	$(\text{ATP-}p_4p_{null}, b_1/b_2, \text{ATP-}p_4p_{null})$	b_1
	$p_4 \times (g_6, g_m)$, $(\text{ATP-}p_{null}, b_2/b_1, \text{ATP-}p_5p_{s_5})$, $(\text{ATP-}p_4p_{null}, b_2/b_1, \text{ATP-}p_3p_{null})$	b_2
l_5	$(\text{ATP-}p_5p_{s_5}, b_1/b_2, \text{ATP-}p_5p_{s_5})$	b_1
	$p_1 \times (g_5, g_m)$, $(\text{ATP-}p_{s_5}, b_2/b_1, \text{ATP-}p_6p_{null})$	b_2
l_6	$(\text{ATP-}p_6p_{null}, b_1/b_2, \text{ATP-}p_6p_{null})$	b_1
	$p_6 \times (g_7, g_m)$, $(\text{ATP-}p_{null}, b_2/b_1, \text{ATP-}p_7p_{s_7})$, $(\text{ATP-}p_6p_{null}, b_2/b_1, \text{ATP-}p_8p_{null})$	b_2
l_7	$(\text{ATP-}p_7p_{s_7}, b_1/b_2, \text{ATP-}p_7p_{s_7})$	b_1
	$p_7 \times (g_1, g_m)$, $(\text{ATP-}p_4p_{null}, b_2/b_1, \text{ATP-}p_4p_{null})$	b_2
l_8	$(\text{ATP-}p_8p_{null}, b_1/b_2, \text{ATP-}p_8p_{null})$	b_1
	$p_8 \times (g_6, g_m)$, $(\text{ATP-}p_{null}, b_2/b_1, \text{ATP-}p_9p_{s_9})$, $(\text{ATP-}p_0p_{null}, b_2/b_1, \text{ATP-}p_0p_{null})$	b_2
l_9	$(\text{ATP-}p_9p_{s_9}, b_1/b_2, \text{ATP-}p_9p_{s_9})$	b_1
	$p_9 \times (g_6, g_m)$, $(\text{ATP-}p_{10}p_{null}, b_2/b_1, \text{ATP-}p_{10}p_{null})$	b_2
l_{10}	$(\text{ATP-}p_{10}p_{null}, b_1/b_2, \text{ATP-}p_{10}p_{null})$	b_1
	$p_{10} \times (g_4, g_m)$, $(\text{ATP-}p_0p_{null}, b_2/b_1, \text{ATP-}p_0p_{null})$, $(\text{ATP-}p_{10}p_{null}, b_2/b_1, \text{ATP-}p_{11}p_{null})$	b_2
l_{11}	$(\text{ATP-}p_{10}p_{null}, b_1/b_2, \text{ATP-}p_{11}p_{null})$	b_1
	$p_{11} \times (g_5, g_m)$, $(\text{ATP-}p_{null}, b_2/b_1, \text{ATP-}p_{12}p_{null})$, $(\text{ATP-}p_{11}p_{null}, b_2/b_1, \text{ATP-}p_{13}p_{null})$	b_2
l_{12}	$(\text{ATP-}p_{12}p_{null}, b_1/b_2, \text{ATP-}p_{12}p_{null})$	b_1
	$p_{12} \times (g_5, g_m)$, $(\text{ATP-}p_{null}, b_2/b_1, \text{ATP-}p_{14}p_{null})$, $(\text{ATP-}p_{12}p_{null}, b_2/b_1, \text{ATP-}p_{15}p_{null})$	b_2
l_{13}	$(\text{ATP-}p_{13}p_{null}, b_1/b_2, \text{ATP-}p_{13}p_{null})$	b_1
	$p_{13} \times (g_2, g_m)$, $(\text{ATP-}p_{null}, b_2/b_1, \text{ATP-}p_{18}p_{null})$, $(\text{ATP-}p_{13}p_{null}, b_2/b_1, \text{ATP-}p_{19}p_{null})$	b_2
l_{14}	$(\text{ATP-}p_{14}p_{null}, b_1/b_2, \text{ATP-}p_{14}p_{null})$	b_1
	$p_{14} \times (g_5, g_m)$, $(\text{ATP-}p_{null}, b_2/b_1, \text{ATP-}p_{16}p_{s_{16}})$, $(\text{ATP-}p_{14}p_{null}, b_2/b_1, \text{ATP-}p_{17}p_{s_{17}})$	b_2
l_{15}	$(\text{ATP-}p_{15}p_{null}, b_1/b_2, \text{ATP-}p_{15}p_{null})$	b_1
	$p_{15} \times (g_3, g_m)$, $(\text{ATP-}p_{null}, b_2/b_1, \text{ATP-}p_{18}p_{null})$, $(\text{ATP-}p_{15}, b_2/b_1, \text{ATP-}p_{20}p_{s_{20}})$	b_2
l_{16}	$(\text{ATP-}p_{16}p_{s_{16}}, b_1/b_2, \text{ATP-}p_{16}p_{s_{16}})$	b_1
	$p_{16} \times (g_4, g_m)$, $(\text{ATP-}p_{s_{16}}, b_2/b_1, \text{ATP-}p_{11}p_{null})$	b_2
l_{17}	$(\text{ATP-}p_{17}p_{s_{17}}, b_1/b_2, \text{ATP-}p_{17}p_{s_{17}})$	b_1
	$p_{17} \times (g_2, g_m)$, $(\text{ATP-}p_{s_{17}}, b_2/b_1, \text{ATP-}p_{21}p_{s_{21}})$	b_2
l_{18}	$(\text{ATP-}p_{18}p_{null}, b_1/b_2, \text{ATP-}p_{18}p_{null})$	b_1
	$p_{18} \times (g_4, g_m)$, $(\text{ATP-}p_{null}, b_2/b_1, \text{ATP-}p_0p_{null})$, $(\text{ATP-}p_{18}p_{null}, b_2/b_1, \text{ATP-}p_hp_{s_h})$	b_2
l_{19}	$(\text{ATP-}p_{19}p_{null}, b_1/b_2, \text{ATP-}p_{19}p_{null})$	b_1
	$p_{19} \times (g_3, g_m)$, $(\text{ATP-}p_{null}, b_2/b_1, \text{ATP-}p_0p_{null})$, $(\text{ATP-}p_{15}, b_2/b_1, \text{ATP-}p_{18}p_{null})$	b_2
l_{20}	$(\text{ATP-}p_{20}p_{s_{20}}, b_1/b_2, \text{ATP-}p_{20}p_{s_{20}})$	b_1
	$p_{20} \times (g_0, g_m)$, $(\text{ATP-}p_{s_{20}}, b_2/b_1, \text{ATP-}p_0p_{null})$	b_2
l_{21}	$(\text{ATP-}p_{21}p_{s_{21}}, b_1/b_2, \text{ATP-}p_{21}p_{s_{21}})$	b_1
	$p_0 \times (g_3, g_m)$, $(\text{ATP-}p_{s_{21}}, b_2/b_1, \text{ATP-}p_{18}p_{null})$	b_2
l_h	$(\text{ATP-}p_hp_{s_h}, b_1/b_2, \text{ATP-}p_hp_{s_h})$	

By using the CRISPR/Cas9 inserting rule, plasmid p_i is consumed, and plasmid p_{s_i} and ATP remain in bacterium b_2. The conjugation rule in bacterium b_2 is designed by the operation of the proceeding instruction l_j. One of the following two cases occurs in bacterium b_2.

- If instruction l_j is an ADD instruction, then bacterium b_2 has the conjugation rule (ATP-$p_{si}, b_2/b_1,$ ATP-$p_j p_{s_j}$). By using the rule, plasmids $p_j p_{s_j}$ and ATP are conjugated to bacterium b_1. In this case, system Π starts to simulate the proceeding ADD instruction l_j.
- If instruction l_j is a SUB instruction, then bacterium b_2 has the conjugation rule (ATP-$p_{si}, b_2/b_1,$ ATP-$p_j p_{null}$), by which plasmids $p_j p_{null}$ and ATP are transmitted to bacterium b_1. In this case, system Π starts to simulate the proceeding SUB instruction l_j.

Therefore, system Π can correctly simulate the ADD instruction of M_u. The system starts from bacterium b_1 having plasmid $p_i p_{s_i}$ and ATP, which are transmitted to bacterium b_2 by the conjugation rule. In bacterium b_2, the number of gene g_r in chromosomal DNA is increased by 1 using the CRISPR/Cas9 gene inserting rule, and plasmids $p_j p_{s_j}$ (if the proceeding instruction l_j is an ADD instruction) or $p_j p_{null}$ (if the proceeding instruction l_j is a SUB instruction) are transmitted to bacterium b_1, which means that system Π starts to simulate instruction l_j.

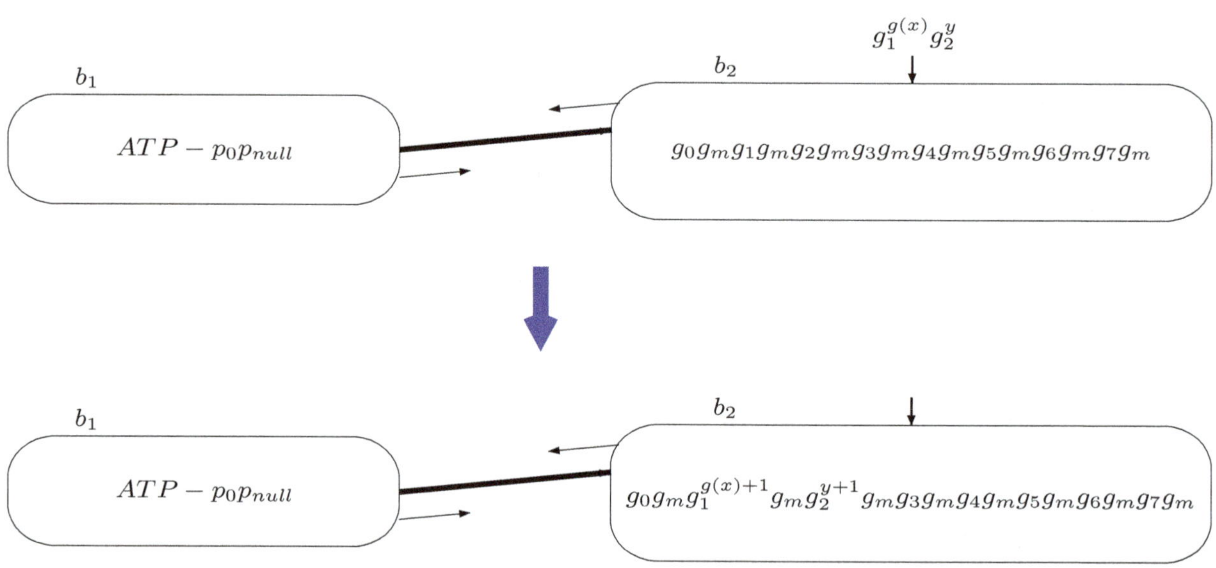

Figure 3. The transition of system Π by reading input information encoded by $g(x)$ copies of genes g_1 and y copies of gene g_2 through input bacterium b_2.

Simulating the SUB instruction: l_i : (SUB(r), l_j, l_k).

We suppose at a certain computation step that system Π has to simulate a SUB instruction l_i : (SUB(r), l_j, l_k). For any SUB instruction l_i, plasmid p_i of the form $p_i = (cas9, gRNA_{g_r}^{delete})$ is associated in system Π. In bacterium b_1, there are plasmids $p_i p_{null}$ and ATP such that the conjugation rule (ATP-$p_i p_{null}, b_1/b_2,$ ATP-$p_i p_{null}$) can be used. In bacterium b_2, it has the following two cases.

- If there is at least one gene g_r existing between neighboring genes g_r and g_m in chromosomal DNA of bacterium b_2 (corresponding to the case that the number stored in register r is $n > 0$), then the CRISPR/Cas9 deleting rule $p_i \times (g_1, g_m)$ is used to delete one copy of gene g_r from chromosomal DNA. This simulates the number stored in register r being decreased by 1. By consuming plasmid p_i, bacterium b_2 retains plasmid p_{null} and ATP such that a conjugation rule (ATP-$p_{null}, b_2/b_1,$ ATP-$p_j p_{s_j}$) or (ATP-$p_{null}, b_2/b_1,$ ATP-$p_j p_{null}$) is used, which depends on whether the proceeding instruction would be an ADD or a SUB instruction. In this way, plasmids $p_j p_{s_j}$ or $p_j p_{null}$) and ATP are transmitted to bacterium b_1. The system starts to simulate instruction l_j.

- If there is no gene g_r existing between neighboring genes g_r and g_m in chromosomal DNA of bacterium b_2 (corresponding to the case that the number stored in register r is 0), then the CRISPR/Cas9 deleting rule $p_i \times (g_1, g_m)$ cannot be used, but a conjugation rule (ATP-$p_i p_{null}, b_2/b_1$, ATP-$p_k p_{s_k}$) or (ATP-$p_i p_{null}, b_2/b_1$, ATP-$p_k p_{null}$) is able to be used. Plasmids ($p_k p_{s_k}$ or $p_k p_{null}$) and ATP are conjugated to bacterium b_1, which means the system starts to simulate instruction l_k.

We note that when plasmids $p_i p_{null}$ are conjugated to bacterium b_2 from bacterium b_1, it may happen that both the CRISPR/Cas9 deleting rule $p_i \times (g_1, g_m)$ and (ATP-$p_i p_{null}, b_2/b_1$, ATP-$p_k p_{s_k}$) (or (ATP-$p_i p_{null}, b_2/b_1$, ATP-$p_k p_{null}$)) can be used. In this case, the CRISPR/Cas9 deleting rule $p_i \times (g_1, g_m)$ will be applied because of the fact that it has priority over the plasmid transferring rule.

The simulation of a SUB instruction is correct: System Π starts from bacterium b_1 having plasmid $p_i p_{null}$ and ATP and ends with plasmid $p_j p_{s_j}$ or $p_j p_{null}$ and ATP (if the number stored in register r is $n > 0$) to start the simulation of instruction l_j; otherwise it ends with plasmid $p_k p_{s_k}$ or $p_k p_{null}$ and ATP (if the number stored in register r is 0) to start the simulation of instruction l_k.

Simulating the halt instruction: l_h : HALT.

When register machine M_u reaches the halt instruction l_h : HALT, the computation of register machine M_u halts. At that moment, bacterium b_1 in system Π holds plasmids $p_h p_{s_h}$ and ATP, and the conjugation rule (ATP-$p_h p_{sh}, b_1/b_2$, ATP-$p_h p_{s_h}$) can be used. By using the rule, plasmids $p_h p_{s_h}$ and ATP are transmitted to bacterium b_2; no gene can be edited by plasmid p_h, and no rule can be used. Hence, the computation of system Π finally halts.

The number of gene g_0 in chromosomal DNA of bacterium b_2 encodes the number stored in register 0 of M_u. If the number stored in register 0 is $n > 0$, then there are $n+1$ copies of gene g_0 in chromosomal DNA of bacterium b_2. The computational result can be obtained by counting the number of gene g_0 in chromosomal DNA of bacterium b_2.

From the above description of system Π and its work, it is clear that system Π can simulate each computation of M_u. We can check that the constructed system Π has

- 2 bacterium for conjugation with each other;
- 22 plasmids p_i for the 22 ADD and SUB instructions with $i = 0, 1, 2, \ldots 21$;
- 9 plasmids p_{s_i} for 9 ADD instructions with $i = 1, 2, 5, 7, 9, 16, 17, 20, 21$;
- 1 plasmid p_{null} for the 13 SUB instructions;
- 2 plasmids p_h and p_{s_h} for the HALT instruction;
- 8 genes g_i for encoding numbers in registers i with $i = 0, 1, 2 \ldots 7$;
- 1 gene g_m for separating gene g_i in chromosomal DNA.

This gives, in total, 2 bacteria, 34 plasmids, and 9 genes.

This concludes the proof. □

3.2. A Small Universal BP System as a Number Generator

In this section, we construct a small universal BP system as a number generator. A BP system Π_u is universal if, given a fixed admissible enumeration of the unary partial recursive functions $(\varphi_0, \varphi_1, \ldots)$, there is a recursive function g such that for each natural number x, whenever we input the number $g(x)$ in Π_u, the set of numbers generated by the system is equal to $\{n \in N | \varphi_x(n) \text{ is defined}\}$. In other words, after introducing the "code" $g(x)$ of the partial recursive function φ_x in the form of $g(x)$ copies of certain genes in chromosomal DNA of the input bacterium, the BP system generates all numbers n for which $\varphi_x(n)$ is defined.

System Π_u has the same topological structure, plasmids, and evolution rules as system Π constructed in Section 3.1, but the input bacterium is b_2 and the output bacterium is b_1. Differently from the universal computing devices considered in Section 3.1, the strategy to simulate a universal register machine as a number generator is as follows.

Step 1. The output bacterium b_1 initially has n copies of gene g_m.

Step 2. System Π_u starts by loading $g(x)$ copies of gene g_1 and n copies of gene g_2 in the input bacterium b_2.

Step 3. The computation of Π_u is activated by using plasmid p_0p_{null} to simulate the register machine M_u from Figure 2, with $g(x)$ stored in register 1, and number n stored in register 2.

If the computation in register machine M_u halts, instruction σ_{l_h} can finally be activated. To simulate register machine M_u reaching the HALT instruction, system Π_u holds plasmids $p_hp_{s_h}$ and transmits them to bacterium b_2. After this, system Π_u halts, as no rule can be used in bacterium b_2. When the system halts, the number of gene g_m in the output bacterium b_1 is the computational result, which is exactly the number n. Hence, the number n can be computed/generated by system Π_u.

The difference between systems Π and Π_u is the loading input information process. The initial configuration and transition of system Π_u by reading input signals encoded by $g(x)$ copies of genes g_1 and n copies of gene g_2 through input bacterium b_2 are shown in Figure 4.

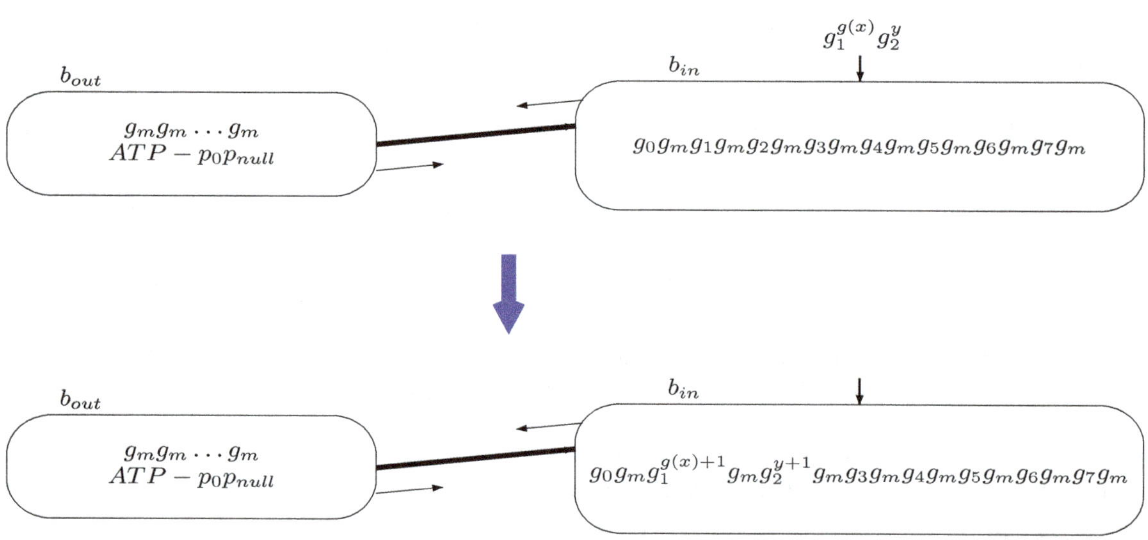

Figure 4. The initial configuration and transition of system Π_u by reading input information encoded by $g(x)$ copies of genes g_1 and n copies of gene g_2 through input bacterium b_2.

We can check that the constructed system Π_u has

- 2 bacterium for the conjugation with each other;
- 22 plasmids p_i for the 22 ADD and SUB instructions with $i = 0, 1, 2, \ldots 21$;
- 9 plasmids p_{s_i} for 9 ADD instructions with $i = 1, 2, 5, 7, 9, 16, 17, 20, 21$;
- 1 plasmid p_{null} for the 13 SUB instructions;
- 2 plasmids p_h and p_{s_h} for the HALT instruction;
- 8 genes g_i for encoding numbers in registers i with $i = 0, 1, 2 \ldots 7$;
- 1 gene g_m for separating gene g_i in chromosomal DNA.

This gives, in total, 2 bacteria, 34 plasmids, and 9 genes.

Therefore, we have the following theorem.

Theorem 2. *There is a Turing universal BP system with 2 bacteria and 34 plasmids that can compute a Turing-computable set of natural numbers.*

4. Conclusions

In this work, we construct two small universal BP systems. Specifically, it is obtained that a BP system with 2 bacteria, 34 plasmids, and 9 genes is universal for both computing recursively enumerable functions and computing/generating a family of sets of natural numbers. It is obtained that 34 plasmids are sufficient for constructing Turing universal BP systems. This provides theoretical support as well as paradigms using a reasonable number of bacteria and plasmids to construct powerful bacterial computers.

Following the research line, finding smaller universal BP systems deserves further research. A possible way to slightly decrease the number of plasmids used in small universal BP systems is using code optimization, exploiting some particularities of the register machine M_u. For example, as considered in [25], for the sequence of two consecutive ADD instructions l_{17}: $(\mathrm{ADD}(2), l_{21})$ and l_{21}: $(\mathrm{ADD}(3), l_{18})$, without any other instruction addressing the label l_{21}, the two ADD modules can be combined. However, a challenging problem regards what the minimum size of a universal BP system is—in other words, what the borderline between universality and non-universality is. Characterization of universality by BP systems is expected. A balance between the number of bacteria and plasmids in universal BP systems can be considered, that is, using more bacteria to reduce the number of plasmids.

It is worth developing the applications of BP systems. Bio-inspiring computing models perform well in computations, particularly in solving computational complex problems in feasible time [34–36]. It is of interest to use BP systems to solve computationally hard problems. Some specific applications using BP systems would be of interest to researchers from biological fields.

In artificial intelligence, there are many bio-inspired algorithms (see, e.g., [37,38]). It is worth designing bacteria-computing-inspired algorithms or introducing bacteria computing operators in classical algorithms. Additionally, it would be meaningful to construct powerful bacterial computers or computing devices in biological labs.

Author Contributions: Conceptualization, X.W. and T.S.; Methodology, X.W.; Software, T.M.; Validation, P.Z., T.S. and X.W.

Acknowledgments: This work was supported by the National Natural Science Foundation of China (Grant Nos. 61502535, 61572522, 61572523, 61672033, and 61672248), PetroChina Innovation Foundation (2016D-5007-0305), Key Research and Development Program of Shandong Province (No. 2017GGX10147), Natural Science Foundation of Shandong Province (No. ZR2017MF004), Talent introduction project of China University of Petroleum (No. 2017010054), Research Project TIN2016-81079-R (AEI/FEDER, Spain-EU) and Grant 2016-T2/TIC-2024 from Talento-Comunidad de Madrid, Project TIN2016-81079-R (MINECO AEI/FEDER, Spain-EU), and the InGEMICS-CM Project (B2017/BMD-3691, FSE/FEDER, Comunidad de Madrid-EU).

References

1. Gitai, Z. The new bacterial cell biology: Moving parts and subcellular architecture. *Cell* **2005**, *120*, 577–586. [CrossRef] [PubMed]
2. Goldstein, E.; Drlica, K. Regulation of bacterial dna supercoiling: Plasmid linking numbers vary with growth temperature. *Proc. Natl. Acad. Sci. USA* **1984**, *81*, 4046–4050. [CrossRef] [PubMed]
3. Summers, D. *The Biology of Plasmids*; John Wiley & Sons: Hoboken, NJ, USA, 2009.
4. Poet, J.L.; Campbell, A.M.; Eckdahl, T.T.; Heyer, L.J. Bacterial computing. *XRDS Crossroads ACM Mag. Stud.* **2010**, *17*, 10–15. [CrossRef]
5. Gupta, V.; Irimia, J.; Pau, I.; Rodríguez-Patón, A. Bioblocks: Programming protocols in biology made easier. *ACS Synth. Biol.* **2017**, *6*, 1230–1232. [CrossRef] [PubMed]

6. Gutierrez, M.; Gregorio-Godoy, P.; del Pulgar, G.P.; Munoz, L.E.; Sáez, S.; Rodríguez-Patón, A. A new improved and extended version of the multicell bacterial simulator gro. *ACS Synth. Biol.* **2017**, 6, 1496–1508. [CrossRef] [PubMed]
7. Adleman, L.M. Molecular computation of solutions to combinatorial problems. *Sciences* **1994**, *266*, 1021–1024. [CrossRef]
8. Carell, T. Molecular computing: Dna as a logic operator. *Nature* **2011**, *469*, 45–46. [CrossRef] [PubMed]
9. Xu, J. Probe machine. *IEEE Trans. Neural Netw. Learn. Syst.* **2016**, *27*, 1405–1416. [CrossRef] [PubMed]
10. Păun, G.; Rozenberg, G.; Salomaa, A. *The Oxford Handbook of Membrane Computing*; Oxford University Press: Oxford, UK, 2010.
11. Păun, G.; Rozenberg, G. A guide to membrane computing. *Theor. Comput. Sci.* **2002**, *287*, 73–100. [CrossRef]
12. Păun, G. Computing with membranes. *J. Comput. Syst. Sci.* **2000**, *61*, 108–143. [CrossRef]
13. Freund, R.; Kari, L., Păun, G. DNA computing based on splicing: The existence of universal computers. *Theory Comput. Syst.* **1999**, *32*, 69–112. [CrossRef]
14. Kari, L.; Păun, G.; Rozenberg, G.; Salomaa, A.; Yu, S. DNA computing, sticker systems, and universality. *Acta Inform.* **1998**, *35*, 401–420. [CrossRef]
15. Martın-Vide, C.; Păun, G.; Pazos, J.; Rodrıguez-Patón, A. Tissue P systems. *Theor. Comput. Sci.* **2003**, *296*, 295–326. [CrossRef]
16. Ionescu, M.; Păun, G.; Yokomori, T. Spiking neural P systems. *Fundam. Inform.* **2006**, *71*, 279–308.
17. Song, T.; Pan, L.; Păun, G. Asynchronous spiking neural p systems with local synchronization. *Inform. Sci.* **2013**, *219*, 197–207. [CrossRef]
18. Ezziane, Z. DNA computing: Applications and challenges. *Nanotechnology* **2005**, *17*, R27. [CrossRef]
19. Chen, X.; Pérez-Jiménez, M.J.; Valencia-Cabrera, L.; Wang, B.; Zeng, X. Computing with viruses. *Theor. Comput. Sci.* **2016**, *623*, 146–159. [CrossRef]
20. Rogozhin, Y. Small universal turing machines. *Theor. Comput. Sci.* **1996**, *168*, 215–240. [CrossRef]
21. Baiocchi, C. Three small universal turing machines. In *Machines, Computations, and Universality*; Springer: Berlin/Heidelberg, Germany, 2001; pp. 1–10.
22. Korec, I. Small universal register machines. *Theor. Comput. Sci.* **1996**, *168*, 267–301. [CrossRef]
23. Iirgen Albert, J.; Culik, K., II. A simple universal cellular automaton and its one-way and totalistic version. *Complex Syst.* **1987**, *1*, 1–16.
24. Kudlek, M.; Yu, R. Small universal circular post machines. *Comput. Sci. J. Mold.* **2001**, *9*, 25.
25. Păun, A.; Păun, G. Small universal spiking neural P systems. *BioSystems* **2007**, *90*, 48–60. [CrossRef] [PubMed]
26. Zhang, X.; Zeng, X.; Pan, L. Smaller universal spiking neural P systems. *Fundam. Inform.* **2008**, *87*, 117–136.
27. Pan, L.; Zeng, X. A note on small universal spiking neural P systems. *Lect. Notes Comput. Sci.* **2010**, *5957*, 436–447.
28. Păun, A.; Sidoroff, M. Sequentiality induced by spike number in SNP systems: Small universal machines. In *Membrane Computing*; Springer: Berlin/Heidelberg, Germany, 2012; pp. 333–345.
29. Song, T.; Jiang, Y.; Shi, X.; Zeng, X. Small universal spiking neural P systems with anti-spikes. *J. Comput. Theor. Nanosci.* **2013**, *10*, 999–1006. [CrossRef]
30. Song, T.; Xu, J.; Pan, L. On the universality and non-universality of spiking neural P systems with rules on synapses. *IEEE Trans. Nanobiosci.* **2015**, *14*, 960–966. [CrossRef] [PubMed]
31. Song, T.; Pan, L. Spiking neural P systems with request rules. *Neurocomputing* **2016**, *193*, 193–200. [CrossRef]
32. Song, T.; Rodríguez-Patón, A.; Gutiérrez, M.; Pan, Z. Computing with bacteria conjugation and crispr/cas9 gene editing operations. *Sci. Rep.* **2018**, sumitted.
33. Minsky, M.L. *Computation: Finite and Infinite Machines*; Prentice-Hall, Inc.: Upper Saddle River, NJ, USA, 1967.
34. Valencia-Cabrera, L.; Orellana-Martín, D.; Martínez-del Amor, M.A.; Riscos-Nú nez, A.; Pérez-Jiménez, M.J. Computational efficiency of minimal cooperation and distribution in polarizationless p systems with active membranes. *Fundam. Inform.* **2017**, *153*, 147–172. [CrossRef]
35. Song, B.; Pérez-Jiménez, M.J.; Pan, L. An efficient time-free solution to qsat problem using p systems with proteins on membranes. *Inf. Comput.* **2017**, *256*, 287–299. [CrossRef]
36. Macías-Ramos, L.F.; Pérez-Jiménez, M.J.; Riscos-Nú nez, A.; Valencia-Cabrera, L. Membrane fission versus cell division: When membrane proliferation is not enough. *Theor. Comput. Sci.* **2015**, *608*, 57–65. [CrossRef]
37. Ma, X.; Sun, F.; Li, H.; He, B. Neural-network-based sliding-mode control for multiple rigid-body attitude tracking with inertial information completely unknown. *Inf. Sci.* **2017**, *400*, 91–104. [CrossRef]

15

Genome-Wide Identification and Comparative Analysis for OPT Family Genes in *Panax ginseng* and Eleven Flowering Plants

He Su [1,†], Yang Chu [2,†], Junqi Bai [1], Lu Gong [1], Juan Huang [1], Wen Xu [1], Jing Zhang [1], Xiaohui Qiu [1], Jiang Xu [2,*] and Zhihai Huang [1,*]

1 The Second Clinical College of Guangzhou University of Chinese Medicine, Guangzhou 510006, China; suhe@cau.edu.cn (H.S.); baijunqi@126.com (J.B.); gonglu0904@126.com (L.G.); juanhuangzi@126.com (J.H.); freexuwen@163.com (W.X.); ginniezj@163.com (J.Z.); qiuxiaohui@gzucm.edu.cn (X.Q.)

2 Institute of Chinese Materia Medica, China Academy of Chinese Medicinal Sciences, Bejing 100700, China; ychu@icmm.ac.cn

* Correspondence: jxu@icmm.ac.cn (J.X.); zhhuang7308@163.com (Z.H.).

† These authors contributed equally to this work.

Abstract: Herb genomics and comparative genomics provide a global platform to explore the genetics and biology of herbs at the genome level. *Panax ginseng* C.A. Meyer is an important medicinal plant for a variety of bioactive chemical compounds of which the biosynthesis may involve transport of a wide range of substrates mediated by oligopeptide transporters (OPT). However, information about the OPT family in the plant kingdom is still limited. Only 17 and 18 OPT genes have been characterized for *Oryza sativa* and *Arabidopsis thaliana*, respectively. Additionally, few comprehensive studies incorporating the phylogeny, gene structure, paralogs evolution, expression profiling, and co-expression network between transcription factors and OPT genes have been reported for ginseng and other species. In the present study, we performed those analyses comprehensively with both online tools and standalone tools. As a result, we identified a total of 268 non-redundant OPT genes from 12 flowering plants of which 37 were from ginseng. These OPT genes were clustered into two distinct clades in which clade-specific motif compositions were considerably conservative. The distribution of OPT paralogs was indicative of segmental duplication and subsequent structural variation. Expression patterns based on two sources of RNA-Sequence datasets suggested that some OPT genes were expressed in both an organ-specific and tissue-specific manner and might be involved in the functional development of plants. Further co-expression analysis of OPT genes and transcription factors indicated 141 positive and 11 negative links, which shows potent regulators for OPT genes. Overall, the data obtained from our study contribute to a better understanding of the complexity of the OPT gene family in ginseng and other flowering plants. This genetic resource will help improve the interpretation on mechanisms of metabolism transportation and signal transduction during plant development for *Panax ginseng*.

Keywords: *Panax ginseng*; oligopeptide transporter; flowering plant; phylogeny; transcription factor

---

1. Introduction

Peptide transportation is a widely observed phenomenon of translocating small peptides across a membrane in a carrier-mediated, energy-dependent manner [1]. Transported peptides are often hydrolyzed and the resulting amino acids are used as substrates for protein synthesis, sources of nitrogen and carbon [2,3], and signals for biological processes such as quorum sensing [4], yeast mating [5], and metal homeostasis regulation [6,7]. There are three distinct protein families related to peptide transportation. The ATP binding cassette (ABC, TC 3.A.1) transporter superfamily is the largest transporter gene family. The members are able to translocate a wide variety of substrates including amino acids, sugars, peptides, proteins, and a large number of hydrophobic compounds and metabolites across extra-cellular and intracellular membranes [8,9]. In contrast to the ABC family, the proton-dependent oligopeptide transporter (PTR, TC 2.A.17) family utilizes a proton gradient other than ATP hydrolysis for dipeptide and tripeptide translocation [10]. The members of PTR proteins have been found in all kingdoms of life except the Archaea [1,11]. PTR also participates in amino acid and nitrate transportation [12]. In addition to dipeptides and tripeptides that are translocated by PTR proteins, tetra-peptides, penta-peptides, and some longer oligopeptides are translocated by a novel protein family known as the oligopeptide transporter (OPT, TC 2.A.67) family [10].

The OPT family is a group of electrochemical potential-driven transporters that catalyze their solutes in an energy-dependent symport manner. CaOPT1 was first cloned from *Candida albicans* (Robin) Berkhout and functional verified in *Schizosaccharomyces pombe* (Lindner) and subsequently defined as OPT but not an ABC or PTR protein by Jeff Becker's laboratory [13–15]. OPTs are suggested to play diverse roles in long-distance sulfur distribution, metal homeostasis, nitrogen mobilization, heavy metal sequestration by transporting glutathione, peptides, and meta-chelates [16]. Phylogenetically, the OPTs can be divided into Oligopeptide Transporter (PT) and Stripe-like (YSL) clades [16,17]. Genes in the YSL clade have been found in Archaea, eubacteria, fungi, and plants but not in animals, which function as metal chelate transporters [6,18,19] consisting of mugeneic acids (MA) or nicotianamine (NA) while genes in the PT clade have only been identified in plants and fungi mediating long-distance metal distribution, nitrogen mobilization, glutathione translocation, and heavy metal sequestration [16,20–25].

In plants, the OPTs may play important roles in plant growth and abiotic and biotic stress responses [13,26]. The OPT member *ZmYS1*, which was first cloned by Curie et al. [6] but re-defined by Yen et al. [27], was proven to mediate the import of Fe-phytosiderophore complexes from soils and long-distance transport of iron-NA complexes [13]. Studies of two *AtOPT3* T-DNA mutants indicated that *AtOPT3* is of importance in both embryo development and iron deficiency signal transduction [1,7]. In addition, *AtOPT3* is found to be expressed in the phloem and functions in long-distance shoot-to-root signaling for Fe/Zn/Mn status. A lack of *AtOPT3* in *Arabidopsis thaliana* (*Arabidopsis*) led to the over-accumulation of cadmium in seeds [23]. Glutathione (GSH) is an essential sulfur-containing tripeptide that performs various important roles in plant processes, including detoxification of xenobiotics, heavy metal transport and resistance, controlling redox status, and long-distance transport of organic sulfur [22]. GSH is a precursor for plants to use to produce phytochelatins (PCs), which is the polymerized form of GSH, by which heavy metals can be transported to a central vacuole for detoxification [28]. AtOPT4 and AtOPT6 from *Arabidopsis* [21,22] and BjGT1 [25] from *Brassica juncea* (*B. juncea*) are all capable of translocating Cd-GSH conjuncts. Moreover, GSH also plays important roles in plant growth and development in response to abiotic and biotic stresses.

The majority of members of OPT proteins seem to contain 16 TMSs including a few of which appear to have 17 TMSs. A homology-based analysis for each TMSs in the OPT family indicated that the 16-TMS proteins might have been generated by three sequential duplications from 2-TMS protein precursors. Additionally, gene fusion might be responsible for the 17-TMS proteins [17]. However, although the OPT proteins have been studied for more than two decades, the majority of studies still focus on model plants such as yeast, *Arabidopsis*, and *Oryza sativa* (rice) [1,21,29,30]. Thanks to the rapid development of whole-genome sequencing techniques, an exponential increase in genome information has provided us with great opportunities to identify more OPT genes in non-model plants and make comparisons among multiple species simultaneously. However, to the best of our knowledge, genome-wide identification of OPT proteins has only been conducted in *Ganoderma lucidum* [31], *Populus trichocarpa*, and *Vitis vinifera* [32]. *Panax ginseng*, which is a Traditional Chinese Medicine, has been used for several millennia and has become more and more popular around the world. It is the most commonly used medicinally species in the *Panax* genera in contrast to the other four species: *Panax quinquefolius*, *Panax vietnamensis*, *Panax japonicus*, and *Panax notoginseng* [33]. Since we finished the genome assembly for *Panax ginseng (P. ginseng)* in our previous report [34], the interest in characterizing OPT genes in *P. ginseng* and comparing it with other genome-assembly-available species has increased.

Although more plant genomes have been mapped in the last decade, studies on the genome-wide identification and comparison of OPT genes among species are still limited. Information on the phylogeny, gene structure, expression patterns, and regulatory networks of OPT genes remains to be discovered. In the present study, we identified OPT genes from *P. ginseng* and 11 flowering plants with the purpose of uncovering the phylogenetic relationships and gene structures of OPT genes in flowering plants as well as investigating the expression profiles and regulators of OPT genes in *P. ginseng*. Our analysis, which combines these types of information, provides new insights into both the structural and functional roles of OPT genes in ginseng and serves as a valuable resource for further study of the roles OPT genes play in plant development and transport of secondary metabolism.

2. Results and Discussion

2.1. Identification of OPT Genes in P. ginseng and 11 Other Flowering Plants

We identified the OPT genes for *P. ginseng* and other species with TransportTP by setting as reference organisms Oryza sativa and *Arabidopsis thaliana* [35]. As a result, a total of 364 OPT candidates were identified in our study (Table 1). Seventeen of the 18 identified genes from *Arabidopsis* were in accordance with the reviewed records deposited in Swiss-Prot (release 2018_10), and, although the other gene *At5g45450*, was not recorded in Swiss-Prot, it was regarded as an OPT gene recorded in GenBank. However, only 12 OPT genes were identified from rice, of which the accession numbers were not in accordance with those records deposited in Swiss-Prot. This might be due to a different version of the rice genome being used. Considering our genome assembly did not scale to a chromosome level, we conducted a manual curation of the 39 OPTs identified from *P. ginseng*. Thereafter, 37 OPT genes were kept for further analysis. In addition, we identified 54 and 26 OPT genes from poplar and grape in our study, respectively, while only 20 and 18 genes were identified by Cao et al. [32]. These results suggested that our identification of OPT genes was accurate and comprehensive.

Table 1. Statistics of OPT genes predicted from *P. ginseng* and 11 flowering plants.

Species	Predicted	De-Redundant	Final
Ginseng	39	37	37
Arabidopsis	18	16	17
Rice	12	10	18
Sorghum	38	26	26
Carrot	25	16	16
Potato	42	29	29
Tomato	23	17	17
Cassava	29	21	21
Clover	31	25	25
Cacao	27	19	19
Poplar	54	28	28
Grape	26	24	23
Total	364	268	276

37 OPT genes were left after manual curation for OPT genes from Ginseng. OPT genes predicted for *Arabidopsis* and rice were replaced with 17 and 18 reviewed OPT genes retrieved from Swiss-Prot. De-redundant indicated only one gene could be kept if there were genes that had 100% similarity to it.

Since we found some OPTs were highly redundant (similar to each other with 100% similarity) within species, we removed the redundant OPTs in order to reduce the subsequent calculation consumption using CD-HIT software [36] by setting the sequence identity threshold to 100%. Lastly, a total of 268 OPT genes were kept. Because a candidate OPT gene named "GSVIVT01007176001" identified from grape contained too many "X"s, we excluded it from further analyses. In order to generate robust results from subsequent studies, we replaced those predicted OPT genes from *Arabidopsis* and rice with reviewed OPT genes retrieved from Swiss-Prot. Furthermore, we introduced two other experimentally verified OPT genes from *B. juncea* and *Zea mays* (*BjGT1* and *Maize_YS_1* respectively [6,25]) into our study. Lastly, 278 OPT genes were used for further analysis (Supplementary File 1).

2.2. Protein Properties of OPT Genes for OPT Genes Identified in P. ginseng and 11 Other Flowering Plants

By examining the properties of OPT genes for each plant species, we found that the number of amino acid residues varied among species. Generally, the number of amino acid residues for OPT genes in *Arabidopsis*, rice, sorghum, and cassava ranged from 552 to 766, which is higher than the rest of those studied species (ranged from 184 to 941, as for *P. ginseng*). The number of residues ranged from 348 to 919 (Table S1, Figure 1D). The distribution of molecular weight for OPT genes was similar to the distribution pattern of residue numbers (Figure 1B). The grand average of hydropathicity (GRAVY) value is a measure of protein hydrophobicity [37]. Our results suggested that GRAVY for those OPT genes mainly ranged from 0.30 to 0.60 (Figure 1A). As OPT genes with the lowest and the highest GRAVY values (0.029 for Potri.017G150620.2.p and 0.87 for PGSC0003DMP400037534) were filtered out for lacking OPT-specific information for further phylogenetic analysis (Figure 2), we expanded the confident range of GRAVY values from 0.329–0.628 to 0.114–0.659 compared with the previous study [32]. In addition, the isoelectric point (pI) of the majority of OPT genes was around 9.0, which suggests that the electrochemical properties of OPT genes might be less varied in the plant kingdom (Figure 1C).

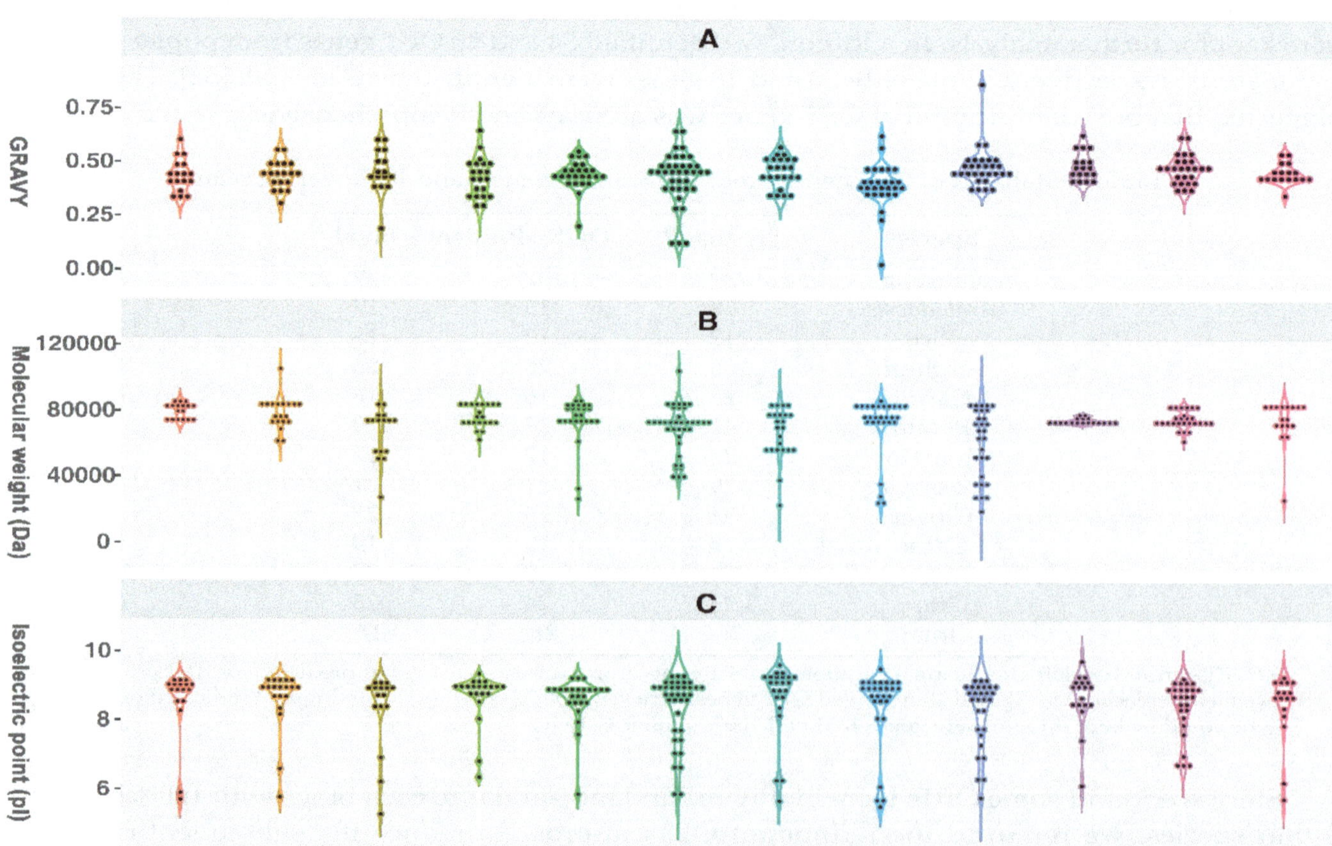

Figure 1. *Cont.*

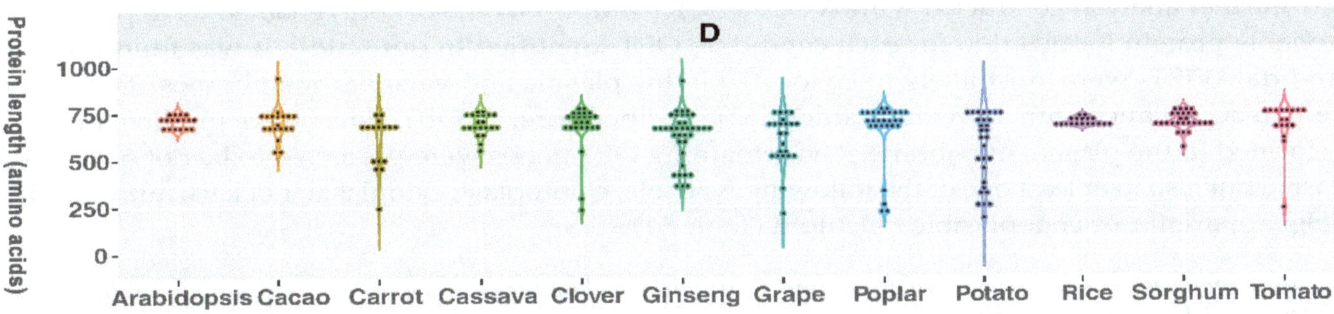

Figure 1. Protein properties for OPT genes identified from *P. ginseng* and 11 flowering plants. (**A**) Grand average of hydropathicity, GRAVY. (**B**) Molecular weight. (**C**), Isoelectric point, pI. (**D**) Number of amino acid residues.

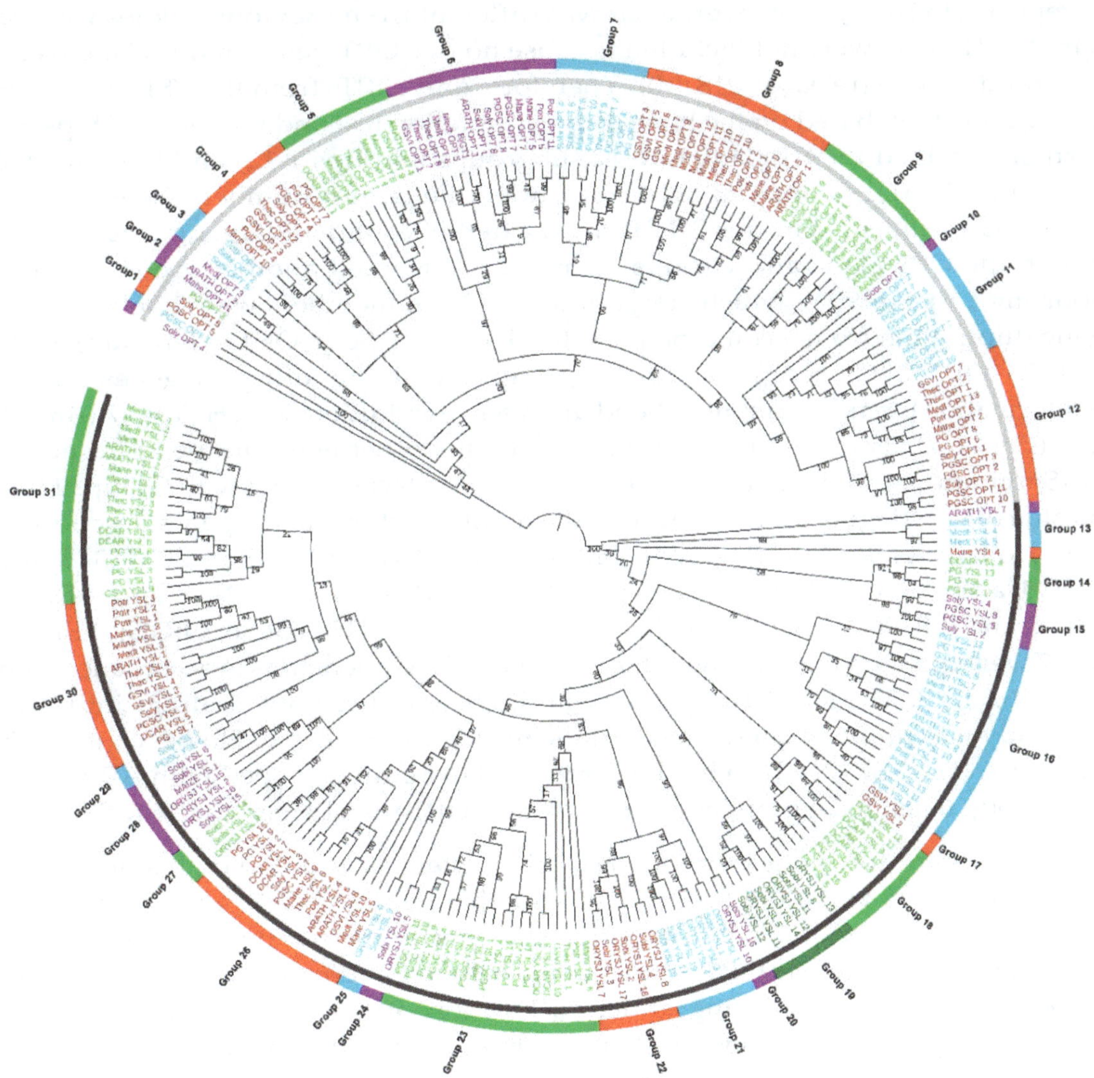

Figure 2. Phylogenetic relationships of OPT genes in *P. ginseng* and 11 other species. Tomato (Soly), Potato (PGSC), Cassava (Mane), *Arabidopsis* (ARATH), Clover (Medt), Poplar (Potr), Grape (GSVI), Cacao (Thec), Sorghum (Sobi), Carrot (DCAR), and Rice (ORYSJ). BjGT1 from *Brassica juncea* and Maize YS1 from *Zea Mays* (maize) are experimentally validated OPT proteins that were retrieved from GenBank database. Light gray in the inner circle indicates the PT clade. Dark gray refers to the YSL clade.

Further analysis conducted with WOLF PSORT (http://woltpsort.org) enabled us to predict the probable protein localization for each candidate OPT identified in our study. It was found that all candidate OPTs were most likely to be located in the plasma and vacuolar membranes. The results were in accordance with a previous study [32]. Furthermore, 17 OPT genes were predicted to only be located in the plasma membrane. The remaining OPT genes were predicted to be not only in the plasma but also in at least one of the following: vacuole, chloroplast, cytoplasm, nucleus, mitochondria, Golgi apparatus, or endoplasmic reticulum (Table S1).

2.3. Phylogenetic Analyses, Classification, and Functional Relatedness of the OPT Genes Identified in P. ginseng and 11 Other Flowering Plants

To unravel the phylogenetic relationships of OPT genes in flowering plants, we conducted a phylogenetic analysis for those genes from 12 flowering plants. All OPT genes were clustered into two major distinct clades known as PT and YSL clade for which the results were consistent with those of previous reports [13,16,32]. However, what was different from previous studies was that the rice OPT genes in the PT clade were not included because no rice OPT genes in this clade were available in the Swiss-Prot database (release 2018_10). Therefore, only OPTs from the YSL clade were used in this phylogenetic analysis. Based on the bootstrap permutation test and the relationships of each OPT gene, we further classified the PT clade into 12 subgroups (Groups 1-12) and the YSL clade into 19 subgroups (Groups 13-31). Groups 23 and 31 included the largest number of members in the YSL clade (each with 19 members). Groups 9-12 formed a highly confident larger group with a bootstrap value of 96% in the OPT clade and Groups 23 and 27-30 formed another group in the YSL clade with a supporting value of 99%, which suggested that those members were likely to have evolved by recent gene duplication from a common ancestor. However, Soly_OPT_4 (Tomato), PGSC_OPT_1 (Potato), and PG_OPT_2 (Ginseng) failed to be grouped with any other PT genes due to a lack of supporting information by maximum likelihood analyses. In addition, ARTH_YSL_7 (*Arabidopsis*) and Mane_YSL_4 (Cassava) also failed to be grouped with any other genes in the YSL clade. Although Sobi_OPT_7 (Sorghum) seemed likely to stand alone, it was in fact grouped with Groups 11 and 12 with a bootstrap value of 93% (Figure 2). Furthermore, the motif structures of the genes described below also supported the group classifications (Figure S1). Moreover, ARATH_OPT_3 and BjGT3 (*Brassica juncea*), ARATH_OPT_1 and ARATH_OPT_5, ARATH_OPT_6 and ARATH_OPT_8 and ARATH_OPT_9, ARATH_YSL_5 and ARATH_YSL_8, ARATH_YSL_4 and ARATH_YSL_6, ORYSJ_YSL_7 (rice) and ORYSJ_YSL_17 were grouped together, with the phylogenetic relationships in accordance with previous study reports [16,30,32]. The consistency of our findings with previous findings indicated that our phylogenetic study was properly conducted and the results were reliable. However, it was interesting to find out that ARATH_YSL_7, which has been reported to be sub-grouped with ARATH_YSL_5 and ARATH_YSL_8 [30], failed to be grouped with any other OPT members in the YSL clade in our study.

Genes with the same functions were often closely related, as found in both a previous study [32] and our study. *BjGT1*, which is the first cloned and characterized OPT gene from *Brassica juncea*, was experimentally validated to be a glutathione transporter mediating cadmium absorption [25]. ARATH_OPT_3, which is another OPT gene that was cloned and characterized in *Arabidopsis*, was reported to be involved in the sensing and translocation of Cd (as well as Fe and Zn) [1,7,23]. These two functionally similar genes were clustered together in our study. In addition, Mazie_YS_1, the first experimentally validated OPT gene responsible for transport of Fe(III)-phytosiderophore chelates, was clustered together with ORYSJ_YSL_15 and ORYSJ_YSL_2 (Figure 2). ORYSJ_YSL_15 has been suggested to be responsible for iron uptake from rhizosphere and for phloem transport of iron by transporting Fe(III)-phytosiderophore chelates while ORYSJ_YSL_2 has been suggested to be responsible for phloem transport of iron by transporting Fe(III)-nicotianamine chelates [38,39]. Furthermore, ARATH_YSL_2 and ARATH_YSL_3 clustered together in Group 31 were both reported to be involved in transport of nicotianamine-chelated metals in the vasculature [40,41]. These results supported the idea that genes with the same functions were closely related. Based on the hypothesis, it would be interesting to test if PG_YSL_1 is involved in iron-transportation since it was clustered together with ARATH_YSL_1 that was found to be involved in transport of iron-nicotianamine chelates [41,42]. Similarly, it would be interesting to test if PG_OPT_1 is akin to ARATH_OPT_6, which was reported to be involved in the transport of glutathione derivatives and metal complexes [21,43,44] and to test whether PG_OPT_9,10 and PG_OPT_11 are involved in increasing plant sensitivity to Cd like ARATH_OPT_7 functions [43].

We identified a total of 45 pairs of paralogs from the phylogenetic analyses (Table S2), which accounted for 11.1% to 70.6% of all OPT candidates in each studied species and shared similar structures within each group (Figure S1). We found that some OPT genes in ginseng were tandemly clustered on the same scaffold (Table S3) and those genes were location-related. For example, PG_OPT_10 and PG_OPT_11 were neighbor paralogs with 1122 bp in between. These genes might be formed by tandemly segmental duplication. PG_YSL_8 and PG_YSL_10 constitute a special tandemly clustered paralogs with a 3214 bp-long shared region. This paralogs pair might be generated by a crossover of chromosome after whole-genome duplication or by gene fusion. It would be interesting to test whether this OPT cluster was functional in further studies. In addition, PG_YSL_4-PG_YSL_5 and PG_YSL_14-PG_YSL_16 formed a special type of gene cluster block in which PG_YSL_4-PG_YSL_14 and PG_YSL_5-PG_YSL_16 were identified as paralogs oriented in the same direction. PG_YSL_18-PG_YSL_19 and PG_YSL_21-PG_YSL_22 constituted another special type of block, in which PG_YSL_18-PG_YSL_21 and PG_YSL_19-PG_YSL_22 were paralogs oriented in opposite directions (Figure 3). From this section of the study, we speculated that both types of cluster blocks were generated from segmental duplication or whole-genome duplication. Since *P. ginseng* is a tetraploid plant, we prefer to believe that genes from those blocks were more likely to be generated by whole-genome duplication. The paralogs blocks arranged in the opposite direction were likely to be generated by subsequent segmental inversion of the chromosome after segmental duplication.

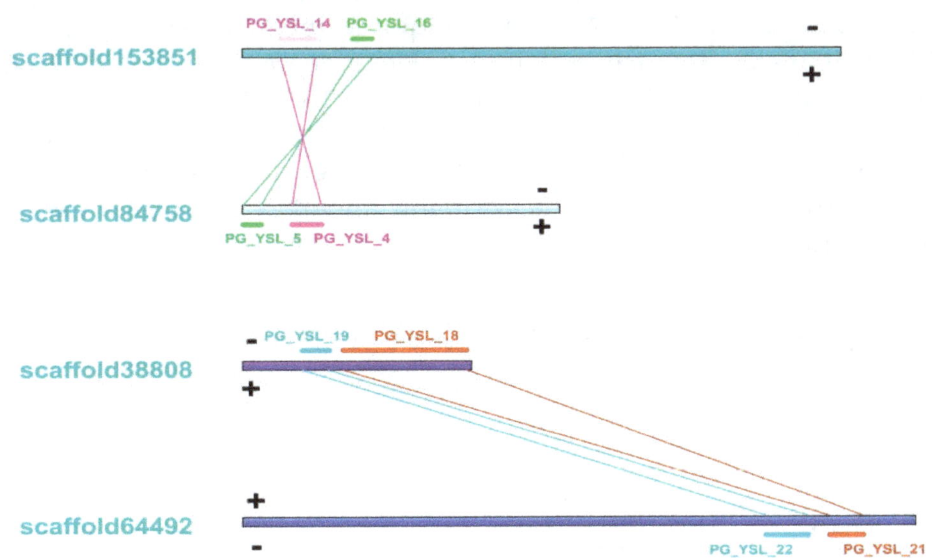

Figure 3. Chromosome locations for two special types of clusters of paralogs blocks. We used the ginseng genome version1 finished by Xu et al. [34] in this study. The scaffold refers to the DNA sequences in the ginseng genome that were generated by bridging non-gapped contigs (assembled with short gun sequencing reads) with mate-pair sequencing reads. A scaffold is equivalent to a chromosome segment.

Ks (synonymous substitution rate) is a widely accepted concept for gene duplication time estimation. In general, the lower *Ks* is, the more recently gene duplication occurred [32]. Since a codon-based alignment of PG_YSL_6 and PG_YSL_17 failed to generate, calculation of *Ks* was excluded from this study (Table S2). Aligned sequences were nearly identical after removing gaps from

Potr_YSL_2/Potr_YSL_3 (poplar), Thec_OPT_7/Thec_OPT_8 (cacao), and Thec_YSL_4/Thec_YSL_5. The estimation of *Ks* for these paralogs also failed. Additionally, *Ks* values for PG_OPT_9/PG_OPT_10 and Sobi_YSL_13/Sobi_YSL_14 were estimated as 0, suggested that they were generated by a very recent duplication event. It was interesting to find that gene duplication of OPT paralogs occurred more recently in the YSL clade than in the PT clade in *P. ginseng*. The phenomenon was similar to grape and clover but contrary to cacao, cassava, and *Arabidopsis*. The duplication event for paralogs occurred more recently in ginseng, potato, poplar, cacao, grape, clover, and sorghum (about 0 to 5 MYA) than in carrot, cassava, *Arabidopsis*, and rice, which indicates that ginseng and other species or their common ancestor might have suffered a high level of gene loss during evolution because of the lack of an older duplication event such as 94.2 MYA for ARATH_OPT_6/ARATH_OPT_9 [45].

2.4. Conserved Domains and Motif Analysis for OPT Genes Identified in P. ginseng and 11 Other Flowering Plants

By searching against the Conserved Domain Database (CDD) [46] with 278 OPT genes, all genes were annotated as OPT genes. However, only 267 were predicted to have specific domains, wherein all the 258 OPT genes used in the phylogeny analysis were covered. Because domain analysis could not provide information about smaller individual motifs and more divergent patterns, we conducted a study of motif analysis with MEME software (Supplementary File 2). As a result, 30 distinct motifs were identified in these genes. Detailed information of those motifs is presented in Supplementary File 3. It is interesting that the motif composition of OPT members in the PT clade is distinct from that in the YSL clade (Figure S1), which was in accordance with the conclusions generated by phylogenetic analysis. In addition, the number of motifs of OPT genes from the PT clade (ranging from 4 to 11, with a median value of 8) was distinct from that of the YSL clade (ranging from 5 to 12, with a median value of 11), which suggests the clade-specific structure of each OPT gene (Figure 4). Furthermore, we found nine motifs (Motif_1,3,6,13,14,19,23,15,29) unique to the PT clade and 10 motifs (Motif_7,8,12,16,17,21,22,24,26,30) unique to the YSL clade, respectively. Six motifs (Motif_10,14,19,23,102,106) were frequently shared by PT clade members (94.4%, 94.4%, 86.9%, 91.6%, 95.3%, and 99.1%, respectively) and 10 motifs (Motif_2,8,9,15,16,18,21,22,26,28) were frequently shared by the YSL clade members (92.0%, 88.7%, 71.3%, 38.7%, 92.0%, 92.0%, 92.7%, 81.3%, 84.7%, 97.3%, and 94.0%, respectively) (Table S4). Those findings might give us new insights into how OPT genes evolved since being separated from their common ancestor and how they functionally diverged during the subsequent evolution process.

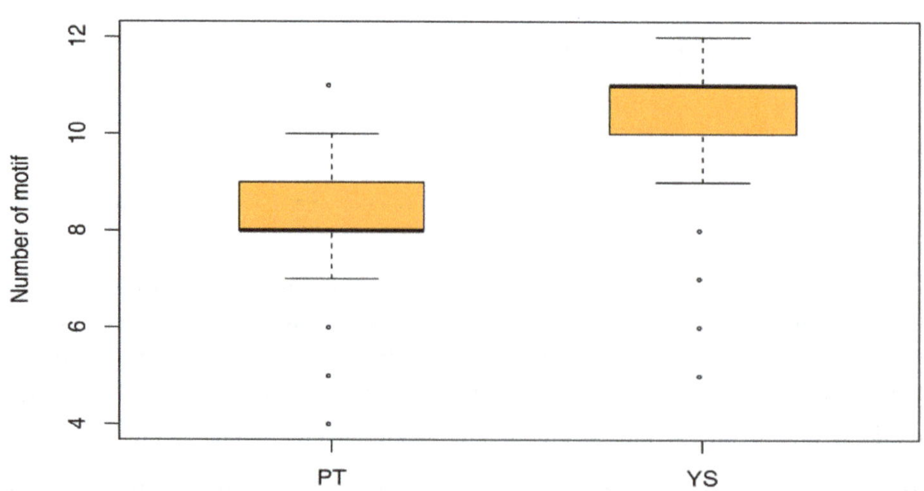

Figure 4. Number of motifs in OPT genes from PT and YS clades. These static results were calculated with xml output of MEME analysis (Supplementary File 5) by our custom R scripts. The boxplot was generated by the built-in function "boxplot" in R.

2.5. Profiling of Expression Patterns for OPT Genes Identified in P. ginseng

In order to examine the expression patterns of the OPT genes in *P. ginseng*, we performed a comprehensive expression analysis by using two sets of RNA-Seq datasets: one from our previous study about *P. ginseng* root [47] and one from a public study about 18 kinds of tissues. In general, genes in the YSL clade were more highly expressed than genes in the PT clade except in the periderm (Figure 5). PG_YSL_2,13 and PG_YSL_15 were expressed evenly in the root with little difference among tissues, which suggests that they might be constitutive OPTs. OPT genes exhibited distinct tissue-specific expression manners. For example, PG_OPT_4,5 and PG_YSL_12 were more likely to be expressed highly in periderm than in the stele or cortex. PG_YSL_11 and PG_YSL_7 had the highest expression in the stele and cortex, respectively, while they were still expressed at a considerably high level in other tissues. The different expression patterns for those OPT genes indicated that a wide range of substrates might be transported in different parts of the plant root.

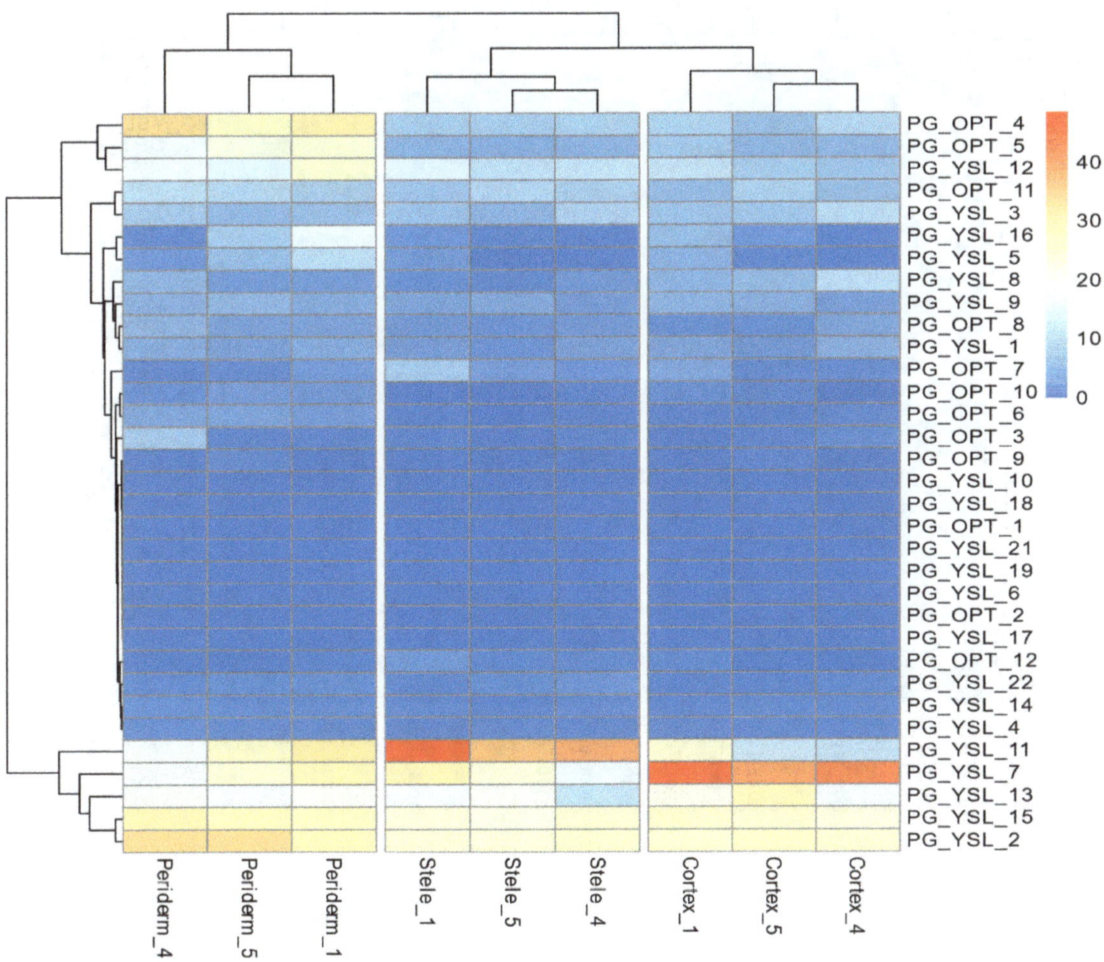

Figure 5. Expression of OPT genes of different tissues in ginseng root. Gene expression of OPT genes was calculated with RNA-Seq data generated by our previous study on *P. ginseng* root. The hierarchical clustered heat map was plotted with 'pheatmap' planted in R package named 'pheatmap'.

Due to the nature of sink tissue of fruit and seeds in plants, the expression characteristics of OPT genes of these tissues are expected to share more common traits than those of other tissues. Based on the expression data, fruit flesh, fruit pedicel, fruit peduncle, and seeds were clustered together, which suggests that the similar expression pattern of those OPT genes might contribute to methods of metabolism relocation. The expression of PG_YSL_1,3 and PG_YSL_7 both peaked in fruit flesh compared with the fruit pedicel, fruit peduncle, and seed, which indicates that lateral transportation might be the most active transportation process during seed development. Additionally, the leaf blade, leaf pedicel, leaf peduncle, and stem, which are physically connected organs forming a complex vascular transportation system in plants, were clustered together by their similar expression pattern, wherein PG_YSL_8 was expressed at the highest level (except for the stem). Moreover, the arm root, fiber root, and leg root were clustered together, and PG_OPT_13 and PG_YSL_16 were highly expressed. It was interesting that PG_YSL_7 was highly expressed in 12-year-old and 25-year-old roots but minimally expressed in five-year-old and 18-year-old roots that were clustered with the main root cortex, the main root epiderm, and the rhizome. Taken together, the expression patterns found in our study and Wang's [48] both suggested that OPT genes were expressed in tissue-specific and location-specific manners by which the transportation and distribution of oligopeptides and their conjugates with metals, signals, etc. were shaped in different ginseng tissues [41] (Figure 6).

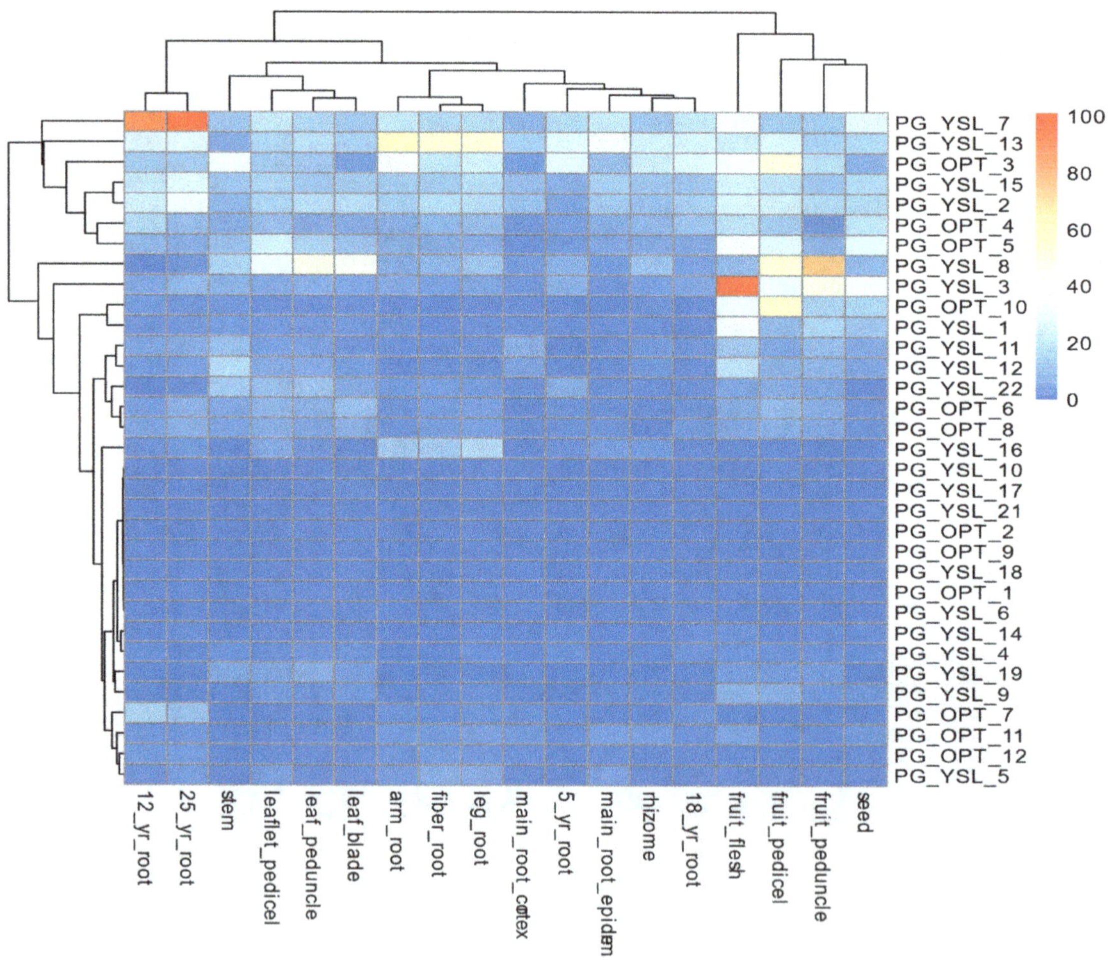

Figure 6. Expression of the OPT genes in 18 organs of ginseng. Gene expression of OPT genes was re-calculated with RNA-seq data from public research by taking our published genome as a reference.

Based on the phylogenetic analysis described above, PG_OPT_4 and PG_OPT_5 were grouped with ARATH_OPT_1 and ARATH_OPT_5, which were proven to be OPT transporters for penta-peptide (KLLLG) in an energy-dependent manner by yeast complementation assay [20,49]. High expression of those genes exclusively in the root suggested that the penta-peptide-related metabolism (metabolism substrates, signal molecules, etc.) transportation might be activated. PG_YSL_12 was identified as another periderm-specific expressed gene found in this study. It was expressed more highly in the stem and fruit flesh than in other tissues (Figure 6). PG_YSL_12 was clustered with ARATH_YSL_5 and ARATH_YSL_8 into one group, which indicates that it might be involved in the transport of nicotianamine-chelated metals (metals-NA) just as ARATH_YSL_2 was in the transport of Fe-NA across the plasma membrane in leaf cells, involving lateral movement of iron away from the xylem [40]. Furthermore, ARATH_YSL_8 might also be involved directly in iron uptake by leaf cells [13].

ARATH_YSL_1 and ARATH_YSL_3 were experimentally verified OPT proteins, which were found to be able to mediate Fe transportation to and from vascular tissues [41]. ARATH_YSL_3 was a sister branch to ARATH_YSL_2, which had been functionally confirmed to be expressed in both roots and shoots and to mediate transport of metal-NA complexes [40], which indicates their functional similarity. ARATH_YSL_1 was clustered into another sister group to a larger group including ARATH_YSL_2 and ARATH_YSL_3, which suggests that members of these two larger group might share some functional similarities. MAIZE_YS_1 (*ZmYS1*) known as a proton-coupled symporter transports iron complexed by plant-derived Fe(III) chelators (phytosiderophores, PS) by scavenging from soil, termed Strategy II [13]. It formed another cluster in the YSL clade with some YSL genes from sorghum and rice with a bootstrap value of 97% (Figure 2). Considering the functional similarity of ARATH_YSL_1,2,3 and *ZmYS1*, OPT genes from groups 27 and 28 and groups 29-31 were suggested to form two sister groups that might be involved in the transport of Fe.

The OPT paralogs were more likely to be generated by segmental tandem duplication rather than transposition [32]. The expression pattern of those duplicated genes may differ if they suffered evolutionary divergence such as neofunctionalization. [50]. No similar expression patterns of duplicated paralogs were identified in the study about poplar and grape [32]. However, we detected five similarly expressed paralogs pairs in this study wherein PG_OPT_4-PG_OPT_5 had similar expression patterns both in our previous study and in Wang's study. PG_YSL_4-PG_YSL_14 and PG_YSL_18-PG_YSL_21 were expressed similarly but with a very low expression level. PG_YSL_6-PG_YSL_8 and PG_YSL_11-PG_YSL_12 were reported to be expressed similarly in Wang's report but not in ours. However, PG_YSL_5/PG_YSL_16 was similarly expressed in our study but not in Wang's study. The similar expression patterns found in our study might be because of the relatively short time has been experienced in *Ginseng* paralogs compared with those paralogs in poplar and grape. On the other hand, the phenomenon that a majority of the identified paralogs in *Ginseng* did not have similar expression patterns, which indicates functional diversificationmight be a result of long-term evolution—adapting to changing environmental conditions after gene duplication.

2.6. Analysis of Co-Expression Network between OPT Genes and Potent Transcription factOr for P. ginseng

The regulation of gene expression in all living cells is dominated by transcriptional initiation, which is regulated by transcription factors, ancillary transcription regulators, and chromatin regulators. Therefore, we conducted an analysis focusing on the co-expression between all transcriptionally modulated genes in the ginseng genome and all transcription factors in order to reveal regulators for OPT genes in *Ginseng*. PlantTFcat is a useful tool for identifying proteins with signature domains specific to 108 major transcription regulators families [51]. We assessed the *Ginseng* genome for identifying those proteins. A total of 5073 distinct genes in the *P. ginseng* genome have been predicted to be transcription factors wherein there are 5457 members (Supplementary File 4). Genes annotated as different transcription factors by PlantTFcat (such as PG39956, annotated as Znf-B, LisH, WD40-like, or PLATZ) were removed from further analysis. The expression values of transcription factors that were mapped by many genes (such as MYB-HB-like, mapped by 334 genes) were determined by the median values of those genes. Lastly, a total of 59 transcription factors and 13 OPT genes were used for network analysis. We used non-parametric Spearman's rank-order correlation for our co-expression analysis due to its robustness for generating biologically relevant gene networks [52].

The matrix of all correlation values for expression values between each pair of transcription factor and OPT gene from a set of nine biological samples is shown in Table S5. At a conservative threshold of $\rho \geq |0.85|$, 141positive and 11 negative correlations involving 13 OPT genes were found (Table S6). The number of transcription factors correlated to an OPT gene ranged from 1 to 27, while it ranged only 1 to 5 for the number of OPT genes correlated with a transcription factor (Tables S7 and S8, Figure 7). For example, PG_OPT_5 and PG_OPT_6 positively correlated with 26 and 27 transcription factors, respectively, while *bHLH* and *WRKY* only correlated with five OPT genes. Our findings suggested that the initiation of transcription of OPT genes might be dominated by a complicated synergetic regulation system consisting of a number of transcription factors. Additionally, transcription factors might act as pleiotropic regulators participating in a variety of transcription regulations for OPT genes. On the other hand, because 19 out of 59 transcription factors were linked to only one OPT gene and two out of 13 OPT genes were linked to only one transcription factor, the results suggested that the transcription of some OPT genes was regulated by specific transcription factors and some transcription factors had specific target genes to regulate.

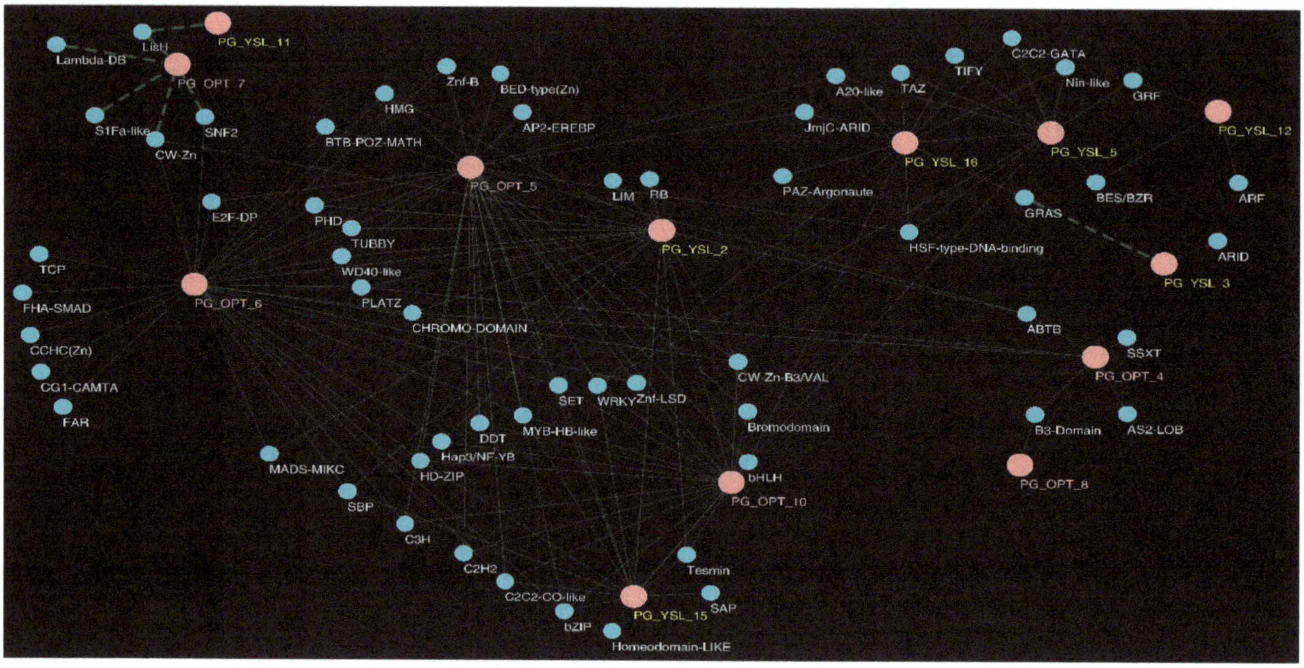

Figure 7. Regulatory gene networks involving transcription master genes and OPT genes. A stringent threshold ($\rho \geq |0.85|$) was used and the visualization was produced in Cytoscape-3.6.1. Nodes for OPT genes are represented by pink circles (yellow labels represent OPT genes in the YSL clade. Pink labels represent OPT genes in PT clade). Nodes for blue circles represent transcriptional regulators. Positive interactions are indicated by green lines and negative interactions are indicated by green dashed lines.

3. Materials and Methods

3.1. Sequence Retrieval and Identification of OPT Genes

We identified OPT genes from *P. ginseng* and 11 other flowering plants by using TransportTP (http://bioinfo3.noble.org/transporter/ [35]). Proteome sequences from 11 flowering plants (*Arabidopsis* TAIR10, rice v7, tomato iTAG2.4, potato v4.03, carrot v2.0, *Manihot esculenta* v6.1, *Medicago truncatula* Mt4.0v1, Poplar v3.1, Grape Genoscope 12X, Cacao v1.1, Sorghum v3.1.1) were retrieved from the Phytozome database (V12.1) [53], wherein the assembly version is followed by each species name. Ginseng proteome sequences were retrieved from http://ginseng.vicp.io:23488/index.php/index/download.html. The proteome sequences for each species were then used for the identification of OPT genes by searching against TransporterTP, setting the E-value threshold to 0.1, and setting *Arabidopsis thaliana* and *Oryza sativa* as the reference organisms.

Subcellular localization of those OPT proteins was predicted with WOLF-PSORT [54,55]. Isoelectric point (pI), molecular weight, and grand average hydropathicity (GRAVY) values were estimated with functions planted in the Peptides package (https://github.com/dosorio/Peptides/) for R.

3.2. Phylogenetic Analysis for OPT Genes

Phylogenetic analysis of these OPT genes was conducted on their conserved domains identified by CDD (Conserved Domain Database, [56,57]) and planted in NCBI with default parameters (50369 PSSMs, e-value of 0.01, maximum number of hits 500). Multiple sequence alignments of those conserved OPT protein were performed with MAFFT v7.158b [58] and followed by manual comparison and refinement. Aligned regions that contain over 50% gaps or ambiguous sites were removed and sequences that contained gaps for more than 50% of the remaining sequence were deleted. Lastly, 258 OPT genes were left for further phylogenetic analysis. In order to select the best evolutionary model for phylogeny reconstruction, we used the function 'modelTest' planted in the R package named 'phangorn' [59] with the parameter 'model' set to 'all' and found 'LG+G+F' was the best model. A maximum likelihood method of phylogenetic analysis based on RAxML v 8.2.9 [60] software was conducted, with the parameter "–bootstop-perms" set to 1000, "-m" set to PROTGAMMALG, and the "-e" set to 0.001. After finishing the reconstruction of the phylogeny of OPT genes for these species, the topology was plotted by the online tool iTOL [61].

3.3. Estimation of Duplication Time for OPT Paralogs

Pairwise alignment of protein sequences of the OPT paralogs was aligned with MAFFT software, and codon-based pairwise alignment of nucleotide sequences were generated by using PAL2NAL [62]. The *Ka* and *Ks* values for paralogous genes were estimated by the program yn00 planted in the PAML package with default parameters [63]. Assuming a molecular clock, the synonymous substitution rates (*Ks*) of the paralogous genes could be regarded as a proxy for time estimation of the segmental duplication events. The approximation of date for duplication events was estimated with the following formula: $T = Ks/2\lambda$, where λ denotes clock-like rates of synonymous substitution. In this study, 1.5×10^{-8} substitutions/synonymous site/year was used for *Arabidopsis*, 6.5×10^{-9} for rice, sorghum, cassava, grape and cacao, 9.1×10^{-9} for poplar [32], 1.08×10^{-8} for clover, 6.68×10^{-9} for *P. ginseng*, 2.69×10^{-9} for potato, and 2.91×10^{-9} for carrot. λ for each species was deduced or collected from previous studies [64,65].

3.4. Analysis of Motif Composition for OPT Genes

Conserved motif analysis for OPT genes in the *P. ginseng* genome was conducted with MEME (http://meme.sdsc.edu). The OPT candidates were run locally with MEME with the following parameters: number of repetitions = any, maximum number of motifs = 30. The other parameters were kept as default values.

3.5. Profiling Expression of OPT Genes for P. ginseng

The gene expression of *P. ginseng* was profiled by RNA-Seq datasets from our previous study [47] and public research [48]. Those datasets could be retrieved from the SRA database by searching BioProject id PRJNA369187 and PRJNA302556. These raw datasets from the SRA database were first converted into FASTQ files by sratoolkit.2.8.0 [66] and then quality controlled by Trimmomatic-0.36 [67]. Lastly, reference-based gene expression of those biological samples was estimated with the HISAT2+StringTie pipeline [68]. FPKM values for each gene were used as gene expression levels. A hierarchical clustered heatmap for OPT genes was plotted with the pheatmap package [69], wherein "manhattan" distance was used for both row-based and column-based clustering.

3.6. Identification of Regulatory Network between OPT Genes and Transcription Factors for P. ginseng

The *P. ginseng* proteome sequence dataset was submitted to the PlantTFcat analysis tool (http://plantgrn.noble.org/PlantTFcat/ [51]) for the identification and classification of transcription factors, chromatin modifiers, and other transcriptional regulators into protein families. Genes that mapped to more than one transcription factors were removed from further analysis. In addition, median values of those genes referring to the same transcription factors were regarded as the transcription factors' expression value. In this study, we used our previous RNA-Seq dataset for construction of the co-expression network between OPT genes and transcription factors. FPKMs of all genes including OPT genes and transcription factors were combined and used for the calculation of Spearman's rank correlation coefficient to predict potential gene regulatory networks. The correlation coefficient (ρ) for each gene pair was calculated by the built-in function "cor" in R, and a threshold of $\rho \geq |0.85|$ was regarded as significant co-expression. Visualization of the network was created in Cytoscape 3.6.1 [70].

4. Conclusions

This study is the first to investigate the chromosomal location, expression profiling, and transcriptional regulation networks of *P. ginseng* OPT genes and provide a comparative genome analysis addressing the phylogeny, gene structure, and paralogs duplication history of the OPT gene family in *P. ginseng* and 11 flowering plants. Chromosomal location analyses revealed that structural variation occurred after segmental duplication, expression profiling, and transcriptional co-expression networks analyses, which indicates that both specific and pleiotropic transcription regulators might be involved in the regulation of OPT genes' expression. Phylogenetic analyses suggested two well-supported clades in the OPT family, which can be further classified into 12 or 19 distinct groups. Motif compositions are conserved in each clade and clade-specific motifs were frequently occupied within each clade. Estimations for paralogs divergence history indicated that the majority of OPT paralogs in *P. ginseng* might have emerged from recent duplications, which was different from the history of *Arabidopsis* or cassava. The study of expression profiles in different organs and tissues of *P. ginseng* has provided insights into possible functional divergence among OPT members and important functional roles in the plant development of some OPT members. These data may provide valuable information for future functional investigations of this gene family.

Author Contributions: Z.H. conceived and designed the research framework. H.C., J.X., and Y.C. prepared the sample and performed the experiments. J.X. and Y.C. provided many important suggestions for data analysis. H.S. analyzed the data. H.S. and J.X. wrote the manuscript. J.B., L.G., J.H., W.X., J.Z., X.Q, and Z.H make revisions to the final manuscript. All authors have read and approved the final manuscript.

References

1. Stacey, M.G.; Koh, S.; Becker, J.; Stacey, G. Atopt3, a member of the oligopeptide transporter family, is essential for embryo development in *Arabidopsis. Plant Cell* **2002**, *14*, 2799–2811. [CrossRef] [PubMed]
2. Perry, J.R.; Basrai, M.A.; Steiner, H.Y.; Naider, F.; Becker, J.M. Isolation and characterization of a saccharomyces cerevisiae peptide transport gene. *Mol. Cell. Biol.* **1994**, *14*, 104–115. [CrossRef] [PubMed]
3. Steiner, H.Y.; Naider, F.; Becker, J.M. The ptr family: A new group of peptide transporters. *Mol. Microbiol.* **1995**, *16*, 825–834. [CrossRef] [PubMed]
4. Swift, S.; Throup, J.P.; Williams, P.; Salmond, G.P.; Stewart, G.S. Quorum sensing: A population-density component in the determination of bacterial phenotype. *Trends Biochem. Sci.* **1996**, *21*, 214–219. [CrossRef]
5. Kuchler, K.; Sterne, R.E.; Thorner, J. Saccharomyces cerevisiae ste6 gene product: A novel pathway for protein export in eukaryotic cells. *Embo J.* **1989**, *8*, 3973–3984. [CrossRef] [PubMed]
6. Curie, C.; Panaviene, Z.; Loulergue, C.; Dellaporta, S.L.; Briat, J.-F.; Walker, E.L. Maize yellow stripe1 encodes a membrane protein directly involved in Fe(III) uptake. *Nature* **2001**, *409*, 346–349. [CrossRef] [PubMed]
7. Stacey, M.G.; Patel, A.; Mcclain, W.E.; Mathieu, M.; Remley, M.; Rogers, E.E.; Gassmann, W.; Blevins, D.G.; Stacey, G. The Arabidopsis atopt3 protein functions in metal homeostasis and movement of iron to developing seeds. *Plant Physiol.* **2008**, *146*, 589–601. [CrossRef] [PubMed]
8. Dean, M.; Hamon, Y.; Chimini, G. The human atp-binding cassette (abc) transporter superfamily. *J. Lipid Res.* **2001**, *42*, 1007–1017. [CrossRef] [PubMed]
9. Higgins, C.F. Abc transporters: From microorganisms to man. *Ann. Rev. Cell Biol.* **1992**, *8*, 67–113. [CrossRef]

10. Hauser, M.; Narita, V.; Donhardt, A.M.; Naider, F.; Becker, J.M. Multiplicity and regulation of genes encoding peptide transporters in saccharomyces cerevisiae. *Mol. Membr. Biol.* **2001**, *18*, 105–112. [CrossRef]
11. Newstead, S. Recent advances in understanding proton coupled peptide transport via the pot family. *Curr. Opin. Struct. Biol.* **2017**, *45*, 17–24. [CrossRef] [PubMed]
12. Williams, L.; Miller, A. Transporters responsible for the uptake and partitioning of nitrogenous solutes. *Ann. Rev. Plant Biol.* **2001**, *52*, 659–688. [CrossRef] [PubMed]
13. Lubkowitz, M. The opt family functions in long-distance peptide and metal transport in plants. In *Genetic Engineering: Principles and Methods*; Setlow, J.K., Ed.; Springer: Boston, MA, USA, 2006; pp. 35–55.
14. Lubkowitz, M.A.; Hauser, L.; Breslav, M.; Naider, F.; Becker, J.M. An oligopeptide transport gene from candida albicans. *Microbiology* **1997**, *143*, 387–396. [CrossRef] [PubMed]
15. Lubkowitz, M.A.; Barnes, D.; Breslav, M.; Burchfield, A.; Naider, F.; Becker, J.M. Schizosaccharomyces pombe isp4 encodes a transporter representing a novel family of oligopeptide transporters. *Mol. Microbiol.* **1998**, *28*, 729–741. [CrossRef] [PubMed]
16. Lubkowitz, M. The oligopeptide transporters: A small gene family with a diverse group of substrates and functions? *Mol. Plant* **2011**, *4*, 407–415. [CrossRef] [PubMed]
17. Gomolplitinant, K.M.; Saier, M., Jr. Evolution of the oligopeptide transporter family. *J. Membr. Biol.* **2011**, *240*, 89. [CrossRef] [PubMed]
18. Feng, S.; Tan, J.; Zhang, Y.; Liang, S.; Xiang, S.; Wang, H.; Chai, T. Isolation and characterization of a novel cadmium-regulated yellow stripe-like transporter (snysl3) in solanum nigrum. *Plant Cell Rep.* **2017**, *36*, 281–296. [CrossRef] [PubMed]
19. Murata, Y.; Ma, J.F.; Yamaji, N.; Ueno, D.; Nomoto, K.; Iwashita, T. A specific transporter for iron(III)-phytosiderophore in barley roots. *Plant J.* **2006**, *46*, 563–572. [CrossRef] [PubMed]
20. Koh, S.; Wiles, A.M.; Sharp, J.S.; Naider, F.R.; Becker, J.M.; Stacey, G. An oligopeptide transporter gene family in arabidopsis. *Plant Physiol.* **2002**, *128*, 21–29. [CrossRef]
21. Wongkaew, A.; Asayama, K.; Kitaiwa, T.; Nakamura, S.-I.; Kojima, K.; Stacey, G.; Sekimoto, H.; Yokoyama, T.; Ohkama-Ohtsu, N. Atopt6 protein functions in long-distance transport of glutathione in arabidopsis thaliana. *Plant Cell Physiol.* **2018**. [CrossRef]
22. Zhang, Z.; Xie, Q.; Jobe, T.O.; Kau, A.R.; Wang, C.; Li, Y.; Qiu, B.; Wang, Q.; Mendoza-Cózatl, D.G.; Schroeder, J.I. Identification of atopt4 as a plant glutathione transporter. *Mol. Plant* **2016**, *9*, 481–484. [CrossRef] [PubMed]
23. Mendoza-Cózatl, D.G.; Xie, Q.; Akmakjian, G.Z.; Jobe, T.O.; Patel, A.; Stacey, M.G.; Song, L.; Demoin, D.W.; Jurisson, S.S.; Stacey, G. Opt3 is a component of the iron-signaling network between leaves and roots and misregulation of opt3 leads to an over-accumulation of cadmium in seeds. *Mol. Plant* **2014**, *7*, 1455–1469. [CrossRef] [PubMed]
24. Vasconcelos, M.W.; Li, G.W.; Lubkowitz, M.A.; Grusak, M.A. Characterization of the pt clade of oligopeptide transporters in rice. *Plant Genome* **2008**, *1*, 77–88. [CrossRef]
25. Bogs, J.; Bourbouloux, A.; Cagnac, O.; Wachter, A.; Rausch, T.; Delrot, S. Functional characterization and expression analysis of a glutathione transporter, bjgt1, from brassica juncea: Evidence for regulation by heavy metal exposure. *Plant Cell Environ.* **2003**, *26*, 1703–1711. [CrossRef]
26. Carole, D.M.; Beno, T.P. Role of glutathione in plant signaling under biotic stress. *Plant Signal. Behav.* **2012**, *7*, 210–212.
27. Yen, M.-R.; Tseng, Y.-H.; Saie, M., Jr. Maize yellow stripe1, an iron-phytosiderophore uptake transporter, is a member of the oligopeptide transporter (opt) family. *Microbiology* **2001**, *147*, 2881–2883. [CrossRef] [PubMed]
28. Cobbett, C.; Goldsbrough, P. Phytochelatins and metallothioneins: Roles in heavy metal detoxification and homeostasis. *Annu. Rev. Plant Biol.* **2003**, *53*, 159–182. [CrossRef]
29. Bourbouloux, A.; Shahi, P.; Chakladar, A.; Delrot, S.; Bachhawat, A.K. Hgt1p, a high affinity glutathione transporter from the yeast saccharomyces cerevisiae. *J. Biol. Chem.* **2000**, *275*, 13259–13265. [CrossRef]
30. Liu, T.; Zeng, J.; Xia, K.; Fan, T.; Li, Y.; Wang, Y.; Xu, X.; Zhang, M. Evolutionary expansion and functional diversification of oligopeptide transporter gene family in rice. *Rice* **2012**, *5*, 1–14. [CrossRef]
31. Xiang, Q.; Shen, K.; Yu, X.; Zhao, K.; Gu, Y.; Zhang, X.; Chen, X.; Chen, Q. Analysis of the oligopeptide transporter gene family in ganoderma lucidum: Structure, phylogeny, and expression patterns. *Genome* **2017**, *60*, 293–302. [CrossRef]

32. Cao, J.; Huang, J.; Yang, Y.; Hu, X. Analyses of the oligopeptide transporter gene family in poplar and grape. *BMC Genom.* **2011**, *12*, 465. [CrossRef] [PubMed]
33. Yun, T.K. Brief introduction of panax ginseng c.A. Meyer. *J. Korean Med. Sci.* **2001**, *16* (Suppl.), S3–S5.
34. Jiang, X.; Yang, C.; Baosheng, L.; Shuiming, X.; Qinggang, Y.; Rui, B.; He, S.; Linlin, D.; Xiwen, L.; Jun, Q. *Panax ginseng* genome examination for ginsenoside biosynthesis. *GigaScience* **2017**, *6*, 1–15.
35. Li, H.; Benedito, V.A.; Udvardi, M.K.; Zhao, P.X. Transporttp: A two-phase classification approach for membrane transporter prediction and characterization. *BMC Bioinform.* **2009**, *10*, 418. [CrossRef] [PubMed]
36. Li, W.; Godzik, A. Cd-hit: A fast program for clustering and comparing large sets of protein or nucleotide sequences. *Bioinformatics* **2006**, *22*, 1658–1659. [CrossRef] [PubMed]
37. Kyte, J.; Doolittle, R.F. A simple method for displaying the hydropathic character of a protein. *J. Mol. Biol.* **1982**, *157*, 105–132. [CrossRef]
38. Shintaro, K.; Haruhiko, I.; Daichi, M.; Michiko, T.; Hiromi, N.; Satoshi, M.; Nishizawa, N.K. Osysl2 is a rice metal-nicotianamine transporter that is regulated by iron and expressed in the phloem. *Plant J.* **2010**, *39*, 415–424.
39. Haruhiko, I.; Takanori, K.; Tomoko, N.; Michiko, T.; Yusuke, K.; Kazumasa, S.; Mikio, N.; Hiromi, N.; Satoshi, M.; Nishizawa, N.K. Rice osysl15 is an iron-regulated iron(III)-deoxymugineic acid transporter expressed in the roots and is essential for iron uptake in early growth of the seedlings. *J. Biol. Chem.* **2009**, *284*, 3470–3479.
40. DiDonido, D., Jr.; Roberts, L.A.; Sanderson, T.; Eisley, R.B.; Walker, E.L. Arabidopsis yellow stripe-like2 (ysl2): A metal-regulated gene encoding a plasma membrane transporter of nicotianamine–metal complexes. *Plant J.* **2004**, *39*, 403–414. [CrossRef]
41. Waters, B.M.; Chu, H.H.; Didonato, R.J.; Roberts, L.A.; Eisley, R.B.; Lahner, B.; Salt, D.E.; Walker, E.L. Mutations in arabidopsis yellow stripe-like1 and yellow stripe-like3 reveal their roles in metal ion homeostasis and loading of metal ions in seeds. *Plant Physiol.* **2006**, *141*, 1446–1458. [CrossRef]
42. Marie, L.J.; Adam, S.; Stéphane, M.; Jean-François, B.; Catherine, C. A loss-of-function mutation in atysl1 reveals its role in iron and nicotianamine seed loading. *Plant J.* **2010**, *44*, 769–782.
43. Cagnac, O.; Bourbouloux, A.; Chakrabarty, D.; Zhang, M.-Y.; Delrot, S. Atopt6 transports glutathione derivatives and is induced by primisulfuron. *Plant Physiol.* **2004**, *135*, 1378–1387. [CrossRef] [PubMed]
44. Pike, S.; Patel, A.; Stacey, G.; Gassmann, W. Arabidopsis opt6 is an oligopeptide transporter with exceptionally broad substrate specificity. *Plant Cell Physiol.* **2009**, *50*, 1923–1932. [CrossRef] [PubMed]
45. Jaillon, O.; Aury, J.M.; Noel, B.; Policriti, A.; Clepet, C.; Casagrande, A.; Choisne, N.; Aubourg, S.; Vitulo, N.; Jubin, C. The grapevine genome sequence suggests ancestral hexaploidization in major angiosperm phyla. *Nature* **2007**, *449*, 463–467. [PubMed]
46. Marchlerbauer, A.; Bo, Y.; Han, L.; He, J.; Lanczycki, C.J.; Lu, S.; Chitsaz, F.; Derbyshire, M.K.; Geer, R.C.; Gonzales, N.R. Cdd/sparcle: Functional classification of proteins via subfamily domain architectures. *Nucleic Acids Res.* **2017**, *45*, D200–D203. [CrossRef] [PubMed]
47. Zhang, J.J.; Su, H.; Zhang, L.; Liao, B.S.; Xiao, S.M.; Dong, L.L.; Hu, Z.G.; Wang, P.; Li, X.W.; Huang, Z.H. Comprehensive characterization for ginsenosides biosynthesis in ginseng root by integration analysis of chemical and transcriptome. *Molecules* **2017**, *22*, 889. [CrossRef]
48. Wang, K.; Jiang, S.; Sun, C.; Lin, Y.; Rui, Y.; Yi, W.; Zhang, M. The spatial and temporal transcriptomic landscapes of ginseng, panax ginseng c. A. Meyer. *Sci. Rep.* **2015**, *5*, 18283. [CrossRef]
49. Osawa, H.; Stacey, G.; Gassmann, W. Scopt1 and atopt4 function as proton-coupled oligopeptide transporters with broad but distinct substrate specificities. *Biochem. J.* **2006**, *393*, 267–275. [CrossRef]
50. Prince, V.E.; Pickett, F.B. Splitting pairs: The diverging fates of duplicated genes. *Nat. Rev. Genet.* **2002**, *3*, 827–837. [CrossRef]
51. Dai, X.; Sinharoy, S.; Udvardi, M.; Zhao, P.X. Planttfcat: An online plant transcription factor and transcriptional regulator categorization and analysis tool. *BMC Bioinform.* **2013**, *14*, 321. [CrossRef]
52. Sapna, K.; Nie, J.; Chen, H.S.; Hao, M.; Ron, S.; Xiang, L.; Lu, M.Z.; Taylor, W.M.; Wei, H. Evaluation of gene association methods for coexpression network construction and biological knowledge discovery. *PLoS ONE* **2012**, *7*, e50411.
53. Goodstein, D.M.; Shu, S.; Russell, H.; Rochak, N.; Hayes, R.D.; Joni, F.; Therese, M.; William, D.; Uffe, H.; Nicholas, P. Phytozome: A comparative platform for green plant genomics. *Nucleic Acids Res.* **2012**, *40*, D1178–D1186. [CrossRef] [PubMed]

54. Yu, N.Y.; Wagner, J.R.; Laird, M.R.; Melli, G.; Lo, R.; Dao, P.; Sahinalp, S.C.; Ester, M.; Foster, L.J.; Brinkman, F.S.L. Psortb 3.0. *Bioinformatics* **2010**, *26*, 1608–1615. [CrossRef] [PubMed]
55. Horton, P.; Park, K.J.; Obayashi, T.; Fujita, N.; Harada, H.; Adamscollier, C.J.; Nakai, K. Wolf psort: Protein localization predictor. *Nucleic Acids Res.* **2007**, *35*, 585–587. [CrossRef] [PubMed]
56. Marchlerbauer, A.; Anderson, J.B.; Chitsaz, F.; Derbyshire, M.K.; Deweesescott, C.; Fong, J.H.; Geer, L.Y.; Geer, R.C.; Gonzales, N.R.; Gwadz, M. Cdd: Specific functional annotation with the conserved domain database. *Nucleic Acids Res.* **2009**, *37*, D205–D210. [CrossRef] [PubMed]
57. Marchler-Bauer, A.; Lu, S.; Anderson, J.B.; Chitsaz, F.; Derbyshire, M.K.; DeWeese-Scott, C.; Fong, J.H.; Geer, L.Y.; Geer, R.C.; Gonzales, N.R.; et al. Cdd: A conserved domain database for the functional annotation of proteins. *Nucleic Acids Res.* **2011**, *39*, D225–D229. [CrossRef]
58. Katoh, K.; Kuma, K.; Toh, H.; Miyata, T. Mafft version 5: Improvement in accuracy of multiple sequence alignment. *Nucleic Acids Res.* **2005**, *33*, 511–518. [CrossRef]
59. Schliep, K.P. Phangorn: Phylogenetic analysis in r. *Bioinformatics* **2011**, *27*, 592–593. [CrossRef]
60. Stamatakis, A. Raxml version 8: A tool for phylogenetic analysis and post-analysis of large phylogenies. *Bioinformatics* **2014**, *30*, 1312–1313. [CrossRef]
61. Letunic, I.; Bork, P. *Interactive Tree of Life (Itol): An Online Tool for Phylogenetic Tree Display and Annotation*; Oxford University Press: Oxford, UK, 2007; pp. 78–82.
62. Suyama, M.; Torrents, D.; Bork, P. Pal2nal: Robust conversion of protein sequence alignments into the corresponding codon alignments. *Nucleic Acids Res.* **2006**, *34*, W609–W612. [CrossRef]
63. Yang, Z. Paml: A program package for phylogenetic analysis by maximum likelihood. *Comput. Appl. Biosci. Cabios* **1997**, *13*, 555–556. [CrossRef] [PubMed]
64. The Potato Genome Sequencing Consortium; Xu, X.; Pan, S.; Cheng, S.; Zhang, B.; Mu, D.; Ni, P.; Zhang, G.; Yang, S.; Li, R.; et al. Genome sequence and analysis of the tuber crop potato. *Nature* **2011**, *475*, 189–195. [CrossRef] [PubMed]
65. Young, N.D.; Debellé, F.; Oldroyd, G.E.D.; Geurts, R.; Cannon, S.B.; Udvardi, M.K.; Benedito, V.A.; Mayer, K.F.X.; Gouzy, J.; Schoof, H.; et al. The medicago genome provides insight into the evolution of rhizobial symbioses. *Nature* **2011**, *480*, 520–524. [CrossRef] [PubMed]
66. Sherry, S. Ncbi sra toolkit technology for next generation sequence data. *Pump Ind. Anal.* **2000**, *3*, 2230–2234.
67. Bolger, A.M.; Lohse, M.; Usadel, B. Trimmomatic: A flexible trimmer for illumina sequence data. *Bioinformatics* **2014**, *30*, 2114–2120. [CrossRef] [PubMed]
68. Pertea, M.; Kim, D.; Pertea, G.M.; Leek, J.T.; Salzberg, S.L. Transcript-level expression analysis of rna-seq experiments with hisat, stringtie and ballgown. *Nat. Protoc.* **2016**, *11*, 1650–1667. [CrossRef] [PubMed]
69. Kolde, R. Pheatmap: Pretty Heatmaps. R Package Version 1.0.8. Available online: https://CRAN.R-project.org/package=pheatmap (accessed on 18 December 2018).
70. Shannon, P.; Markiel, A.; Ozier, O.; Baliga, N.S.; Wang, J.T.; Ramage, D.; Amin, N.; Schwikowski, B.; Ideker, T. Cytoscape: A software environment for integrated models of biomolecular interaction networks. *Genome Res.* **2003**, *13*, 2498–2504. [CrossRef] [PubMed]

16

To Decipher the *Mycoplasma hominis* Proteins Targeting into the Endoplasmic Reticulum and their Implications in Prostate Cancer Etiology using Next-Generation Sequencing Data

Mohammed Zakariah [1,†], Shahanavaj Khan [2,3,*,†], Anis Ahmad Chaudhary [4], Christian Rolfo [5], Mohamed Maher Ben Ismail [6] and Yousef Ajami Alotaibi [6]

1 Research Center, College of Computer and Information Science, King Saud University, Riyadh 11451, Saudi Arabia; mzakariah@ksu.edu.sa
2 Nanomedicine & Biotechnology Research Unit, Department of Pharmaceutics, College of Pharmacy, P.O. Box 2457, King Saud University, Riyadh 11451, Saudi Arabia
3 Department of Bioscience, Shri Ram Group of College (SRGC), Muzaffarnagar 251002, UP, India
4 Department of Pharmacology, College of Medicine, Al-Imam Mohammad Ibn Saud Islamic University, Riyadh 11451, Saudi Arabia; anis.chaudhary@gmail.com
5 Phase I-Early Clinical Trials Unit, Oncology Department, Antwerp University Hospital, "Centre for Oncological Research (CORE)", 2650 Edegem, Belgium; Christian.Rolfo@uza.be
6 Computer Science Department, College of Computer and Information Sciences, King Saud University, Riyadh 11451, Saudi Arabia; mbenismail@ksu.edu.sa; (M.M.B.I.); yaalotaibi@ksu.edu.sa (Y.A.A.)
* Correspondence: khan.shahanavaj@gmail.com.
† These authors contributed equally to this work.
Academic Editors: Xiangxiang Zeng, Alfonso Rodríguez-Patón and Quan Zou

Abstract: Cancer was initially considered a genetic disease. However, recent studies have revealed the connection between bacterial infections and growth of different types of cancer. The enteroinvasive strain of *Mycoplasma hominis* alters the normal behavior of host cells that may result in the growth of prostate cancer. The role of *M. hominis* in the growth and development of prostate cancer still remains unclear. The infection may regulate several factors that influence prostate cancer growth in susceptible individuals. The aim of this study was to predict *M. hominis* proteins targeted into the endoplasmic reticulum (ER) of the host cell, and their potential role in the induction of prostate cancer. From the whole proteome of *M. hominis*, 19 proteins were predicted to be targeted into the ER of host cells. The results of our study predict that several proteins of *M. hominis* may be targeted to the host cell ER, and possibly alter the normal pattern of protein folding. These predicted proteins can modify the normal function of the host cell. Thus, the intercellular infection of *M. hominis* in host cells may serve as a potential factor in prostate cancer etiology.

Keywords: prostate cancer; *Mycoplasma hominis*; endoplasmic reticulum; systems biology; protein targeting

1. Introduction

Bacterial infection is recognized to play a significant role in the progression and advancement of various forms of cancers, including prostate, lung, gastric, and colon cancer [1–3]. Recent study showed that prostate gland restrains a plethora of different strains of bacteria [4]. Bacterial dysbiosis, inflammation, and other factors are associated with the growth of prostate cancer, although the exact mechanisms involved in growth of cancer due to bacterial infection are not very clear. Mycoplasmas are bacteria that lack cell walls, and are causative agents of various diseases related to respiratory and urogenital tract among humans [5,6]. The dominant types of mycoplasmas found in the urogenital system of humans include *Mycoplasma genitalium*, *Ureaplasma urealyticum*, and *M. hominis*. The relationship between mycoplasmas and the human population was first detected in the 1960s [7]. Different studies have highlighted the connection between *M. hominis* infection and prostate cancer advancement [8–11]. This association was further supported by numerous studies on *M. hominis* infection and higher classification of prostate cancer. The cell cycle signal cascade, including DNA repair mechanisms and apoptosis, may be altered following mycoplasma infection [8]. Although a meta-analysis report has suggested a suspicious role of *M. hominis* in the growth of prostate cancer [12], various recent studies instead confirmed the involvement of different bacteria in the progression of different types of cancer [13–15]. Pathogenic bacteria and their subcellular constituents interact with several types of host cell receptors that may change the expression of various genes. These enigmatic alterations may affect the normal control and regulation of host cells [8,16]. Chronic inflammation and chronic infections (or both) are the cause of 20% of different forms of cancers in humans.

Induction of *pro-inflammatory cytokines* and reactive *oxygen species*, regulated by *chronic inflammation*, may promote nitration and chlorination of nucleic acids and proteins. *M. hominis* infection serves as a factor promoting the growth and development of prostate cancer, but several other causes are linked with the growth of prostate cancer [17–19]. Apart from chronic inflammation and mutations, different cyclomodulins have been associated with the growth of prostate cancer by the disturbance of homeostasis in *M. hominis*-infected cells. Some specific strains of bacteria have ability to produce different toxins known as cyclomodulins, that interfere in the host cell cycle, which suggests a potential association of different pathogenic bacteria with different type of cancers [20]. It has supposed that cyclomodulins have the capability to affect the usual growth cycle of the host cell, and are expected to grow as etiological aspects for *M. hominis*-mediated prostate cancer [8]. *M. hominis* strain is considered a usual Gram-negative pathogen. It is known to multiply and inhabit intracellularly during the progression of prostate cancer [21,22]. As *M. hominis* is colonized in the urogenital tract that comprises the prostate, *M. hominis* infection leads to some precise effects in the progression of prostate cancer. The mass collection of genomes for *M. hominis* revealed 715,165 base pairs and a G + C content of 26.94%.

Various strains of different bacteria are involved in the intracellular infection and duplication in specific host cells, wherein bacterial pathogens change the usual functioning of cells through the localization of their own proteins in different components of host cells, such as endoplasmic reticulum (ER), Golgi complex, mitochondria, nucleus, plasma membrane, and cytoplasm [13,21,22]. The complete genome of ATCC 27545 strain of *M. hominis* contains 563 open reading frames (ORFs) and encodes different enzymes and proteins. *M. hominis* has the capacity to naturally undergo intercellular replication, allowing the localization of numerous proteins within the host. The targeted proteins work as component of the host cell proteome. Hence, it is likely that several *M. hominis* proteins may possibly get localized within the host cell, due to the availability of signature sequence and evolutionary relatedness of proteins targeted within the cellular compartments of host cells. ER is an main compartment involved in proper protein folding, post-translation modification, translocation, and regulation of cellular homeostasis [23,24]. The unfolded protein response (UPR) is a highly conserved evolutionarily adaptive response that disrupts the ER physiology. UPR has been shown to be altered by different viruses and plays various roles during bacterial infection [25,26]. Both UPR and ER stress activation are involved in the growth and progression of various types of cancers [27]. The whole proteome of *M. hominis* may disturb the normal behavior of infected host cells and get involved in the development of prostate cancer. The main objective of the current work was to predict the protein localization of *M. hominis* in host cells and evaluate their role in the etiology of prostate cancer. We focused on ER proteins using bioinformatics tools and techniques and explored protein localization of *M. hominis* in the ER of host cells. We investigated the possible implication and relations of *M. hominis* proteome in the etiology of prostate cancer.

2. Results

2.1. Selection of the Whole Proteome Database of M. hominis

The whole protein sequence of ATCC-27545 strain of *M. hominis* was collected from UniProt, which has maximum proteins (563) with respect to other available strains. The unique selection of UniProt database was associated with its specific characteristics, including opulent, entirely classified, comprehensive, and accurately annotated sequences.

2.2. Prediction of the Subcellular Localization by cNLS Mapper

Nineteen proteins of *M. hominis* were predicted to be targeted in the ER with different NLSs (Figure 1). Figure 1 shows that among 19 ER targeting proteins, 15, 03, and 01 protein exhibit 0–3.0, 3.0–5.0, and 5.0–8.0 monopartite NLSs cutoff values respectively. Furthermore, the bipartite NLS was observed in 19 proteins and 06, 11, and 02 proteins exhibited 0–3.0, 3.0–5.0, and 5.0–8.0 cutoff values for NLSs, respectively. Different *M. hominis* proteins in a particular host cell may change the usual functioning of the host cell and promote the process of growth and development of cancer. cNLS (classical nuclear localization signals) mapper worked on amino acid sequence patterns, executed by three easy rules according to the NLSs classification [28]; these three rules are principally clusters of K and R basic amino acids and spaces between the clusters. cNLS mapper predicted nuclear localization signals for eukaryotic cells in *M. hominis* proteins. The literature of cNLS mapper has showed that the proteins with cut off value 8–10, 7–8, 3–5, and 1–2, were analyzed as targeted to the nucleus, targeted to both nucleus, partially targeted to nucleus and cytoplasm, and targeted to the cytoplasm, respectively. On the existence of multiple NLS sequence in same proteins, the elevated cut off value was documented.

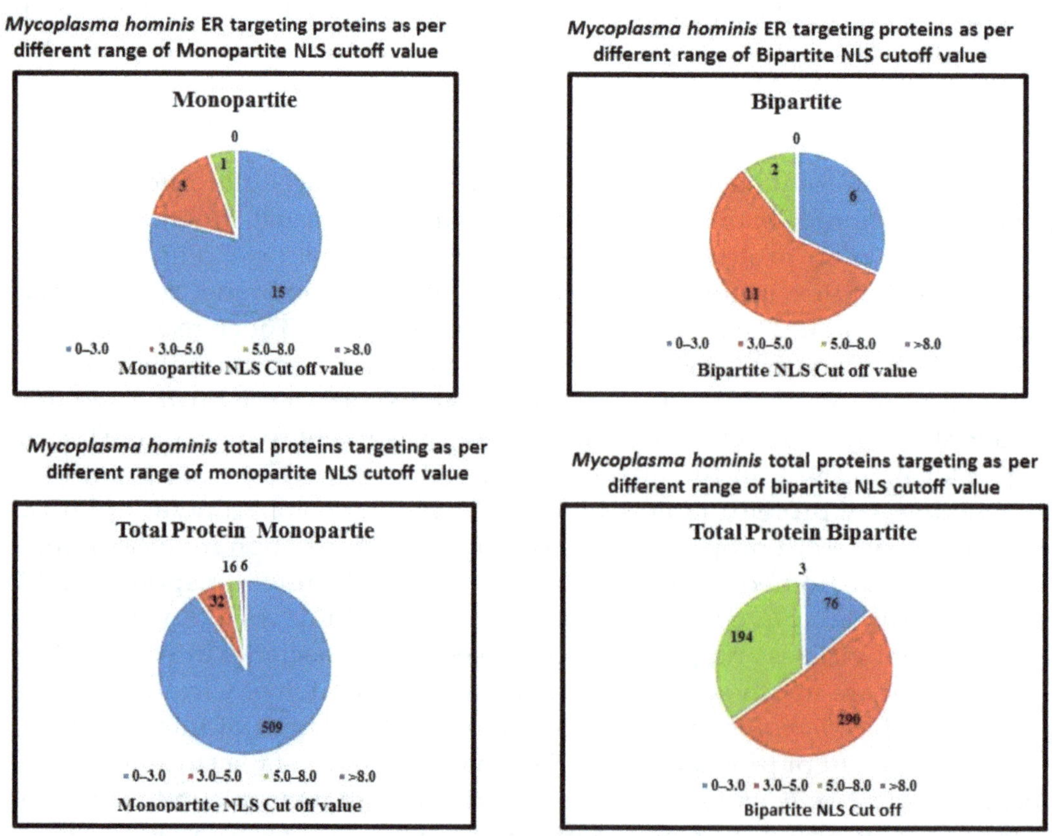

Figure 1. In silico analyses of the *Mycoplasma hominis* proteins localization in endoplasmic reticulum of host cells and their relationship with various NLS values.

2.3. Prediction of ER Localization in M. hominis Proteins Using Hum-Mploc 3.0

From the whole proteins (563) of *M. hominis*, only 19 were observed to be targeted into the ER of the host cell, as analyzed using Hum-mPLoc 3.0. The increase in cutoff values of NLS in monopartite and bipartite resulted in a decrease in ER localization (except 3.0–5.0 cutoff value of bipartite) (Table 1). Similarly, the increase in the molecular weight resulted in the decrease in the ER protein targeting of *M. hominis* (except for the range of 0–20 kDa). Proteins of 20–40 kDa molecular weights were observed mainly localized in the ER of the host cell (Table 2). Moreover, the outcome of values of isoelectric point (pI) failed to illustrate any constant pattern for ER localization of different proteins of *M. hominis* (Table 3). However, it has demonstrated that the bimodal character (pI and subcellular localization) in bacteria is likely to be a common property of proteomes, and is connected with the requirement of various pI values depending on subcellular protein localization [29].

Table 1. Computational prediction of *M. hominis* proteins targeted to the endoplasmic reticulum (ER) of host cells and their relation to all proteins with a similar NLS.

NLS	NLS Cutoff	Number of Proteins Targeting ER	Total Number of Proteins in This Range	Percentage
Monopartite NLS	0–3.0	15	509	2.94
	3.0–5.0	3	32	9.3
	5.0–8.0	1	16	6.25
	>8.0	0	6	0
Bipartite NLS	0–3.0	6	76	7.89
	3.0–5.0	11	290	3.79
	5.0–8.0	2	194	1.03
	>8.0	0	3	0

Table 2. Computational prediction of *Mycoplasma hominis* proteins targeted to ER of host cells and their relation to all proteins with similar molecular weight.

Molecular Weight	Number of Proteins Targeting to ER	Total Number of Proteins	Percentage
0–20 kDa	3	128	2.34
20–40 kDa	10	205	4.87
40–60 kDa	4	109	3.66
60–80 kDa	1	61	1.63
>80 kDa	1	60	1.66

Table 3. Computational prediction of *Mycoplasma hominis* proteins targeting to ER of host cells and their relation to all proteins with similar pI value.

Range of pI Value	Number of Proteins Targeting to ER	Total Number of Proteins	Percentage
3.0–5.0	0	25	0
5.0–6.0	1	105	0.95
6.0–7.0	0	68	0
7.0–8.0	0	35	0
8.0–9.0	2	109	1.83
9.0–10.0	12	181	6.62
10.0–11.0	4	33	12.12
12.0–13.0	0	7	0

The protein localization patterns of *M. hominis* in the ER of the host cell with diverse NLS values are shown in Figure 1. The patterns of protein localization in *M. hominis* for various ranges of molecular weights are shown in Figure 2. In the Figure 3, we illustrate the protein localization patterns of *M. hominis* in the ER of the host cell at various ranges of pI values. Various ER-targeting proteins are thought to be involved in different biochemical pathways. The details of the outcome of the ER-targeting *M. hominis* proteins with different roles are shown in Table 4. Various localized proteins could interfere in the regular growth behavior of host cells, thus leading to the alteration in the regular functioning of the host cell biochemical pathways. We have suggested a possible association of ER-targeting proteins of *M. hominis* in prostate cancer etiology. We have recently predicted the possible impacts of nucleus-, mitochondria-, and cytoplasm-targeting *M. hominis* proteins on the carcinogenesis of prostate cancer [10,30].

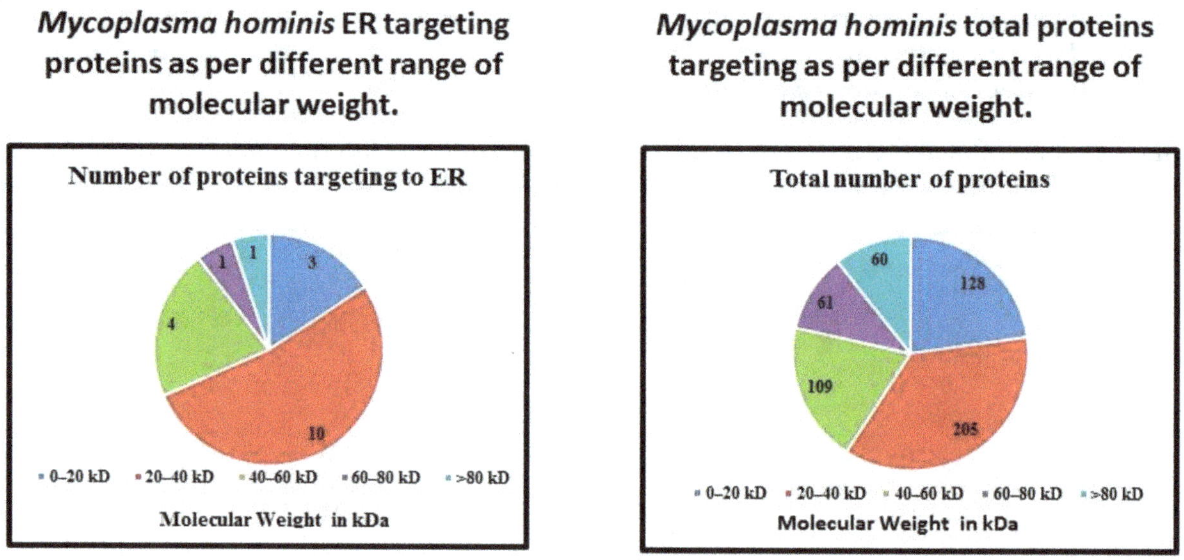

Figure 2. In silico analyses of the *Mycoplasma hominis* proteins localization in endoplasmic reticulum of host cells and their relation to proteins with different ranges of molecular weight.

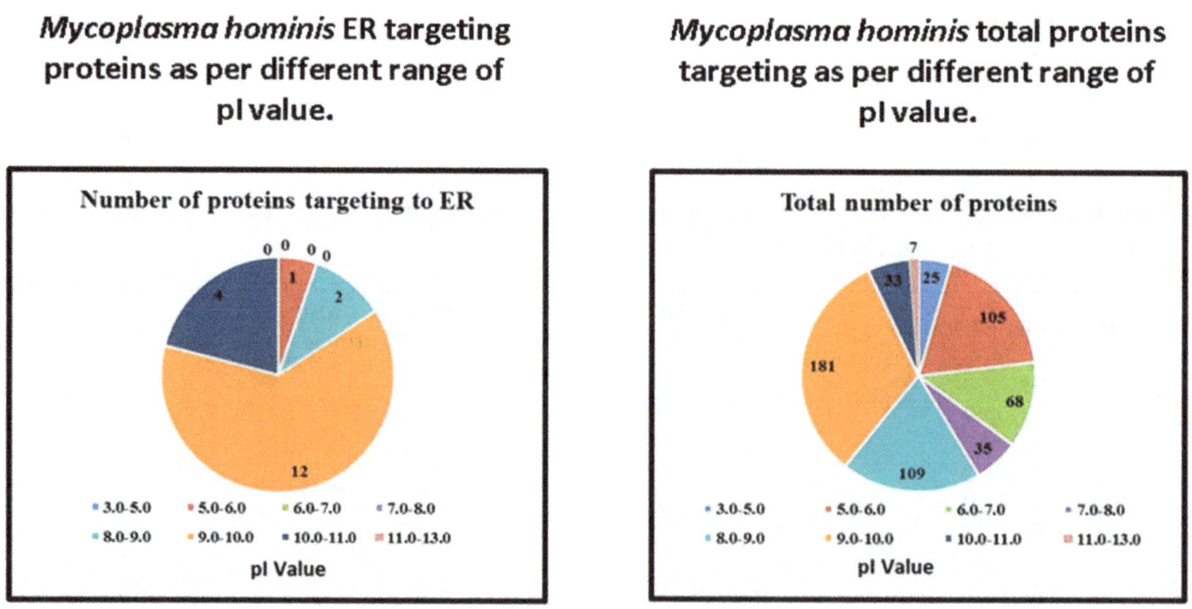

Figure 3. In silico analyses of the *Mycoplasma hominis* proteins localized in endoplasmic reticulum of host cells and their relation to proteins with different ranges of pI values.

Table 4. Details of *Mycoplasma hominis* proteins targeted to ER of host cells with their functions.

Accession Number	Protein Name	Function in Bacteria	Protein Existence	pI	Molecular Weight	NLS Mapper		Hum-mPLoc 3.0
						Monopartite	Bipartite	
A0A097NT54	Uncharacterized protein	Unknown	Protein predicted	10.17	13,450	0	22	Endoplasmic reticulum
A0A097NSU8	Uncharacterized protein	Unknown	Protein predicted	5.45	36,963	0	3.8	Endoplasmic reticulum
A0A097NTC6	Lipoprotein signal peptidase (EC 3.4.23.36)	Aspartic-type endopeptidase activity.	Protein inferred from homology	8.44	23,620	0	3.3	Endoplasmic reticulum
A0A097NTZ8	Cation transporting P-type ATPase (EC 3.6.3.8)	ATP binding, calcium-transporting ATPase activity, and metal ion binding activity.	Protein inferred from homology	8.56	107,148	2	6.1	Endoplasmic reticulum
A0A097NT87	Protein translocase subunit SecY	Intracellular protein transmembrane transport activity, protein transport activity hrough the Sec complex	Protein inferred from homology	9.51	54,926	6	3.9	Endoplasmic reticulum
A0A097NTV1	Membrane protein	Unknown	Protein predicted	9.54	25,294	0	5.5	Endoplasmic reticulum
A0A097NTF0	Prolipoprotein diacylglyceryl transferase (EC 2.4.99.-)	Phosphatidylglycerol-prolipoprotein diacylglyceryl transferase activity and involved in lipoprotein biosynthetic process.	Protein inferred from homology	9.56	35,944	0	3.8	Endoplasmic reticulum
A0A097NSJ5	Potassium transporter KtrB	Cation transmembrane transporter activity.	Protein predicted	9.6	58,434	0	3	Endoplasmic reticulum
A0A097NT10	ComEC/Rec2-related protein	Unknown	Protein predicted	9.65	54,059	0	4.4	Endoplasmic reticulum
A0A097NSI4	Uncharacterized protein	Unknown	Protein predicted	9.66	38,283	0	3.3	Endoplasmic reticulum
A0A097NSM2	Membrane protein	Unknown	Protein predicted	9.72	33,519	0	3.4	Endoplasmic reticulum
A0A097NSJ8	1-acyl-sn-glycerol-3-phosphate acyltransferase (EC 2.3.1.51)	1-acylglycerol-3-phosphate *O*-acyltransferase activity.	Protein predicted	9.79	31,034	0	4.7	Endoplasmic reticulum
A0A097NT14	Cobalt ABC transporter permease	Unknown	Protein predicted	9.85	35,998	5	4.4	Endoplasmic reticulum
A0A097NTJ8	ABC transporter permease	Transporter activity.	Protein inferred from homology	9.86	65,590	0	2.9	Endoplasmic reticulum
A0A097NTE8	Uncharacterized protein	Unknown	Protein predicted	9.9	57,701	0	5	Endoplasmic reticulum
A0A097NSN9	Uncharacterized protein	Unknown	Protein predicted	9.97	14,493	0	2.6	Endoplasmic reticulum
A0A097NSG1	Uncharacterized protein	Unknown	Protein predicted	10.1	17,661	4	2.6	Endoplasmic reticulum
A0A097NTR2	Membrane protein	Transporter activity.	Protein predicted	10.19	33,456	4	3.2	Endoplasmic reticulum
A0A097NT45	Membrane protein	Unknown	Protein predicted	10.47	20,769	0	2.5	Endoplasmic reticulum

3. Discussion

Prostate cancer is the frequently identified cancer and sixth leading reason of cancer-related death globally. The frequency of prostate disease has relentlessly expanded in Asian nations, including China, India, and Malaysia [19]. Cancer was initially considered as a genetic disease. Nevertheless, various studies have revealed the association between bacterial infections and progression of different cancer types. The infection may regulate different factors that influence prostate cancer progress in susceptible individuals. It has been reported that protein molecules of host cells face chronic and acute challenges for the maintenance of their integrity [31]. Proteostasis/protein homeostasis allows health and proper growth of eukaryotic cells. The proteostasis depends upon a complex network of proteases molecular, proteins, chaperones, and different regulatory factors [32]. Inadequacies in protein homeostasis or proteostasis direct to several cardiovascular, neurodegenerative, metabolic, and oncological disorders [31]. Different factors, such as chemical toxins, exogenous proteins, drugs, and environmental factors compromise proteostasis, which stimulates proteome stress, and cause different disorders [33,34]. Proteostasis in humans is regulated through quality control (QC) network of approximately 800 different proteins [35,36]. The ER is an important subcellular compartment responsible for the regulation of protein folding, post-translation modification, and protein translocation. Interruption in the environment of ER by pathological agents may cause alterations in glycosylation, DNA damage, nutrient deprivation, oxidative stress, calcium depletion, and energy fluctuation/disturbance, thereby resulting in ER stress and consequent accretion of misfolded or unfolded proteins in the ER lumen. These host cells must overcome perturbations of ER functions and stress for their survival. If ER stress is left unresolved, it may disturb the normal functioning of apoptosis [37]. The disturbance in apoptotic regulatory protein Bcl-2 function results in the increased transcription of p53 unregulated modulator of apoptosis (PUMA), Bcl2-like11 (BIM), BH3-only proteins, and NADPH oxidase activator (NOXA). ER stress promotes the interactions between Bax and PUMA, resulting in the release of cytochrome *c* and apoptosis activation by caspase-dependent modulation of p53 proteins. Many *M. hominis* proteins outmaneuver the host cell machinery, and possibly change the normal behavior of host cells, which may promote the growth of prostate cancer. The involvement of *M. hominis* in the progression of prostate cancer is not yet clear. The rationale of this study was to predict *M. hominis* proteins that are targeted into the ER of the host cell, and assess their potential roles in the growth and progression of prostate cancer.

Of the whole proteome, 19 proteins of *M. hominis* were expected to be targeted to the ER of host cells. These predicted proteins of *M. hominis* may modify the normal function of the host cell. The study of protein targeting into the host cell is very important to detect the progression and development of cancer, especially if the cancer growth is associated with the intracellular bacterial infection. The targeting of different bacterial proteins into host cell compartments, such as cytoplasm, mitochondria, and nucleus, has an important effect on the etiology of different cancer types [14,15,38]. The host cell gets affected by various types of bacterial proteins, which alter the usual development and normal behavior of the host cell. Infection is considered as a possible factor in the progression and development of different types of cancer, especially when it is linked with chronic inflammation that leads to cancer progression in various cases [39,40]. Several novel techniques could be developed to detect and treat cancer based on the study of chronic infection related to cancer and their mechanisms and activities that promote cancer [40]. The potential involvement of infection of *M. hominis* in prostate cancer and its evolution and progression depend on the estimation of its protein targeting into various sections of host cells. The protein-targeting ability of *M. hominis* may lead to several consequences related to prostate cancer etiology. Various *M. hominis* proteins targeted into host cells may disturb the behavior and functioning of infected cells [3,9].

Numerous advanced techniques have been developed for the analysis of protein localization; however, these may be inefficient, owing to their high cost and time-consuming protocols [41]. Several bioinformatics tools have been developed to calculate the subcellular targeting of proteins, thereby offering several advantages for investigational procedures [42–44]. The tools developed basically focus on the identity/alignment search or recognition of a particular sequence motif essential for particular protein localization [44]. The research work of protein targeting is also essential to infer the different functions of bacterial proteins within the host cell. The possible role of targeted proteins can be analyzed on the evidence of their relationships with different host proteins, whose role is already revealed. This information will act as a starting point for upcoming wet lab experiments.

Although nuclear-targeting proteins play a crucial role by controlling the normal functioning of host cells, other cell organelle-targeting proteins are also involved in the regulation of the normal behavior of host cells. At present, different predictors are developed for the prediction of specific motifs in protein sequences. Six NLS classes have been classified, wherein nuclear import proteins are transferred through α/β pathway of importin. The well-known dimer of α and β importin is present in the nuclear import receptor, wherein α importin works as a possible adapter protein and binds to cNLS that is identified either twice (bipartite NLS) or once (monopartite NLS), with highly basic stretches of amino acids [45]. The potential activity of NLS sequences changes within the identical class of NLS with altered sequence of NLS [46]. Hence, in the current study we particularly employed NLS mapper predictor in the present study to analyze NLS activity as an alternative to NLS sequence based on the contribution of every residue of amino acid in the NLSs. This could lead to more accurate prediction of results [44]. NLS predictor senses the activity of an NLS as a separate protein sequence, rather than the whole structural sequence of a particular native protein.

Hum-mPLoc 3.0 was used in the current in silico study for the analysis of *M. hominis* proteins targeted into the host subcellular compartment. Hum-mPLoc 3.0 tool is based on different Support Vector Machine (SVMs) method systematized in a decision tree, and predicts 12 different human subcellular localization, including ER, nucleus, mitochondrion, cytoplasm, Golgi apparatus, centriole, cytoskeleton, endosome, peroxisome, lysosome, extracellular region, and plasma membrane [47]. The tRNA threonylcarbamoyladenosine biosynthesis protein TsaE, membrane protein, lipoprotein signal peptidase, 1-acyl-*sn*-glycerol-3-phosphate acyltransferase, prolipoprotein diacylglyceryl transferase, cobalt ABC transporter permease, ComEC/Rec2-related protein, protein translocase subunit SecY, and potassium transporter KtrB are predicted to be targeted into the ER of the host cell, resulting in the alteration in the normal pattern of protein folding in ER.

4. Materials and Methods

4.1. Proteins and Prediction Analysis

The whole proteome of *M. hominis* comprising 563 proteins was selected for the prediction of proteins that are targeted into host cells. Computational predictions were used for the analysis of proteins targeted into the ER of host cells. Complete data were collected after the prediction analysis to predict implications of ER-targeting proteins in prostate cancer etiology.

4.2. Choice of Protein Database for M. hominis

We used the UniProt database (www.uniprot.org) to select the specific strain of *M. hominis*. The UniProt database is considered as a complete resource for the sequence of proteins and data annotation. This database was prepared with the combination of PIR database, Swiss-Prot, and TrEMBL activities [48], and is a collection of huge data with respect to *M. hominis* and its subcellular location, as described in Swiss-Prot/TrEMBL or PIR-PSD [49,50]. This database has two proteomes related to *M. hominis* strains, namely, ATCC-23114/PG21 and ATCC-27545 [51]. All *M. hominis* proteins of ATCC-27545 strain were used for the in silico analysis of the subcellular proteins targeted into host cells by using predictor cNLS mapper (Tsuruoka, Japan). Furthermore, Hum-mPLoc 3.0 predictor (Shanghai, China) was implicated to predict the possible location of ER-targeting *M. hominis* proteins in host cells.

4.3. Prediction of the Subcellular Localization by cNLS Mapper

Proteins targeted in various organelles of host cells were predicted using cNLS mapper [44]. Whole protein sequence of *M. hominis* was utilized for the analysis of monopartite and bipartite NLSs. The precise cNLS mapper cutoff values were 8–10, 1–2, 5–3, and 7–8, that were used to predict the localization of proteins in nucleus, cytoplasm, both cytoplasm and nucleus, and partially nucleus, respectively, as defined in the literature of cNLS. The values of monopartite and bipartite NLSs (basic amino acid stretches) were analyzed using the cNLS mapper in whole protein sequence of *M. hominis*. These basic stretches mediate binding of NLS to importin-α transport receptor, and this complex binds to importin β, through which a specific protein localizes to the nucleus. cNLS mapper detects contribution of each residue in NLS and predicts NLS activity, which is suggested to give more accurate prediction performance [44].

4.4. Prediction of ER Localization in M. hominis Proteins Using Hum-Mploc 3.0

Hum-mPLoc 3.0 was utilized to determine *M. hominis* protein localization in the ER and covered about 12 different human subcellular compartments. The protein sequences have showed multiview complementary features, i.e., peptide-based functional domains, context vocabulary annotation-based gene ontology, and amino acid residue-based statistical features, as most of the existing predictors to determine the subcellular targeting of human proteins are limited with unique location site. To prevail this barrier, Hum-mPLoc, a new ensemble classifier [47], was established and used for cases with multiple location sites. The predictor Hum-mPLoc is accessible generously by researchers from the web server at http://202.120.37.186/bioinf/hum-multi. This predictor has been employed to predict various human protein entries in Swiss-Prot database which do not have subcellular location annotations or are interpreted as "uncertain." Hum-mPLoc predictor may analyze the possible targeting of specific protein in three kingdoms, namely, plants, fungi, and animals. In the current research work, we determined proteins targeting in the ER of host cells using *M. hominis* proteins sequence as the query.

5. Conclusions

Protein homeostasis/proteostasis, is very crucial for viability and health of cells [33]. Alteration in proteostasis has been connected with the growth of many disorders, including cancer, which is considered as the most challenging disease of the current era [31]. Among different virulence aspects, various bacterial protein toxins that are somehow associated to the progression of various types of cancer have been the possible targets or markers for the management of cancer. The possible connection between *M. hominis* infection and risk of prostate cancer has gained attention in the past few years, but no detailed information is available. Here we predicted the connection between *M. hominis* infection and prostate cancer, and found that the intercellular *M. hominis* infection in host cells acts as a potential factor in prostate cancer etiology, owing to the accumulation of several *M. hominis* proteins in the ER of host cells. ER is an important organelle of eukaryotic cells involved in the regulation of secretory pathways, and release and storage of calcium. Misfolded proteins in ER cause ER stress through their accumulation and stimulation of UPR. ER stress and UPR activation are associated with the progression of different types of diseases in human, including different types of cancer. The present research work paves way for the analysis of the potential involvement of specific strain of *M. hominis* in prostate cancer.

Author Contributions: The project work was conceived by S.K. and Y.A.A., M.Z. performed bioinformatics analyses along with S.K., M.M.B.I., C.R. and A.A.C. were involved in protein sequencing. All authors were helped in writing and critical review of manuscript and approved the final draft of manuscript.

Acknowledgments: This project was supported by the Research Groups Program (Research Group number RG-1439-033), Deanship of Scientific Research, King Saud University, Riyadh, Saudi Arabia.

References

1. Eaton, K.; Yang, W. Registered report: Intestinal inflammation targets cancer-inducing activity of the microbiota. *eLife* **2015**, *4*, e04186. [CrossRef] [PubMed]
2. Arthur, J.C.; Perez-Chanona, E.; Muhlbauer, M.; Tomkovich, S.; Uronis, J.M.; Fan, T.J.; Campbell, B.J.; Abujamel, T.; Dogan, B.; Rogers, A.B.; et al. Intestinal inflammation targets cancer-inducing activity of the microbiota. *Science* **2012**, *338*, 120–123. [CrossRef] [PubMed]
3. Barykova, Y.A.; Logunov, D.Y.; Shmarov, M.M.; Vinarov, A.Z.; Fiev, D.N.; Vinarova, N.A.; Rakovskaya, I.V.; Baker, P.S.; Shyshynova, I.; Stephenson, A.J.; et al. Association of *Mycoplasma hominis* infection with prostate cancer. *Oncotarget* **2011**, *2*, 289–297. [CrossRef] [PubMed]
4. Cavarretta, I.; Ferrarese, R.; Cazzaniga, W.; Saita, D.; Luciano, R.; Ceresola, E.R.; Locatelli, I.; Visconti, L.; Lavorgna, G.; Briganti, A.; et al. The Microbiome of the Prostate Tumor Microenvironment. *Eur. Urol.* **2017**, *72*, 625–631. [CrossRef] [PubMed]
5. Huang, S.; Li, J.Y.; Wu, J.; Meng, L.; Shou, C.C. Mycoplasma infections and different human carcinomas. *World J. Gastroenterol.* **2001**, *7*, 266–269. [CrossRef] [PubMed]
6. Razin, S.; Yogev, D.; Naot, Y. Molecular biology and pathogenicity of mycoplasmas. *Microbiol. Mol. Biol. Rev.* **1998**, *62*, 1094–1156. [PubMed]
7. Cimolai, N. Do mycoplasmas cause human cancer? *Can. J. Microbiol.* **2001**, *47*, 691–697. [CrossRef] [PubMed]
8. Feng, S.H.; Tsai, S.; Rodriguez, J.; Lo, S.C. Mycoplasmal infections prevent apoptosis and induce malignant transformation of interleukin-3-dependent 32D hematopoietic cells. *Mol. Cell. Biol.* **1999**, *19*, 7995–8002. [CrossRef] [PubMed]
9. Barykova Iu, A.; Shmarov, M.M.; Logunov, D.; Verkhovskaia, L.V.; Aliaev Iu, G.; Fiev, D.N.; Vinarov, A.Z.; Vinarova, N.A.; Rakovskaia, I.V.; Naroditskii, B.S.; et al. Identification of Mycoplasma in patients with suspected prostate cancer. *Zhurnal Mikrobiol. Epidemiol. Immunobiol.* **2010**, *4*, 81–85.
10. Khan, S.; Zakariah, M.; Rolfo, C.; Robrecht, L.; Palaniappan, S. Prediction of *Mycoplasma hominis* proteins targeting in mitochondria and cytoplasm of host cells and their implication in prostate cancer etiology. *Oncotarget* **2017**, *8*, 30830–30843. [CrossRef] [PubMed]
11. Lo, S.C.; Tsai, S. Mycoplasmas and human prostate cancer: An exciting but cautionary note. *Oncotarget* **2011**, *2*, 352–355. [PubMed]
12. Caini, S.; Gandini, S.; Dudas, M.; Bremer, V.; Severi, E.; Gherasim, A. Sexually transmitted infections and prostate cancer risk: A systematic review and meta-analysis. *Cancer Epidemiol.* **2014**, *38*, 329–338. [CrossRef] [PubMed]
13. Gagnaire, A.; Nadel, B.; Raoult, D.; Neefjes, J.; Gorvel, J.P. Collateral damage: Insights into bacterial mechanisms that predispose host cells to cancer. *Nat. Rev. Microbiol.* **2017**, *15*, 109–128. [CrossRef] [PubMed]
14. Khan, S.; Imran, A.; Khan, A.A.; Abul Kalam, M.; Alshamsan, A. Systems Biology Approaches for the Prediction of Possible Role of Chlamydia pneumoniae Proteins in the Etiology of Lung Cancer. *PLoS ONE* **2016**, *11*, e0148530. [CrossRef] [PubMed]
15. Khan, S. Potential role of Escherichia coli DNA mismatch repair proteins in colon cancer. *Crit. Rev. Oncol. Hematol.* **2015**, *96*, 475–482. [CrossRef] [PubMed]
16. Tsai, S.; Wear, D.J.; Shih, J.W.; Lo, S.C. Mycoplasmas and oncogenesis: Persistent infection and multistage malignant transformation. *Proc. Natl. Acad. Sci. USA* **1995**, *92*, 10197–10201. [CrossRef] [PubMed]
17. De Marzo, A.M.; Platz, E.A.; Sutcliffe, S.; Xu, J.; Gronberg, H.; Drake, C.G.; Nakai, Y.; Isaacs, W.B.; Nelson, W.G. Inflammation in prostate carcinogenesis. *Nat. Rev. Cancer* **2007**, *7*, 256–269. [CrossRef] [PubMed]
18. Sfanos, K.S.; De Marzo, A.M. Prostate cancer and inflammation: The evidence. *Histopathology* **2012**, *60*, 199–215. [CrossRef] [PubMed]
19. Ito, K. Prostate cancer in Asian men. *Nat. Rev. Urol.* **2014**, *11*, 197–212. [CrossRef] [PubMed]
20. Buc, E.; Dubois, D.; Sauvanet, P.; Raisch, J.; Delmas, J.; Darfeuille-Michaud, A.; Pezet, D.; Bonnet, R. High prevalence of mucosa-associated, *E. coli* producing cyclomodulin and genotoxin in colon cancer. *PLoS ONE* **2013**, *8*, e56964. [CrossRef] [PubMed]
21. Dessi, D.; Delogu, G.; Emonte, E.; Catania, M.R.; Fiori, P.L.; Rappelli, P. Long-term survival and intracellular replication of *Mycoplasma hominis* in Trichomonas vaginalis cells: Potential role of the protozoon in transmitting bacterial infection. *Infect. Immun.* **2005**, *73*, 1180–1186. [CrossRef] [PubMed]

22. Namiki, K.; Goodison, S.; Porvasnik, S.; Allan, R.W.; Iczkowski, K.A.; Urbanek, C.; Reyes, L.; Sakamoto, N.; Rosser, C.J. Persistent exposure to Mycoplasma induces malignant transformation of human prostate cells. *PLoS ONE* **2009**, *4*, e6872. [CrossRef] [PubMed]
23. Yadav, R.K.; Chae, S.W.; Kim, H.R.; Chae, H.J. Endoplasmic reticulum stress and cancer. *J. Cancer Prev.* **2014**, *19*, 75–88. [CrossRef] [PubMed]
24. Pluquet, O.; Pourtier, A.; Abbadie, C. The unfolded protein response and cellular senescence. A review in the theme: Cellular mechanisms of endoplasmic reticulum stress signaling in health and disease. *Am. J. Physiol. Cell Physiol.* **2015**, *308*, C415–C425. [CrossRef] [PubMed]
25. Dumartin, L.; Alrawashdeh, W.; Trabulo, S.M.; Radon, T.P.; Steiger, K.; Feakins, R.M.; di Magliano, M.P.; Heeschen, C.; Esposito, I.; Lemoine, N.R.; et al. ER stress protein AGR2 precedes and is involved in the regulation of pancreatic cancer initiation. *Oncogene* **2016**, *36*, 3094–3103. [CrossRef] [PubMed]
26. Celli, J.; Tsolis, R.M. Bacteria, the endoplasmic reticulum and the unfolded protein response: Friends or foes? *Nat. Rev. Microbiol.* **2015**, *13*, 71–82. [CrossRef] [PubMed]
27. Wang, M.; Kaufman, R.J. The impact of the endoplasmic reticulum protein-folding environment on cancer development. *Nat. Rev. Cancer* **2014**, *14*, 581–597. [CrossRef] [PubMed]
28. Hicks, G.R.; Raikhel, N.V. Protein import into the nucleus: An integrated view. *Annu. Rev. Cell Dev. Biol.* **1995**, *11*, 155–188. [CrossRef] [PubMed]
29. Schwartz, R.; Ting, C.S.; King, J. Whole proteome pI values correlate with subcellular localizations of proteins for organisms within the three domains of life. *Genome Res.* **2001**, *11*, 703–709. [CrossRef] [PubMed]
30. Khan, S.; Zakariah, M.; Palaniappan, S. Computational prediction of *Mycoplasma hominis* proteins targeting in nucleus of host cell and their implication in prostate cancer etiology. *Tumour Biol.* **2016**, *37*, 10805–10813. [CrossRef] [PubMed]
31. Balch, W.E.; Morimoto, R.I.; Dillin, A.; Kelly, J.W. Adapting proteostasis for disease intervention. *Science* **2008**, *319*, 916–919. [CrossRef] [PubMed]
32. Gupta, R.; Kasturi, P.; Bracher, A.; Loew, C.; Zheng, M.; Villella, A.; Garza, D.; Hartl, F.U.; Raychaudhuri, S. Firefly luciferase mutants as sensors of proteome stress. *Nat. Methods* **2011**, *8*, 879–884. [CrossRef] [PubMed]
33. Liu, Y.; Miao, K.; Li, Y.; Fares, M.; Chen, S.; Zhang, X. A HaloTag-Based Multicolor Fluorogenic Sensor Visualizes and Quantifies Proteome Stress in Live Cells Using Solvatochromic and Molecular Rotor-Based Fluorophores. *Biochemistry* **2018**. [CrossRef] [PubMed]
34. Liu, Y.; Fares, M.; Dunham, N.P.; Gao, Z.; Miao, K.; Jiang, X.; Bollinger, S.S.; Boal, A.K.; Zhang, X. AgHalo: A Facile Fluorogenic Sensor to Detect Drug-Induced Proteome Stress. *Angew. Chem. Int. Ed. Engl.* **2018**, *56*, 8672–8676. [CrossRef] [PubMed]
35. Hartl, F.U.; Bracher, A.; Hayer-Hartl, M. Molecular chaperones in protein folding and proteostasis. *Nature* **2011**, *475*, 324–332. [CrossRef] [PubMed]
36. Wood, R.J.; Ormsby, A.R.; Radwan, M.; Cox, D.; Sharma, A.; Vopel, T.; Ebbinghaus, S.; Oliveberg, M.; Reid, G.E.; Dickson, A.; et al. A biosensor-based framework to measure latent proteostasis capacity. *Nat. Commun.* **2018**, *9*, 287. [CrossRef] [PubMed]
37. Tabas, I.; Ron, D. Integrating the mechanisms of apoptosis induced by endoplasmic reticulum stress. *Nat. Cell Biol.* **2011**, *13*, 184–190. [CrossRef] [PubMed]
38. Khan, A.A. In silico prediction of escherichia coli proteins targeting the host cell nucleus, with special reference to their role in colon cancer etiology. *J. Comput. Biol.* **2014**, *21*, 466–475. [CrossRef] [PubMed]
39. Poutahidis, T.; Cappelle, K.; Levkovich, T.; Lee, C.W.; Doulberis, M.; Ge, Z.; Fox, J.G.; Horwitz, B.H.; Erdman, S.E. Pathogenic intestinal bacteria enhance prostate cancer development via systemic activation of immune cells in mice. *PLoS ONE* **2013**, *8*, e73933. [CrossRef] [PubMed]
40. Karin, M.; Lawrence, T.; Nizet, V. Innate immunity gone awry: Linking microbial infections to chronic inflammation and cancer. *Cell* **2006**, *124*, 823–835. [CrossRef] [PubMed]
41. Huh, W.K.; Falvo, J.V.; Gerke, L.C.; Carroll, A.S.; Howson, R.W.; Weissman, J.S.; O'Shea, E.K. Global analysis of protein localization in budding yeast. *Nature* **2003**, *425*, 686–691. [CrossRef] [PubMed]
42. Zhou, H.; Yang, Y.; Shen, H.B. Hum-mPLoc 3.0: Prediction enhancement of human protein subcellular localization through modeling the hidden correlations of gene ontology and functional domain features. *Bioinformatics* **2017**, *33*, 843–853. [CrossRef] [PubMed]
43. Kumar, M.; Raghava, G.P. Prediction of nuclear proteins using SVM and HMM models. *BMC Bioinform.* **2009**, *10*, 22. [CrossRef] [PubMed]

44. Kosugi, S.; Hasebe, M.; Tomita, M.; Yanagawa, H. Systematic identification of cell cycle-dependent yeast nucleocytoplasmic shuttling proteins by prediction of composite motifs. *Proc. Natl. Acad. Sci. USA* **2009**, *106*, 10171–10176. [CrossRef] [PubMed]
45. Lange, A.; Mills, R.E.; Lange, C.J.; Stewart, M.; Devine, S.E.; Corbett, A.H. Classical nuclear localization signals: Definition, function, and interaction with importin alpha. *J. Biol. Chem.* **2007**, *282*, 5101–5105. [CrossRef] [PubMed]
46. Kosugi, S.; Hasebe, M.; Entani, T.; Takayama, S.; Tomita, M.; Yanagawa, H. Design of peptide inhibitors for the importin alpha/beta nuclear import pathway by activity-based profiling. *Chem. Biol.* **2008**, *15*, 940–949. [CrossRef] [PubMed]
47. Shen, H.B.; Chou, K.C. Hum-mPLoc: An ensemble classifier for large-scale human protein subcellular location prediction by incorporating samples with multiple sites. *Biochem. Biophys. Res. Commun.* **2007**, *355*, 1006–1011. [CrossRef] [PubMed]
48. Apweiler, R.; Bairoch, A.; Wu, C.H.; Barker, W.C.; Boeckmann, B.; Ferro, S.; Gasteiger, E.; Huang, H.; Lopez, R.; Magrane, M.; et al. UniProt: The Universal Protein knowledgebase. *Nucleic Acids Res.* **2004**, *32*, D115–D119. [CrossRef] [PubMed]
49. Consortium, U. Activities at the Universal Protein Resource (UniProt). *Nucleic Acids Res.* **2014**, *42*, D191–D198.
50. Boeckmann, B.; Bairoch, A.; Apweiler, R.; Blatter, M.C.; Estreicher, A.; Gasteiger, E.; Martin, M.J.; Michoud, K.; O'Donovan, C.; Phan, I.; et al. The SWISS-PROT protein knowledgebase and its supplement TrEMBL in 2003. *Nucleic Acids Res.* **2003**, *31*, 365–370. [CrossRef] [PubMed]
51. Pereyre, S.; Sirand-Pugnet, P.; Beven, L.; Charron, A.; Renaudin, H.; Barre, A.; Avenaud, P.; Jacob, D.; Couloux, A.; Barbe, V.; et al. Life on arginine for *Mycoplasma hominis*: Clues from its minimal genome and comparison with other human urogenital mycoplasmas. *PLoS Genet.* **2009**, *5*, e1000677. [CrossRef] [PubMed]

Permissions

All chapters in this book were first published in MDPI; hereby published with permission under the Creative Commons Attribution License or equivalent. Every chapter published in this book has been scrutinized by our experts. Their significance has been extensively debated. The topics covered herein carry significant findings which will fuel the growth of the discipline. They may even be implemented as practical applications or may be referred to as a beginning point for another development.

The contributors of this book come from diverse backgrounds, making this book a truly international effort. This book will bring forth new frontiers with its revolutionizing research information and detailed analysis of the nascent developments around the world.

We would like to thank all the contributing authors for lending their expertise to make the book truly unique. They have played a crucial role in the development of this book. Without their invaluable contributions this book wouldn't have been possible. They have made vital efforts to compile up to date information on the varied aspects of this subject to make this book a valuable addition to the collection of many professionals and students.

This book was conceptualized with the vision of imparting up-to-date information and advanced data in this field. To ensure the same, a matchless editorial board was set up. Every individual on the board went through rigorous rounds of assessment to prove their worth. After which they invested a large part of their time researching and compiling the most relevant data for our readers.

The editorial board has been involved in producing this book since its inception. They have spent rigorous hours researching and exploring the diverse topics which have resulted in the successful publishing of this book. They have passed on their knowledge of decades through this book. To expedite this challenging task, the publisher supported the team at every step. A small team of assistant editors was also appointed to further simplify the editing procedure and attain best results for the readers.

Apart from the editorial board, the designing team has also invested a significant amount of their time in understanding the subject and creating the most relevant covers. They scrutinized every image to scout for the most suitable representation of the subject and create an appropriate cover for the book.

The publishing team has been an ardent support to the editorial, designing and production team. Their endless efforts to recruit the best for this project, has resulted in the accomplishment of this book. They are a veteran in the field of academics and their pool of knowledge is as vast as their experience in printing. Their expertise and guidance has proved useful at every step. Their uncompromising quality standards have made this book an exceptional effort. Their encouragement from time to time has been an inspiration for everyone.

The publisher and the editorial board hope that this book will prove to be a valuable piece of knowledge for researchers, students, practitioners and scholars across the globe.

List of Contributors

Abdul Wahab and Omid Mahmoudi
Department of Electronics and Information Engineering, Jeonbuk National University, Jeonju 54896, Korea

Jeehong Kim
Department of New & Renewable Energy, VISION College of Jeonju, Jeonju 55069, Korea

Kil To Chong
Department of Electronics Engineering, Jeonbuk National University, Jeonju 54896, Korea
Advance Electronics & Information Research Center, Jeonbuk National University, Jeonju 54896, Korea

Francesca Tessaro and Leonardo Scapozza
Pharmaceutical Biochemistry, School of Pharmaceutical Sciences, University of Geneva CMU, Rue Michel-Servet 1, 1211 Geneva 4, Switzerland
Institute of Pharmaceutical Sciences of Western Switzerland, University of Geneva, 1211 Geneva, Switzerland

Pengmian Feng
Innovative Institute of Chinese Medicine and Pharmacy, Chengdu University of Traditional Chinese Medicine, Chengdu 611730, China
School of Public Health, North China University of Science and Technology, Tangshan 063000, China

Zhaochun Xu
Key Laboratory for Neuro-Information of Ministry of Education, School of Life Science and Technology, Center for Informational Biology, University of Electronic Science and Technology of China, Chengdu 610054, China
Computer Department, Jingdezhen Ceramic Institute, Jingdezhen 333403, China

Hui Yang, Hao Lv and Hui Ding
Key Laboratory for Neuro-Information of Ministry of Education, School of Life Science and Technology, Center for Informational Biology, University of Electronic Science and Technology of China, Chengdu 610054, China

Li Liu
Laboratory of Theoretical Biophysics, School of Physical Science and Technology, Inner Mongolia University, Hohhot 010021, China

Baofang Hu
School of Information Science and Engineering, Shandong Normal University, Jinan 250014, China
School of Data and Computer Science, Shandong Women's University, Jinan 250014, China
Shandong Provincial Key Laboratory for Distributed Computer Software Novel Technology, Shandong Normal University, Jinan 250014, China

Hong Wang, Lutong Wang and Weihua Yuan
School of Information Science and Engineering, Shandong Normal University, Jinan 250014, China
Shandong Provincial Key Laboratory for Distributed Computer Software Novel Technology, Shandong Normal University, Jinan 250014, China

Yinghan Hong
School of Computer Science and Technology, Guangdong University of Technology, Guangzhou 510006, China
School of Physics and Electronic Engineering, Hanshan Normal University, Chaozhou 521041, China

Zhifeng Hao
School of Computer Science and Technology, Guangdong University of Technology, Guangzhou 510006, China
School of Mathematics and Big Data, Foshan University, Foshan 528000, China

Guizhen Mai
School of Computer Science and Technology, Guangdong University of Technology, Guangzhou 510006, China

Han Huang
School of Software Engineering, South China University of Technology, Guangzhou 510006, China

Arun Kumar Sangaiah
School of Computing Science and Engineering, Vellore Institute of Technology, Vellore-632014, Tamil Nadu, India

Bin Wang, Shihua Zhou and Xuedong Zheng
Key Laboratory of Advanced Design and Intelligent Computing, Dalian University, Ministry of Education, Dalian 116622, China

Yingjie Xie
Applied Technology College, Dalian Ocean University, Dalian 116300, China

Changjun Zhou
College of Mathematics, Physics and Information Engineering, Zhejiang Normal University, Jinhua 321004, China

Ke Han
School of Computer and Information Engineering, Harbin University of Commerce, Harbin 150028, China

Miao Wang and Lei Zhang
Life sciences and Environmental Sciences Development Center, Harbin University of Commerce, Harbin 150010, China

Chunyu Wang
School of Computer Science and Technology, Harbin Institute of Technology, Harbin 150001, China

Chun Chen and Ming-Han Lee
Department of Chemistry, National Dong Hwa University, Shoufeng, Hualien 97401, Taiwan

Ching-Feng Weng
Department of Life Science and Institute of Biotechnology, National Dong Hwa University, Shoufeng, Hualien 97401, Taiwan

Max K. Leong
Department of Chemistry, National Dong Hwa University, Shoufeng, Hualien 97401, Taiwan
Department of Life Science and Institute of Biotechnology, National Dong Hwa University, Shoufeng, Hualien 97401, Taiwan

Yuting Xing, Chengkun Wu, Xi Yang, Wei Wang and En Zhu
School of Computer Science, National University of Defense Technology, Changsha, Hunan 410073, China

Jianping Yin
School of Computer Science and Network Security, Dongguan University of Technology, Dongguan, Guangdong 523808, China

Jianying Yuan and Dequan Guo
The Postdoctoral Station at Xihua University Based on Collaboration Innovation Center of Sichuan Automotive Key Parts, Xihua University, Chengdu 610039, China
School of Control Engineering, Chengdu University of Information Technology, Chengdu 610225, China
School of Aeronautics and Astronautics, University of Electronic Science and Technology, Chengdu 610054, China

Gexiang Zhang
Robotics Research Center, Xihua University, Chengdu 610039, China
School of Electrical Engineering, Southwest Jiaotong University, Chengdu 610031, China

Prithwineel Paul
School of Electrical Engineering, Southwest Jiaotong University, Chengdu 610031, China

Ming Zhu
School of Control Engineering, Chengdu University of Information Technology, Chengdu 610225, China

Qiang Yang
Robotics Research Center, Xihua University, Chengdu 610039, China

Xiaoke Ping, Tengyue Wang, Na Lin, Feifei Di, Yangyang Li, Hongju Jian, Kun Lu, Jiana Li, Xinfu Xu and Liezhao Liu
College of Agronomy and Biotechnology, Chongqing Engineering Research Center for Rapeseed, Academy of Agricultural Sciences, State Cultivation Base of Crop Stress Biology for Southern Mountainous Land, Southwest University, Chongqing 400715, China

Hao Wang
Hybrid Rapeseed Research Center of Shanxi Province, Shanxi Rapeseed Branch of National Centre for Oil Crops Genetic Improvement, Yangling 712100, China

Buwen Cao
College of Information and Electronic Engineering, Hunan City University, Yiyang 413000, China
College of Computer Science and Electronic Engineering, Hunan University, Changsha 410082, China

Shuguang Deng and Hua Qin
College of Information and Electronic Engineering, Hunan City University, Yiyang 413000, China

Pingjian Ding
College of Computer Science and Electronic Engineering, Hunan University, Changsha 410082, China

Shaopeng Chen
College of Mathematics and Computer Science, Hunan Normal University, Changsha 410081, China

Guanghui Li
College of Computer Science and Electronic Engineering, Hunan University, Changsha 410082, China
School of Information Engineering, East China Jiaotong University, Nanchang 330013, China

Xun Wang and Tongmao Ma
College of Computer and Communication Engineering, China University of Petroleum, Qingdao 266580, China

Pan Zheng
Department of Accounting and Information Systems, University of Canterbury, Christchurch 8041, New Zealand

Tao Song
College of Computer and Communication Engineering, China University of Petroleum, Qingdao 266580, China
Departamento de Inteligencia Artificial, Universidad Politcnica de Madrid (UPM), Campus de Montegancedo, 28660 Boadilla del Monte, Spain

He Su, Junqi Bai, Lu Gong, Juan Huang, Wen Xu, Jing Zhang, Xiaohui Qiu and Zhihai Huang
The Second Clinical College of Guangzhou University of Chinese Medicine, Guangzhou 510006, China

Yang Chu and Jiang Xu
Institute of Chinese Materia Medica, China Academy of Chinese Medicinal Sciences, Bejing 100700, China

Mohammed Zakariah
Research Center, College of Computer and Information Science, King Saud University, Riyadh 11451, Saudi Arabia

Shahanavaj Khan
Nanomedicine & Biotechnology Research Unit, Department of Pharmaceutics, College of Pharmacy, King Saud University, Riyadh 11451, Saudi Arabia
Department of Bioscience, Shri Ram Group of College (SRGC), Muzaffarnagar 251002, UP, India

Anis Ahmad Chaudhary
Department of Pharmacology, College of Medicine, Al-Imam Mohammad Ibn Saud Islamic University, Riyadh 11451, Saudi Arabia

Christian Rolfo
Phase I-Early Clinical Trials Unit, Oncology Department, Antwerp University Hospital, "Centre for Oncological Research (CORE)", 2650 Edegem, Belgium

Mohamed Maher Ben Ismail and Yousef Ajami Alotaibi
Computer Science Department, College of Computer and Information Sciences, King Saud University, Riyadh 11451, Saudi Arabia

Index

www.ingramcontent.com/pod-product-compliance
Lightning Source LLC
LaVergne TN
LVHW060526210926
848485LV00005B/1156